Will to Live

Will to Live

AIDS THERAPIES AND
THE POLITICS OF SURVIVAL

João Biehl

Photographs by Torben Eskerod

PRINCETON UNIVERSITY PRESS

PRINCETON AND OXFORD

Copyright © 2007 by Princeton University Press

Published by Princeton University Press, 41 William Street, Princeton,
New Jersey 08540

In the United Kingdom: Princeton University Press, 6 Oxford Street,
Woodstock, Oxfordshire OX20 1TW

Second printing, and first paperback printing, 2009
Paperback ISBN: 978-0-691-14385-9

The cloth edition of this book has been cataloged as follows

Library of Congress Control Number: 2007934333
ISBN: 978-0-691-13008-8

British Library Cataloging-in-Publication Data is available

Publication of this book has been aided by Princeton University's Committee on
Research in the Humanities and Social Sciences

This book has been composed in Sabon and Futura

Printed on acid-free paper. ∞

press.princeton.edu

Printed in Canada

10 9 8 7 6 5 4 3 2

For my parents

And from this indigent river,
this blood-mud that meanders
with its almost static march
through sclerosis and cement

and from the people who stagnate
in the river's mucus,
entire lives rotting
one by one to death,

you can learn that the human being
is always the best measure,
and that the measure of the human
is not death but life.

—*João Cabral de Melo Neto,*
Education by Stone

Contents

Introduction

A New World of Health

The Right to a Nonprojected Future

In his book *A Bias for Hope*, economist Albert O. Hirschman (1971) challenges social scientists to move beyond categorical prejudgments, beyond the sole search for general laws and orderly sequences of what is required for wider social and political transformation. Having in mind the Latin American countries in which he worked (including Brazil), Hirschman challenges us, instead, to engage the *unexpected*.

The study of how beliefs, attitudes, and values are refashioned and molded by "more or less accidentally undertaken practices," Hirschman argues, "widens the limits of what is or is perceived to be possible, be it at the cost of lowering our ability, real or imaginary, to discern the probable" (p. 28). At stake is helping "to defend the right to a nonprojected future as one of the truly inalienable rights of every person and nation; and to set the stage for conceptions of change to which the inventiveness of history and a 'passion for the possible' are admitted as vital actors" (p. 37).

This book addresses the crucial question of what happens when such luminous prospects of social science are politically and technologically operationalized. Brazil has, against all odds, invented a public way of treating AIDS. In 1996, it became the first developing country to adopt an official policy that universalized access to antiretroviral drugs (ARVs), about five years before global policy discussions moved from a framework that focused solely on prevention to one that incorporated universal treatment. Some 200,000 Brazilians are currently taking ARVs that are paid for by the government, and this policy is widely touted as a model for stemming the AIDS crisis in the developing world. This lifesaving policy came into existence through an unexpected alliance of activists, government reformers, development agencies and the pharmaceutical industry. *Will to Live* moves between a social analysis of the institutional practices shaping the Brazilian response to AIDS and the stories and lives of people affected by it.

HIV/AIDS is the first major epidemic of present-day globalization. Of more than 40 million people estimated to be HIV-infected worldwide, 95 percent live in middle- or low-income countries, causing life expectancy to drop dramatically in those countries worst hit. In late 2003, with only about 400,000 people receiving treatment, the World Health Organization (WHO) and the Joint United Nations Programme on HIV/AIDS (UNAIDS) announced their goal of having 3 million HIV-positive people on antiretroviral therapy by 2005. The results have been mixed, but by any account Brazil has been a leader in the effort to universalize access to treatment. By the end of 2004, the number of people on ARVs had increased to 700,000 globally—in the developing world, this figure stood at 300,000, of which half lived in Brazil (UNAIDS 2004). And when the deadline arrived at the end of 2005, with an estimated 6.5 million people requiring treatment, 1.2 million were on ARVs—encouraging, but still short of the target (UNAIDS 2006). Brazil, with less than 3 percent of the world's HIV/AIDS cases, still accounted for nearly 15 percent of people on ARVs.

Throughout this book, I examine the value systems and the political and economic factors underlying the Brazilian AIDS policy, and identify the novel power arrangements (both national and global) that are crystallized in the policy, in its articulation and implementation. As I probe the policy's social and medical reach, particularly in impoverished urban settings where AIDS is spreading most rapidly, I also inquire into the micro-politics and desires that invest ARVs, making survival possible.[1] I draw from research I carried out over the past ten years among people working in state, corporate, scientific, and nongovernmental institutions, and also from fieldwork among marginalized AIDS patients and grassroots care services.

To understand the radically different world of AIDS post-treatment access I had to move in time and space, back and forth between a difficult analysis of how the afflicted understand themselves—born of careful ethnographic work and long-term conversations I was privileged to have—and a more experience-distant investigation into how therapeutics mix with activism and political economy: locally, nationally, and globally.[2] Fieldwork allows us to see these various actors and forces at work, reminding us that there is no short cut to understanding the multiplicities of reality and the practical articulations through which technologically extended life happens. Ethnography remains, in my view, a vital social scientific antidote to what Hirschman identifies as "compulsive and mindless theorizing." As he writes, "Quick theoretical fix has taken its place in our culture alongside the quick technical fix" (1970, p. 329).

Although much of my research stands within the traditional boundaries of ethnography (charting the lives of individuals and institutions over time through open-ended interviews and participant observation), I also make use of alternative forms of evidence—some of them quantitative—developed in collaboration with researchers and practitioners from other disciplines, including epidemiology. Yet during my fieldwork, I often found myself returning from what I "[saw] and heard with blood-shot eyes and pierced eardrums."[3] Where words and numbers fell short, I teamed up with photographer Torben Eskerod; his photographs, interspersed throughout these pages, highlight the plight and singularity of the abandoned AIDS patients with whom I worked.

Examining this constellation of evidence from an anthropological perspective sheds light on how scientific and technological developments, medicine, and political-economic institutions do their work over time and across cultures. Biotechnological innovations engender unlikely coalitions that both expose the inadequacies of reigning public health paradigms and act to reform, if to a limited extent, global values and mechanisms (of drug pricing and types and scope of philanthropic and humanitarian interventions, for example). Mediated by an activist state, these therapeutic coalitions also expose national contradictions and bring about novel institutions, modes of life, and inequalities. Brazil's response to AIDS thus provides a unique opportunity both to apprehend shifting public-private involvements in a neoliberal landscape and to assess their immediate and long-term effects.

Some of the questions that guided my ethnographic and social epidemiological investigation include: Which public health values and political and technological practices make this therapeutic policy possible, and what guarantees its sustainability? How has the AIDS policy become a kind of public good, emblematic of the state's universal reach, even though it is not enjoyed by all citizens? What networks of care emerge around the distribution of lifesaving drugs? How do the poorest understand and negotiate medical services? How do their lifestyles and social support systems influence treatment adherence? What happens to poverty as these individual sufferers engage the pharmaceutical control of AIDS? What do these struggles over drug access and survival say about the state of human rights, politics, and equity on the ground and globally? Which forms of health are sufficient to liberate life, wherever it is confined?

Universal Access to Lifesaving Therapies

Brazil is the epicenter of the HIV/AIDS epidemic in South America and accounts for 57 percent of all AIDS cases in Latin America and the Caribbean.[4] AIDS was first reported in Brazil in 1980, and through mid-2002, the Ministry of Health had reported nearly 240,000 cumulative cases. HIV prevalence in Brazil is higher than in most of its neighbors, although this is in part due to more accurate reporting. At the end of 2001, an estimated 610,000 individuals were living with HIV/AIDS (an adult prevalence of 0.7 percent).

Social epidemiological studies show considerable heterogeneity in HIV infection rates, with large numbers infected among vulnerable populations and a fast-growing number of heterosexual transmissions. In 1998, 18 percent of sex workers tested in São Paulo were HIV-positive, and in certain areas of the country, intravenous drug users contribute to almost 50 percent of all AIDS cases. Since 1998, the death rate from AIDS has steadily declined, an achievement attributed to the country's AIDS policy (Okie 2006; Dourado et al. 2006).

In the Brazilian AIDS world, the vital actors with a passion for the possible were not just professional politicians. Throughout the 1990s, a range of different groups and institutions—activists and local nongovernmental organizations (NGOs), central and regional governments, and grassroots organizations, along with development agencies such as the World Bank—came together, helping to address what was earlier perceived to be a hopeless situation. This combination of social organization and education, political will (at various levels of government), and international cooperation made it possible for Brazil to overcome AIDS denial and to respond to an imminent crisis in a timely and efficient way.

Social mobilization forced the government to democratize its operations further. AIDS activists and progressive health professionals migrated into state institutions and actively participated in policy making.

They showed creativity in the design of prevention work and audacity in solving the problem of access to AIDS treatment. In their view, the prices pharmaceutical companies had set for ARVs and the protection they received from intellectual property rights laws and the World Trade Organization (WTO) had artificially put these therapies out of reach of the global poor. After framing the demand for free and universal access to ARVs as a human right, in accordance with the country's constitutional right to health, activists lobbied for specific legislation to make the drugs universally available.

The Brazilian government was able to reduce treatment costs by reverse-engineering drugs and promoting the production of generics in both public- and private-sector laboratories. Had an infrastructure for the production of generics not been in place, the story being told today would probably be different. For its part, the Health Ministry also negotiated substantial drug price reductions from pharmaceutical companies by threatening to issue compulsory licenses for patented drugs. Media campaigns publicized these actions, generating strong national and international support.

The result—a policy of biotechnology for the people—has dramatically improved the quality of life of the patients covered. According to the Health Ministry, both AIDS mortality and the use of AIDS-related hospital services have subsequently fallen by 70 percent (MS 2002). Known for its stark socioeconomic inequalities and its perpetual failure to develop to its full potential, Brazil has faced down AIDS, at last becoming "the country of the future" that idealist Stefan Zweig (1941) imagined in the 1940s.

The Brazilian treatment rollout has become an inspiration for international activism and a challenge for the governments of other poor countries devastated by the AIDS pandemic.[5] This policy challenges the perception that treating AIDS in resource-poor settings is economically unfeasible, and it calls our attention to the ways in which biotechnology can be integrated into public policy even in the absence of an optimal health infrastructure.[6] It likewise opens up the political and moral debate over delivering life-extending drugs to countries where patients are poor and institutions have limited capacity, as well as the debate over the immediate and long-term medical implications of doing so.[7]

By 2000, the Brazilian national AIDS program had been named by UNAIDS as the best in the developing world, and in 2003 it received the $1 million Gates Award for Global Health. Brazil is now sharing its know-how in a range of ways. It has taken on a leadership role at the WHO's AIDS program and it is supporting international networks aimed at facilitating treatment access and technological cooperation on HIV/AIDS. In the past years, the Brazilian government has also been leading developing nations in WTO deliberations over a flexible balance between patent rights and public health needs.

We are still far from achieving international justice in the realm of AIDS, but the Brazilian response has at least helped to expose the failures of reigning paradigms that promote public-private partnerships for the resolution of social problems. Brazil's national response has also shown the limits of international development agencies when confronted with the need to act directly on behalf of the poorest. Practically speaking, Brazil opened channels for horizontal south-south collaborations and devised political mechanisms (as fleeting and fragile as they may be) for poor countries to level out some of the pervasive structural inequalities that destine their populations to disease and ill health.

A Political Economy of Pharmaceuticals

Although a compacted and all-encompassing sovereignty is hard to locate in today's geopolitical order, states do not necessarily weaken amid economic globalization.[8] But they do reform and reconfigure themselves, developing new strengths and novel articulations with populations. Brazil's response to AIDS "is a microcosm of a new state-society partnership," Fernando Henrique Cardoso, Brazil's former president (1995–2002) and the country's most prominent sociologist, stated in an interview with me in May 2003: "I always said that we needed to have a porous state so that society could have room for action in it, and that's what happened with AIDS."

Cardoso had no qualms about extrapolating, using the AIDS policy as evidence of the "success" of his state reform agenda—a state open to civil society, decentralized, fostering partnerships for the delivery of services, efficient, ethical, and, if activated, with a universal reach. "Government and social movement practically fused. Brazilian society now organizes itself and acts on its own behalf." From this perspective, the state appears through its model policies.

As with all things political and economic, the reality underlying the AIDS policy is convoluted, dynamic, and filled with gaps. The politicians involved in the making of the AIDS policy were consciously engaged in projects to reform the relationship between the state and society, as well as the scope of governance, as Brazil molded itself to a global market economy. One of this book's central arguments is that on the other side of the signifier *model policy* stands a new political economy of pharmaceuticals, with international and national particularities. As NGO activism converged with state policy making, and as the public health paradigm shifted from prevention to treatment access, political rights have moved toward biologically based rights.

Neoliberal governmentality has taken a new shape. Rather than actively seeking areas of need to address, the new market-oriented state selectively recognizes the claims of organized interest groups that "represent" civil society, leaving out broader public needs for life-sustaining assistance—in the domains of housing, economic security, and so forth. To be "seen" by the state, people have to join these groups and engage in lobbying and lawmaking.

Ethnography helps to uncover the circumstances and contradictions that are inherent to this novel form of therapeutic mobilization, already abstracted in Cardoso's articulation of a "mobilized Brazilian society" and a "porous and activist state." Ethnography complicates. It is a way of grounding and dissecting such abstractions, illuminating the contingency, multiple interests, and unevenness of the political game that is under way.

Given the increasingly global frames of disease control, the way a state deals with AIDS reveals its statecraft: in the Brazilian case, engagement with—and submission to—the forces of globalization. Just a few months before approving the AIDS treatment law in November 1996, the Brazilian government had given in to industry pressures to enshrine strong patent protections in law. Brazil was at the forefront of the developing countries that supported the creation of the WTO, and it had signed the Trade-Related Aspects of Intellectual Property Rights treaty (TRIPS). Parallel to the new patent legislation, pharmaceutical imports to Brazil have increased substantially. Currently, Brazil is the eleventh largest pharmaceutical market in the world.

As the AIDS policy unfolded, Brazil attracted new investments, leading to novel public-private cooperation over access to medical technologies. While Brazil experimented with new modes of regulating markets for lifesaving treatments, pharmaceutical companies took the conflicts over drug pricing and the relaxation of patent laws at the WTO as opportunities both to negotiate broader market access in Brazil and to open up unforeseen AIDS markets in other countries. The industry has also been able to expand clinical research in Brazil, now run in partnership with public health institutions. American pharmaceutical companies have at the same time successfully downplayed the WTO as they lobbied for strict bilateral and regional trade agreements that made local production of generic drugs unviable.

Global markets are incorporated via medical commodities. This process is mediated by development organizations and has crucial implications for the nature and scope of national and local public health interventions. Magic-bullet approaches (i.e., delivery of technology regardless of health care infrastructure) are increasingly the norm. The Brazilian AIDS policy was aligned with a pharmaceutically focused form of health delivery that was being put into practice as part of the government's vision of cost-effective social actions (involving the decentralization and rationalization of assistance amid the dismantling of public health institutions). In recent years, Brazil has seen an incremental change in the concept of public health, now understood less as prevention and clinical care and more as access to medicines—what I call the *pharmaceuticalization of public health*.

The medical accountability at stake in this innovative policy has drastic implications for Brazil's 50 million urban poor, either indigent or making their living through informal and marginal economies. Despite the allegedly universal reach of the AIDS policy, poor AIDS patients have not been explicitly targeted for specific governmental policies related to housing, employment, or economic security. The urban poor gain some public attention during political elections—even then only in the most general terms—and through the limited aid of international agencies. Through AIDS, however, new fields of exchange and possibility have emerged.

Medicines, as I argue throughout this book, have become key elements in the state's arsenal of action. As AIDS activism migrated into state institutions, and as the state played an increasingly activist role in the international politics of drug pricing, AIDS became, in many ways, the "country's disease." In May 2007, for example, Brazil broke the patent of an AIDS drug (Efavirenz, produced by Merck) for the first time—a step recently taken by Thailand—and authorized the import of a generic version from India. Activists worldwide hailed this sovereign decision as a landmark in struggles over the sustainability of countrywide treatment rollouts. Yet, while new pharmaceutical markets have opened, and ARVs have been made universally available (the state is *actually* present through the dispensation of medicines), it is up to individuals and communities to take on locally the roles of medical and political institutions.

This pharmaceuticalization of governance and citizenship, obviously efficacious in the treatment of AIDS, nonetheless crystallizes new inequalities.[9] My ethnography illuminates how this medical intervention—funded and organized by the state alongside international institutions and produced by the pharmaceutical industry—has resulted in effective treatment for working-class and middle-class Brazilians, meanwhile leaving those in the marginalized underclass by the wayside. These individuals cope by using survival strategies that require extraordinary effort and self-transformation.

Persistent Inequalities

Just as the complex Brazilian response to AIDS must be understood within the wider context of the country's democratization and the restructuring of both state and market, so too must it be seen in light of its interaction with local worlds and the subsequent refiguring of personal lives and values.[10]

I was in the coastal city of Salvador (the capital of the northeastern state of Bahia) conducting fieldwork when ARVs began to be widely available in early 1997. For the previous two years I had been charting the local politics of AIDS and documenting life with AIDS among the homeless and the residents of Caasah, a grassroots health service.

Considered by many the "African heart of Brazil," Salvador has an estimated population of 2.5 million and is a center of international tourism. The capital of the country until 1763, it was the entry point for millions of slaves brought from West Africa. Bahia, the largest state in the northeast region of Brazil, has a population of some 12.5 million.[11] Forty-one percent of Bahia's families live below the country's poverty line, and the top income quintile holds 69.5 percent of the wealth in the state. With about 70 percent of the total AIDS cases of the state, Salvador lies at the center of Bahia's AIDS epidemic.

Local epidemiologists and public health officers in the late 1990s had claimed that AIDS incidence was on the decline in both the city and the region, ostensibly in line with the country's successful control policy. But the AIDS reality I saw in the streets of downtown Salvador contradicted this profile. A large number of AIDS sufferers remained epidemiologically and medically unaccounted for, thereafter dying in abandonment. Meanwhile, community-run initiatives triaged care for some of the poorest and sickest.

A central concern of my ethnography has been to produce alternative epidemiological evidence and to generate some form of visibility and accountability for the abandoned subjects with AIDS.[12] As anthropologist-physician Paul Farmer has shown in the context of AIDS in Haiti and the United States, inequalities of power, ranging from poverty to racial and gender discrimination, determine who is at risk for HIV infection and who has access to what services (1992, 1999, 2003). By working closely with those who deliver care to the neediest and by attending to and documenting these patients' voices and experiences, one can identify and weigh the social factors promoting HIV transmission. One can also illuminate variations in the course of disease and in the value systems that lie within medical infrastructures. How, I wondered, would the ARV rollout fare in that context of multiple scarcities and ineffective regional politics? How would the most vulnerable transform a death sentence into a chronic disease? Which social experimentation could make such medical transformation possible?

Here, Hirschman's "right to a nonprojected future" begs for enactment and institutionalization. Caasah, a focal point of my research, was founded in 1992, when a group of homeless AIDS patients, former prostitutes, transvestites, and drug users squatted in an abandoned maternity ward in the outskirts of Salvador. "Caasah had no government," recalled Celeste Gomes, Caasah's director. "They did whatever they wanted in here. Everybody had sex with everybody, they were using drugs. There were fights with knives and broken bottles, and police officials were threatening to kick us out."

Soon, perhaps surprisingly, Caasah became an NGO and began to receive funding from a World Bank loan disbursed through the Brazilian government. By 1994, eviction threats had ceased and the service had gathered resources for basic maintenance. Caasah had formalized partnerships with municipal and provincial Health Divisions, buttressed by strategic exchanges with hospitals and AIDS NGOs.

Throughout the country, other "houses of support" (*casas de apoio*) like Caasah mediate the relationship between AIDS patients and the haphazard, limited public health care infrastructure. They address the paradox that medication is available, but public institutions are barely functioning. By 2000, at least one hundred of the country's five hundred registered AIDS NGOs were houses of support. However, in order to belong to these makeshift institutions of care, people must break with their old habits, communities, and routines as they forge new biographies.

By the mid-1990s, the unruly patients in Caasah had been evicted, and a smaller version of the group began to undergo an intense process of resocialization mediated by psychologists and nurses. Eighty "out-patients" remained eligible for monthly food aid. Patients who wanted to stay in the institution had to change their antisocial behaviors and adhere to medical treatments. Caasah now had a reasonably well-equipped infirmary post, with a triage room and a pharmacy. Religious groups visited the place on a regular basis and many residents adopted religion as an alternative value system.

As Celeste put it, "With time, we domesticated them. They had no knowledge whatsoever, and we changed this doomed sense of 'I will die.' Today they feel normal, like us, they can do any activity, they just have to care not to develop the disease. We showed them the importance of using medication. Now they have this conscience, and they fight for their lives."

Caasah's residents and administrators constituted a viable public that effectively sustained itself in novel interactions with governmental institutions and local AIDS services. In this "AIDS-friendly environment," people did not have to worry about the stigma that came with having AIDS "on the outside," and there was scheduled routine and an infrastructure that made it easier to integrate drug regimens into the everyday. At least for some, this unvarnished public—as desperate as it was creative—came to shape not only adherence to the ARVs but another chance at life.

•••

To document this particular public, to do justice to the singularity of its many lives, photographer Torben Eskerod joined me in the field in March 1997. With a simple chair and a black cloth against a wall, we improvised a photography studio outside Caasah's main building. Torben photographed each person as he or she wished to be portrayed, and I recorded their stories, past and present.

When we returned in December 2001, things had changed dramatically. Caasah had been relocated to a new state-funded building (though it remained an NGO). With treatment regimens available, functional residents had been asked to move out, and Caasah had been redesigned as a short-term care facility for ill patients (a "house of passage," *casa de passagem*) and a shelter for HIV-positive orphans. The hospice now had a team that worked directly with local hospitals and admitted the patients that "fit into the institution and its norms," in the words of Celeste. Disturbingly, there was no systematic effort to track these patients and their treatment actively once they left.

At the state hospital I learned of a triage system for AIDS patients, of which Caasah is part. "Homeless AIDS patients remain outside the system," one of the hospital's social workers told me. "Doctors say that they do not put these patients on ARVs for there is no guarantee that they will continue the treatment. They are concerned about the development of viral resistance to medication." The hospital's leading infectious disease specialist confirmed that "in theory, obviously, the doctor cannot withhold ARVs from drug users and homeless patients . . . but the fact is that the homeless patient does not return for routine ambulatory checkups. So what I do is tell the patient that he has to come back. If he returns and demonstrates a strong will, we begin treatment. . . . But they never, or rarely, come back."

We looked for our former collaborators and tracked down those who had left Caasah. Some had died; others had survived, married, and had children. As Torben took their portraits once again, they told us about all sorts of financial pressures, battles over discrimination, and the difficulty of obtaining access to quality health care. They told us about their *will to live.*

The patients photographed by Torben intimately engage us, their faces and words relating personal travails and the larger issues surrounding AIDS treatment and social inequality. Their very presence, brought so close to us through Torben's lens, establishes an alternative register of engagement and meaning that animates this book: How do these subjects both reflect one another and differ among themselves? What makes them visible or invisible in their neighborhoods? What is their place in a nation's order and in new medical regimes? How do we relate large-scale institutions and forces to local politics and personal trajectories? What is the staying power of these subjects' interior force of life? What might their stories, standing alone and taken collectively, suggest through their concatenation? Each dimension merits a closer look.

In this ethnographic work, double takes were both literal and figurative. Our 1997 work redoubled when we returned to Caasah four-and-a-half years later, providing us with a distinctive longitudinal perspective. And comparing these different moments in time—then and now—in turn opened a critical space for examining *what happens in the meantime.* Our methodology thus operated in dialogic, open-ended, and reflexive fashion: moving back and forth, across time and space, to offer a distinctive understanding of private and public becoming in the face of death and AIDS therapies.

Lives

"Take me to my father's house"

Edileusa came out of the building and into the backyard, where Torben and I had arranged the outdoor studio. She confronted us with a request: "I want to leave. I was waiting for you to come and take me away. God sent you."

Abruptly, she sat in the chair in front of the camera. "You can see how I look into it."

Edileusa had been in Caasah's triage room for the past twelve days, and this was her first day out. She had worked as a prostitute from an early age. The administrators and nurses knew very little else about her past or about the progression of her disease.

Edileusa continued: "But now I am cured. I am God's daughter. I had an ulcer. It is healed. Virgin Mary, it is so good to be free, a free person."

She remembered being expelled from home. "They were ashamed of me. They said that I had stained their lives." But now "my blood is clean. I took all exams; their results are in the infirmary. . . . See, I don't have marks on my body."

Edileusa kept staring straight into the camera. "Nobody likes me here. My things are ready. Put me in your car, and I will tell you where to leave me."

Where to leave you?

"Take me to my father's house."

She looked down. I tried to explain to her that we could not take her away, that she was recovering, that we were doing a work of . . .

"Then it is already done," she declared, and she left.

The next day, I looked for Edileusa in her room, but she was not there; she was back in the triage room. Two weeks later, she died.

Edileusa, 1997

"Today is another world"

"My name is Luis Cardoso dos Santos. I am thirty-six years old. Do you want me to close my eyes? You know, even when I am asleep, I sense when someone is coming into the room. My unconscious is very special. It makes me foresee things."

Luis was brought up by his mother, a *mãe de santo* (a priestess in Candomblé, a traditional Afro-Brazilian religion): "At the age of nine my head was shaved, and I was also initiated as a child-saint." Luis did not finish elementary school: "I had to work. I sewed clothing and later, I will not deny it, I smuggled goods from Paraguay."

Luis had lived in Caasah since 1995. "As you might have noticed, *I was* a homosexual—not a crazy one, though. Today *I am* a patient. I work for Caasah." The administration hired Luis as an office assistant, and he also helped to take care of Tiquinho, a fifteen-year-old hemophiliac boy who had grown up in Caasah.

"For two years, I had been weak, going to the ambulatory services, but the doctors never found out what I had. Finally, they drew my blood for an AIDS test. With the test result, the world closed for me. My family and friends discriminated against me. For them, AIDS was a crime. A doctor sent me here," remembered Luis. Caasah's nursing team helped Luis "to disembark from that death trip."

Luis loved to spread the word about his revival: "Soon I saw that one Luis had died and that another had emerged. Day after day, I feel better. I always try to bring the other patients a friendly word; we all need a friendly word. I will not give death any chance to come near me anymore. If death wants me, it must search for me."

•••

"Today is another world," Luis told us, as he looked at the portrait that Torben had made of him in 1997. "I have nothing to say against

22

the antiretrovirals. I am under Dr. Nanci's care, a very well-known and respected doctor. Celeste and the psychologists motivated me a lot. But I don't live here anymore, and I must take care of myself. I got used to the medication. Medication is me now."

A disability pension and the salary he was earning as Caasah's office assistant allowed Luis to rent a shack with a friend, to eat well, and to save a little, because, as he put it, "I want to have my own corner."

"I don't stop to analyze. What is past does not interest me. I came to face AIDS as a routine, like jumping from bed, brushing my teeth. Medication is a breakfast, and, as a breadwinner, I must go to work. . . . I can be well and then all of a sudden one day fall ill. I must adapt to this movement and face every specific problem. Today AIDS is both struggle and hope."

Luis insinuated he was dating. He also proudly told us that he had adopted an AIDS orphan in Caasah and was paying for the boy's grandmother to take care of him.

"I go step by step, pushing life forward with my will."

Luis, 1997

Luis, 2001

"If I only had thought then the way I think now"

Rose's left hand was atrophied, and she limped. "It is all from drug use. See my hands? I hit the nerves, and they had to cut two fingers." She did anything to get drugs: "I robbed and prostituted myself. I was crazy. I went to the street, to a bar, left with a client, did his game, and drugged myself with the money."

Rose and other healthy patients in Caasah repeatedly pointed to the marks on their bodies as images of past misdeeds, as if they were now in another place, seeing and judging their past selves from a photographic distance. "Ah, now I see. . . . If I only had thought then the way I think now."

Rose grew up in the interior and was expelled from home at the age of thirteen, after she became pregnant. She moved into Pelourinho's red-light district (in Salvador's colonial compound). By the early 1990s, as she saw her friends dying, she realized that "the party was over." By the end of 1993, Rose learned that she was both pregnant and HIV-positive. A physician who did volunteer work among prostitutes arranged Rose's move to Caasah. The father of one of Rose's babies was a founding member of Caasah.

One by one, Rose gave up her four children for adoption (the youngest was adopted by Naiara, Caasah's vice president). "What else could I have done? I couldn't give them a house. I would like my children to know that they were not raised by others because of abandonment, but because of my lack of experience—and also because I thought that I would not live much longer."

Rose has lived longer than she expected. For four years, she has officially been off illegal substances. She has remained asymptomatic, has become literate, and has learned to make handicrafts. At the time of the photograph, she was involved with Jorge Ramos, another resident, and

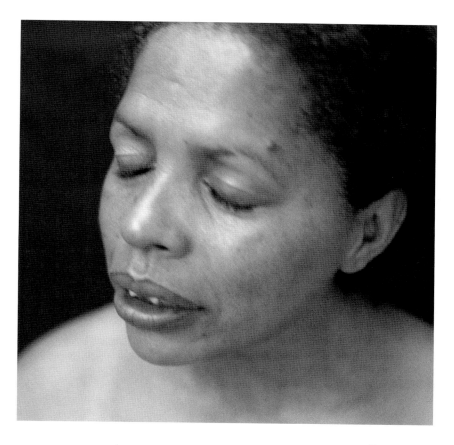

Rose, 1997

was beginning to take ARVs. "I take life in here as if it were a family, the family I did not have," she stated.

•••

"Welcome to the end of the world," Rose said jokingly, as we entered her brick shack, located at the lower end of a muddy hill in the outskirts of Salvador. "I am a new creature, you can write my story." She told us, "I am sold on the antiretrovirals. I am part of this multitude that will do whatever is necessary to guarantee our right to these therapies. I am proud of Brazil."

Caasah helped Rose to get the shack from the government, and she was living there with her one-year-old daughter. She had also taken in her teenage son, who had been under the custody of Professor Carlos, Caasah's nurse. "I am always struggling to pay the bills and raise my children, for I am mother and father."

Rose wept as she recalled how Jorge had died before the girl was born. She had done all that was medically possible. "Jessica got AZT, but the last exam showed that she is still seropositive." Rose knew that the child's HIV status could change until she reached the age of two, "and I pray every day for this to happen."

That same week we met with Rose's doctor. She expressed great respect for the way Rose takes care of herself and navigates the meager and largely unequal networks of care.

"I love Dr. Nanci," Rose told us. "She is a blessing to me and my family. She supports those who want to live. I never had another doctor, and I never had this problem of having to change treatment."

Rose was proud to be "a good patient, but not a fanatic one," she added. "I drink a beer and have some fun on the weekends, but I know my limits, what my body can take."

Yes, "people are still dying with AIDS in the streets," she stated, "but I am no longer there."

Rose, 2001

"Why will I think about the future?"

Nerivaldo stumbled over the chair in front of the camera. Scars from drug injections marked his arms, neck, and forehead. "If all was so simple," he murmured, as he took off his shirt for the portrait. His hormone-induced breasts had almost disappeared. Nerivaldo had begun injecting drugs at a very young age, and he turned tricks as a transvestite until he suffered a stroke three years ago. "First it was the foot; it got a little diseased. Then it was the hand, and I was paralyzed. I recall the doctor saying, 'It is all over.'"

Nerivaldo's face shifted, and his eyes opened and closed erratically, making it impossible to get a clear, still shot. "I have a cataract. It is very difficult to see." Torben then moved the camera closer, as Nerivaldo fought his blindness.

Now in his late twenties, Nerivaldo grew up in the streets. His mother died when he was eight years old. "I have a father, but I do not know where he is." Nerivaldo had lived in Caasah for a few months in 1995, but he did not comply with its strict discipline and medical routine, so he left. He wanted to live his life on his own, familiar terms: "I wanted to enjoy myself as much as I could, to play with my friends out there."

Sometime later, one of Caasah's leaders found him begging in a church, suffering from wounds infected with maggots. Nerivaldo was offered a last chance. He praises the regular care he gets now: "They bandage me up, give me medicine. No one has ever done that for me."

But how is it to live in Caasah?

"It is a depression." Depression springs from the experience of Caasah, an experience in which people are simultaneously well and sick: "We see some dying and others recovering. It's all there is." A resigned understanding of his place in history binds Nerivaldo to this reality. "I am already an aidético [a person with AIDS]. Why will I think about the future?"

30

Nerivaldo, 1997

Nerivaldo related a dream from the night before. "I was first talking alone, to myself. But then I was talking to myself with another person." When asked if he remembers a face, he answers, "My mother. My mother is dead. She died giving birth to my sister. So there is no way I can see my mother. I said, 'Mother, mother.' She responded, 'My son, my son.' Only this. That's how I speak in the dream. I see nothing."

•••

We were told that Nerivaldo had left Caasah again only a few months after the 1997 photograph was taken. He lived with his lover in an abandoned building in downtown Salvador. He fell ill again and found refuge in a makeshift asylum, Casa de Mãe Preta, where he died in 1998.

"A child is what I wanted most in life"

"I have the need to talk, to speak all truths," Evangivaldo proclaimed. "I have this sad psychosis in my head, but nobody comes in here." He spoke to us through a door from his quarantined room.

A former street inhabitant, Evangivaldo was being treated at the AIDS unit of the state hospital. Despite his extremely contagious, crusted Norwegian scabies, a social worker had sent him to Caasah with a "satisfactory discharge." Itches and scabies soon spread throughout Caasah. Healthier residents had developed a strong contempt for anyone who posed a danger to their immune systems, and they urged the administration to send Evangivaldo back to the streets.

Evangivaldo asserted that his greatest fear was not AIDS, but rather what he termed "AIDS citizens": "There are people in here who think that they are superior because of the color of their skin, or because they have a doctor who likes them, or because they have a better health condition. The other day, a guy hit me on the back. But I am not someone who creates confusion. Who knows, the guy could even do something bad to me; he could kill me. They do not do medical autopsies here."

As we were ending this photographic work, Evangivaldo asked to be photographed. He sat in the chair. "I feel a bit different now that I have someone to talk to. As if I were sleeping . . . I dream with birds, trees, and myself at the river shore. The time I lived in the interior, cleared the land, planted manioc, made it into flour, and cut wood. I was left alone, the only child, and came to Salvador in 1980. I carried tons of flour on my back to buy a pair of sandals.

"Later, my work was to take prostitutes to the ships that docked here. Greeks, Filipinos, Koreans, Africans, Chileans, Europeans, I saw all those orgies. . . . An Italian man once gave me gifts. Women were never so tender to me. I am sure he did not infect me. I felt this desire—how can I say it?—it was love. He told me of those landscapes he had seen. He never

Evangivaldo, 1997

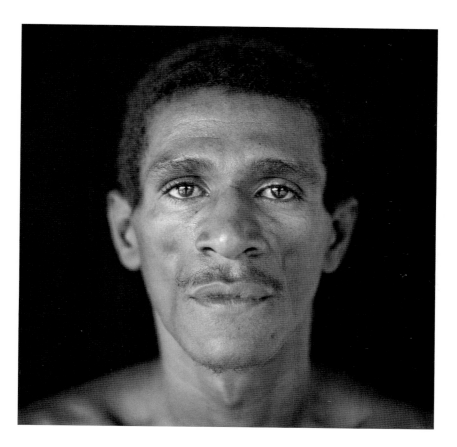

Evangivaldo, 2001

returned. Now I think that had I traveled away, none of this would have happened; rather, other novelties."

•••

"What a joy you give me by coming back," beamed Evangivaldo. His face was barely recognizable. But the aesthetic side effects of ARVs were the least of his concerns. "Today I woke up anguished. We had no gas to cook. I hope you can help me."

For the past three days we had tried in vain to reach him. We had a wrong address and no telephone. "I already had to move four times, because people discovered that Fátima and I have AIDS." The couple had met at Caasah and now had a two-year-old daughter. Juliana was born HIV-positive, but after eighteen months her HIV status changed. "She is healthy now. . . . A child is what I wanted most in life. Juliana fulfilled my desire, a dream I had. I thought I would die without being a father, but now I have a fruit of the earth."

Still on antipsychotic medication and fighting stigma, Evangivaldo explained that although he and Fátima took ARVs, the couple had no way to pay for medication to treat opportunistic diseases, and they were now badly in debt at the local drugstore—a common story, according to Caasah's director.

"It is the financial part of life that tortures me," Evangivaldo said. "When I see them with no food, it makes me ill. But when I find a job and there is nothing lacking at home, then for me it is another life and it is all good."

We sat under a tree in Caasah's backyard, and Evangivaldo looked at his 1997 photograph: "This work was important to me, it marked my history. Then I thought that I would not live. So many of those who thought they were the big guys of Caasah are already dead. My politics is to see things humanly. The one who is strong now must help the weaker."

having children; restarting
the forbidden fruit of AIDS,

"To have HIV . . . is like not having money"

When I first saw Valquirene in October 1996, she walked around the compound with a cloth over her head, and whenever she was in her room she threatened to throw herself out the window. Her family had left her in a psychiatric institution, where she had been given antipsychotic medication, and finally she had been dropped off, "like a carcass," at Caasah. "The doctors and my mother made me crazy," she told me.

Caasah's nurses eased her away from her chemical dependence. In early 1997, she became the first patient at Caasah to be treated with the ARVs recently made available by the government. Three months later, she had acquired, in her own words, "twenty-two pounds and a new body," and was ready to move out of Caasah.

Valquirene, wearing a black cap, approached us with some hesitation. She said that she had lived through so much that she too wanted a portrait made. As she sat, I told her that she could put her head down or show only her back, if she preferred. Instead, she took off her cap and used it as a mask. Her eyes were strikingly blue. She later told us, "I look like my maternal grandfather. He was German and married a black woman. My father's parents were Portuguese and Indian."

With the new therapies, she says, "I have no fear of living. I am only afraid of dying because of my son. I want him to study, to be a great man—then I can die." Valquirene was learning to do handicrafts to maintain herself and planned to study computing. "These days, you need the English language." She had also sought the advice of an activist lawyer, who was helping her to regain custody of her two-and-a-half-year-old son.

"I think that I am a different person now and that it is not such a bad thing to have HIV. It's like not having money. And, in Brazil, everybody experiences that."

●●●

Valquirene regained her health and the custody of her child. She never returned to Caasah for further assistance. Former residents at Caasah reported that she had found a boyfriend, also HIV-positive, and moved with him to the interior, where she ran a food stand. Some said that Valquirene had taken the child with her, while others said that she had left the child under the care of her mother.

Valquirene, 1997

"Too much medication"

"My image will burn the negative. I am too ugly. . . . What are you doing?"

"Measuring the light," Torben answered.

"Sometimes the eyes say something and the face says something else."

Soraia closed her eyes.

"My head is airy now, too much medication. I forget the room I am in. Sometimes I put something in one place, but then it is not there; I put it somewhere else."

It was Soraia's second day out of the triage room. Twenty-two years old, she had migrated from a rural area in the south five years earlier. Before going to the hospital and then to Caasah, Soraia worked the streets of upper-class Pituba Beach as a prostitute. Because of her white skin, she was a prized commodity.

Soraia believes that she was contaminated with HIV by madmen. "They took me by force, held me down, and I could not react." Later, she mentioned having to leave her home in the south because "my brothers beat me up. . . . My parents are both dead."

Only in the hospital did she find out that she had AIDS. "I thought my life was over. I took a piece of glass and began cutting my wrist, but the nurse found me. I did not sleep, I did not eat; I only wept in the hospital."

Soraia was renting a house with two other prostitutes and their four children. But when she left the hospital, her housemates kicked her out, saying that she would contaminate their children. "I had many things. I had a refrigerator. They took everything, even my clothes.

"It is not good to be imprisoned, but I must get used to being in Caasah. You know, all these pills are in fact a kind of drug. Sometimes I feel doped, I get crazy, but then it passes. It is good to take them, right? To cure, to be well.

Soraia, 1997

"Last night, I dreamed about my family. I came home at night and wanted to sleep. My sister told me that I should not go to bed, that I was no longer allowed to live there. Then I went to the street, and all of a sudden I was ambushed. A nurse then woke me up to take medication. . . . I don't know anything else. I lost the sense of the future. Now I have nothing to lose."

•••

Former residents told us that Soraia had left Caasah in 1997 and that she had resumed prostitution for a while. But for several years thereafter, she had not been seen.

"A beautiful place"

Tiquinho, a child with hemophilia, grew up in Caasah. A transfusion of HIV-tainted blood infected him at a young age, and the mayor of his tiny town, Rio Leal, forced him to leave. Now fifteen years old, he goes back to the interior once a year, during Christmas, to visit his mother and eight siblings.

Tiquinho was mostly silent in our meetings. Now and then, he whispered a few comments: "When they told me I had AIDS, it made me nervous. But I began to understand that what I have is not the end of the world. I am not a person who demands things. Sometimes when I am in pain, it takes a long time for people to know what I am feeling."

One day, he drew a house with a Brazilian flag on it: green for forests, blue for the sky, yellow for gold, white for peace, full of stars, and without the slogan "order and progress." Tiquinho is illiterate. On another piece of paper, he drew a smiling babylike creature, big and in diapers, with a woman coming out of the head, her mouth open wide. He would like to date, he said: "I want a girlfriend . . . to talk to."

Brief, simple, and sincere, Tiquinho also spoke of death: "At first, I was afraid of dying. But I do not fear it anymore. I think that when I die, I will go to a beautiful place."

•••

Tiquinho continued to live in Caasah. A born-again Christian, he had begun to participate actively in his church's activities and was learning how to read and write. Caasah's director did not believe that Tiquinho should spend his whole life there. She was making arrangements for Tiquinho to learn to self-medicate so that he could live with his family, who want him to return.

Tiquinho, 1997

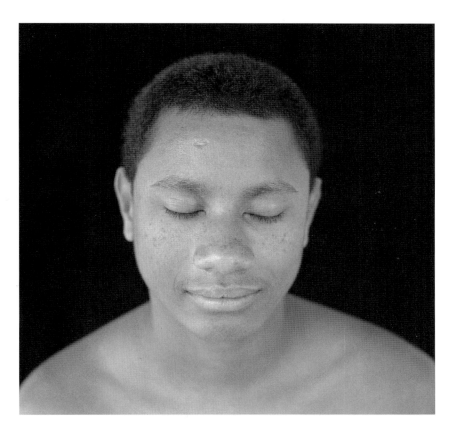

Tiquinho, 2001

The Politics of Survival

What most interests me as an anthropologist is the process of return-ing to the field. Repeatedly returning, one begins to grasp what happens in the meantime—the events and practices that enable wider social and political change, alongside those that debilitate societies and individuals, dooming them to stasis and intractability. In such returns, entanglements and intricacies are revealed. We witness the very temporality of politics, technology, money, and survival. The ethnographer demarcates previ-ously uncharted landscapes and tracks people moving through them. By addressing complicated transformations of institutions and lives in contexts of adversity, ethnography is uniquely qualified to confront and humanize the ways problems and policies are framed and interventions carried out.

Through unexpected political bodies such as Caasah, the question of social inclusion and health is worked out in different ways and lev-els. Operating within everyday violence and alongside AIDS therapies, there is an intense institutional and human maneuvering, a pulse of sorts. Treatment adherence is shaped in connectivity. Channels of communica-tion linking the individual, the state, and the medical institution are con-stantly renegotiated, and newly found personal identities cannot be taken for granted.

I find Gilles Deleuze's insights on the "subjectivity of milieus" helpful to think through the maps people draw as they traverse old and new pub-lic formations. There is never a moment in which we are not plunged into actual milieus. The settings and situations we move through are made of qualities, substances, powers, and events. And the maps we make of these dynamic trajectories and of the intensities we experience in the process "are essential to psychic activity," the philosopher writes (1997). Such "constellations of affects" are especially relevant to individual and group becoming vis-à-vis AIDS therapies: "From one map to the next, it is not a matter of searching for an origin, but of evaluating *displacements*. It is no longer an unconscious of commemoration but one of mobilization, an unconscious whose objects take flight rather than remaining buried in the ground" (p. 63).

At the margins, both the institutional and the pharmacological matters surrounding AIDS treatment undergo considerable flux. And AIDS survivors themselves live in a state of flux, simultaneously acknowledging and disguising their condition while they participate in what I call *local economies of salvation*. Now using ARVs and living on their own, Caasah's former residents face the daily challenge of translating medical investments into social capital and wage-earning power. They live between moments, between spaces, scavenging for resources. Through all these circuits, subjectivity is refigured as a *will to live.* In turn, this interior force of life—articulated at the intersection of social death, biotechnology, and the unevenness of public services—also refigures the terms of care and citizenship.

The political economy of AIDS, spanning both national and international institutions, creates an environment within which individuals and local AIDS organizations are codependent and must recraft positions and possibilities with every exchange. Their transactions are legitimated by a humanitarian and pharmaceutical discourse of lifesaving and civic empowerment. In adhering to drug regimens and making new and productive lives for themselves, patients are—in this discourse—saved. However, merely guaranteeing existence in such dire contexts, amid the dismantling of institutions of care, involves a constant calculus that goes well beyond numbers of pills and the timing of their intake.

The political grounds of existence have been increasingly individualized and atomized, and poor patients rarely become activists. Even as they search for employment, AIDS survivors work hard to remain eligible for whatever the state's paternalistic politics have made available—renewal of disability benefits, free bus vouchers, and additional medication at local health posts, for instance. Being adopted by a doctor and becoming a model patient (by complying with treatment in spite of a miserable situation) greatly facilitates this. And this material calculus becomes all the more important as patients form new families and resume a life considered normal, which was previously impossible for them.

This is not an inclusive form of care or of citizenship. Many are left out, saddled with other categorizations, such as drug addict, prostitute, beggar, or thief. Burdened by these labels, it is difficult for individuals to self-identify or to be identified as AIDS victims deserving of treatment and capable of adherence. To get that to which they are legally entitled, these individuals must not only identify themselves as belonging to the class of those served but also constantly seek out services. To retain services, furthermore, they must behave in particular ways. As a result, they largely remain part of the underground economy and constitute a hidden AIDS epidemic.

An ethnographic analysis of these linkages between macro-level policies, AIDS therapies, and care (or lack thereof) broadens our understanding of the forms that AIDS is taking and of what determines health outcomes among the most vulnerable. It concomitantly considers the organizational contexts that overdetermine inclusion and exclusion and the highly specific cultural contexts in which people assign value to their health status and lives. As life extension is now also possible for those who have been historically neglected, ethnography charts the trajectories that determine this mobilization. In noticing and documenting this *micro-politics of survival*, ethnography illuminates the paths through which people become the physicians of themselves and of their immediate worlds amid the growing tension between health as a public and a private good.[13]

By keeping these interrelated aspects in view—activism and political economy, pharmaceuticals and public health, population and individual, medicine and subjectivity—one organizes a more effective discussion of changing political cultures and ethics in an urgent time. *Will to Live* enters precisely into this sort of discussion. As I chart the trajectories leading to pharmaceutical forms of governance and personhood, I also attempt to open space for people who are missing in official data, policy decisions, and accounts. All this said, it is encouraging that with regard to AIDS, discussions about Brazil have shifted from social death to the maintenance and advance of a life-extending policy already well under way.

Chapter One

Pharmaceutical Governance

Globalization and Statecraft

"Brazil's response to HIV/AIDS is a valuable case study of innovative interactions between state and civil society, prevention and care, economic imperatives and ethical values, large-scale action and targeted programs," Fernando Henrique Cardoso, Brazil's eloquent former president, told me in an interview in May 2003. I met with Cardoso in Princeton, at the Institute for Advanced Study, where he was participating in a meeting of the board of trustees. After leaving the presidency, Cardoso had been traveling the international lecture circuit and had taken a professorship at Brown University.

The AIDS program was one of the most successful social initiatives consolidated during his administration. In years past, Cardoso has used the program as evidence of the transformation of the state and of social policy that he spearheaded.[1] "I always said that the government must open itself to society. This must be a fluid relationship. Social movements ought to be part of the design and implementation of social programs. That's how the neediest can be reached. AIDS is the greatest example."

The new state-society synergy reflected in the country's AIDS policy has developed in the wake of Brazil's democratization and the state's attempt to position itself strategically in the context of globalization, Cardoso argued: "We cannot do politics as if globalization did not exist. Globalization is not in the future; it is here now. I did not want it, but there is no other alternative. This is structural. This new phase of capitalism limits all states, of course, including the United States, but it also opens up new perspectives for states. The old producing state had no ways to capitalize or compete. You must have competition in a market that is global and not local. As we privatized and broke monopolies, we also had to create new agencies and rules to oversee the market, for you cannot allow the state not to have voice in these areas."[2]

Brazil had registered one of the highest rates of gross domestic product (GDP) growth in the world from the beginning of the century until 1980, but from then on the economy had practically stood at a standstill, with hyperinflation and a stream of exchange devaluations that finally declined under the Cardoso administration. In the previous fifty years, the Brazilian state had increasingly intervened in the production of goods and services, but this was no longer resulting in growth. By 1990, Brazil had the largest foreign debt in the world: $112.5 billion. The country's transition from a twenty-year military dictatorship to a democratic regime occurred under the weak, still centralized, and clientelistic government of José Sarney (1985–89) and later the neoliberalizing administration of Fernando Collor de Mello, who was impeached in 1992 under allegations of corruption and abuse of power. Itamar Franco became interim president, followed by Cardoso. In 2003, Luis Inácio Lula da Silva from the Workers' Party (Partido dos Trabalhadores, PT) became Brazil's new president (he has been reelected for a second term).

Cardoso said that both he and the new president "in the end say the same thing," that is, "that globalization is asymmetric and that it does not eliminate the differences imposed on nations. We have to take concrete steps toward decreasing this asymmetry, mainly at the trade level so that we can have access to markets, and also to control financing mechanisms." He made the case that Lula's government was basically following the same "ultra-orthodox" economic line of his administration—"there is no other alternative"—but that, "surprisingly," the new government lagged in social program innovation: "The proposals they have are centralized, very vague, mismanaged, and don't match with what Brazil already is." Cardoso was proud of the ways the AIDS program—with its multisectoral partnerships and high-tech delivery capacity—had pushed the envelope of what was governmentally possible.[3]

In this chapter, I discuss the history of the Brazilian AIDS policy against the background of economic globalization, paying particular attention to the role of science and technology in the design and implementation of responses.[4] I balance Cardoso's account with that of other policy makers, AIDS activists, scientists and health professionals, and representatives of the pharmaceutical industry. As I highlight changes the Brazilian AIDS policy underwent over time, I explore how this policy has become paradigmatic of new state-society-market interactions.

The AIDS policy is a product of intense negotiations. Its final form responds to diverse economic and political interests, as well as to the alternatives crafted by citizens. How do these key public- and private-sector actors interact, and how do they theorize their actions? What are the factual grounds of their rhetoric of agency and partnership? How do they measure the success of their lifesaving goals, and how do they conceive social inclusion and ethics? How does governance through model therapeutic policies affect statecraft and politics writ large?

The Social Science of a Transforming Regime

The AIDS policy evolved in a paradoxical space, caught between a downsizing of central government and the desire of politicians to create, in Cardoso's own words, "new rules for the political game." During our May 2003 conversation, Cardoso insisted that, even though the external environment today leaves less room for governments to shape national economies, states do not necessarily "disaggregate."[5] Rather, he spoke of the state as contingent in nature and of his own political efforts to find new ways, beyond protectionism, to frame the rules of accountability and to find a state voice vis-à-vis the market. "One cannot judge a priori or simplistically whether globalization is good or bad. One must see and decide in practice what is good and what is bad about it. We had to open the economy, keep a strict budget, and stop using inflation as a tool for capitalization. The state guarantees competition and also pushes for changes in the local productive basis."

Instead of decrying the generalized vulnerability that comes with the free flow of capital, Cardoso emphasized the country's overall gain from economic globalization, particularly regarding its technological infrastructure: "The products you make must meet global standards. The car can no longer be a cart, as former president Collor used to say. It must be for export. So you must remodel your production. The same product you sell outside, you sell inside. The producer complains for he will have to invest more, but the consumer wins with that."

Speaking as someone who "successfully" shepherded Brazil toward the global market, Cardoso insisted that the state retains sovereign power: "It is a question of responsibility. Nobody orders you to do these reforms. To say that the World Bank or the International Monetary Fund [IMF] forces you to limit public spending, for example, is a phantasmagoria of those who don't know how things work. These institutions bring experiences to the negotiating table, and a country like Brazil has enough weight to accept them or not."

As I interviewed Cardoso and other policy makers, I was struck by the ethos of power and innovation they conveyed and by their extensive use of social scientific idioms to describe their work and vision. While speaking of political alternatives under the constraints of global trade rules, they frequently mentioned terms such as *contingency*, *assemblages*, *governmentality*, *flexibility*, and *scales*—terms by now familiar to many social scientists studying contemporary politics.[6] They had politicized these concepts, and this discursive practice needs to be critically assessed.

The rhetoric of state agency and the abstractions that Cardoso articulated—a mobilized civil society and activism within the state—are part of a new political discourse. This performative language belongs to a public sphere strongly influenced by social scientists, as well as by politicians who do not want to take responsibility for their decisions to conform to the norms of globalization. For example, Cardoso makes no specific reference to the measures his administration took to open the economy, such as changes in intellectual property legislation and the privatization of state industries. This political discourse does not acknowledge the economic factors and value systems that are built into policy making today.

I asked Cardoso how he saw the reform of the state in light of the country's historical social inequality. He spoke bitterly against what he called "anachronistic revolutionary paradigms," as if sensing a criticism in my question: "There is no real way to break with all that is here. So you must evolve, right? The overall progress you get from evolutionary social policies is slow; wealth distribution does not happen rapidly, it happens over time. So you must improve and make social policies addressing popular interests. And this is what we were doing."

Thus, in his "pragmatic" approach to globalization, Cardoso articulates a market concept of society. Citizens are consumers and have "interests" rather than "needs." The government does not actively search out particular problems or areas of need to attend to—that is the work of mobilized interest groups. "There has never before been so much NGO action within the government as has occurred in the past ten years. In all our social programs there was some kind of social movement involved."

During his two administrations, centralized decision making, clientelism, and corruption—as he saw it—had been replaced by joint state and grassroots activities informed by public opinion, particularly in the fields of education and health. According to Cardoso, these elements of cooperation and nongovernmental involvement are key for maximizing the state's regulatory power and equity in the face of the market's agency in resource allocation and benefits. The work of NGOs and their international counterparts gave voice to specific mobilized communities and helped to consolidate his idea of public actions that were "wider and more efficacious than state action."

In these conditions, lawmaking is the main arena of state action—and putting new laws into practice is an activist matter. Cardoso lauds the signing of the AIDS treatment law in November 1996 and "even the agrarian reform law. They said nothing would pass." In mobilizing for a law and approving it, the state realizes its social contract. In Cardoso's vision, specific policies and legislation replace a wider social contract. In practice, people have to engage with lawmaking and jurisprudence to be seen by the state; as a result, the implementation of the law becomes subject to a whole range of exclusionary dynamics related to economic considerations and specific social pressure.[7]

In fact, empowered by the national AIDS program, activists successfully forced the government to draft two additional legal articles that would allow compulsory licensing of patented drugs in a public health crisis, and this legislation created a venue for state activism vis-à-vis the pharmaceutical industry. To make the antiretroviral treatment rollout economically feasible, the government had to invest in the production of generic drugs and engage in political battles over pricing with major pharmaceutical corporations. Besides successfully bargaining down the price of drugs essential to the AIDS cocktail, Brazil also won a 2001 confrontation with the U.S. government over patent legislation. All these initiatives created an international dialogue on intellectual property and medicines, and in the process, Brazil helped to constitute a southern trade bloc at the World Trade Organization (WTO) aimed at creating a worldwide system of drug price differentiation.

For Cardoso, the AIDS policy is thus emblematic of how state-society partnerships can actually create mechanisms to facilitate a more equitable international situation: "The idea that nothing can be done because rich countries are stronger is generally true, but not always. You can fight and, in the process, gain some advantages. You must penetrate all international spheres, try to influence and branch out. . . . The question of solidarity must be continuously addressed." Brazil's struggle for drug price reduction, he says, "shows that under certain conditions you can gain international support to change things. All the nongovernmental work, global public opinion, changes in legislation, and struggles over patents are evidence of new forms of governmentality in action . . . thereby engineering something else, producing a new world."

In what follows, I describe the practices of activists and the state, as well as the unexpected events that led to this innovative policy amid political and economic restructuring. The voices of these various players, at times discordant, highlight the strengths, weaknesses, and controversies surrounding Brazil's AIDS policy. The sustainability of the "model policy" is not a given. It has to be constantly negotiated in the marketplace. Its continuity lies largely in the capacity of the new government to grasp fully the pharmaceutical modus operandi it inherited from the Cardoso administration, to navigate it intelligently, and to manage it toward equity.

AIDS, Democratization, and Human Rights

HIV/AIDS emerged in Brazil in the early 1980s, concurrent with the demise of the military state. Its growth coincided with the country's democratization amid a ruined economic and social welfare system (Czeresnia 1995; Galvão 2000; Parker and Daniel 1991; Parker et al. 1994). The economy's "miracle years" were over. During the military years, Brazil had experienced fast industrial growth based on a policy of import substitution. But the combination of the 1982 onset of the international debt crisis and a fall in raw material prices forced sharp reductions in imports, public expenditures, and private investments in the country, as in all of Latin America (Adelman 1988). Income fell, unemployment rose, and the public health care system and sanitation services were left to deteriorate.

Living conditions worsened for those left behind by development plans and affected by epidemics supposedly once eradicated, such as dengue fever and malaria. In this context of economic decline, a progressive multisectoral mobilization—including political parties, universities, labor unions, and NGOs—united around democracy in Brazil, culminating in the drafting of a new constitution in 1988.

The struggle for a universal health care system brought the general population into the debates over democratization, and political rights have indeed become equated, for many, with medical rights. "Health," the new constitution reads, "is a right of every individual and a duty of the state, guaranteed by social and economic policies that seek to reduce the risk of disease and other injuries, and by universal and equal access to services designed to promote, protect, and restore health" (Constitution of the Federative Republic of Brazil 1988). The constitution granted all citizens the right to procure free medical assistance from public services and from government-reimbursed private providers. The principles of universality, equity, and integrality in health services were supposed to guide the new unified Brazilian health care system, known as Sistema Único de Saúde, or SUS (Fleury 1996). In practice, however, the right to health care would have to find ways to be realized amid decentralization and fiscal austerity. In 1989, for example, the federal government spent $83 on health per person, but in 1993, this amount plunged to only $37.[8] The private nonprofit and for-profit health care sector would end up delivering the bulk of medical services, including government-subsidized inpatient care. Today, high-cost treatments tend to dominate funding at the expense of health promotion and disease prevention programs.

In this section and the next, I examine the ways activists mobilized and successfully put HIV/AIDS on the political agenda. Operating at the level of identity politics, translocal solidarities, and civic participation, these activists established AIDS as a problem of social justice and demanded that the state fulfill its constitutionally mandated biopolitical obligations. They made their biographies public, carried out prevention and assistance work, and moved from micro- to macro-interventions. According to sociologist Herbert de Souza (Betinho), this movement from "anti to pro" does not imply reducing individualities "to a nameless condition without particularity, but to personalizing and to diversifying the public" (1991, p. 9). Participation in an emergent global public sphere and in policy making also implied a professionalization of activism, and heated debates ensued over the representativeness and political autonomy of AIDS activists, as well as over the scope and reach of interventions.

Epidemiological surveillance services registered the first HIV/AIDS cases in 1982: seven homosexual or bisexual men (later, one HIV/AIDS case from 1980 was found in São Paulo). In 1984, 71 percent of all HIV/AIDS cases were among men who had sex with men; injecting drug users and hemophiliacs were also affected. The virus was most prevalent in urban centers—as of 1985, 89 percent of the reported cases came from São Paulo and Rio de Janeiro (Castilho and Chequer 1996). Over the following two decades, this epidemiological profile rapidly and dramatically changed (Bastos and Barcellos 1995, 1996; Cassano, Frias, and Valente 2000).

For example, in May 2000 the homosexual/bisexual mode of transmission accounted for less than 30 percent of the total number of AIDS infections registered since the beginning of the epidemic; transmission through intravenous drug use accounted for 20 percent (MS 2002). By the late 1980s and early 1990s, heterosexual transmission had become predominant, and the number of women infected grew considerably. In 1985, there were twenty-five men for every woman with HIV/AIDS; by 1990, the ratio had reached 6:1, and in 2000 it arrived at 2:1. The feminization of the epidemic also led to a gradual growth of mother-to-child HIV transmission. In 1990, vertical transmission was responsible for 47 percent of HIV infections among children; in 2002, this number had risen to 90 percent.

The epidemic has also rapidly spread among the poor and otherwise disadvantaged. In 1985, for example, 79 percent of the reported HIV/AIDS cases involved individuals who had at least a high school education; by 2000, 73.8 percent were illiterate or had only finished elementary school (Fonseca et al. 2000; see also Cassano et al. 2000).

By 1985, still early in the progression of the virus, all five regions of the country had already reported HIV/AIDS cases. The Health Ministry and the media, however, kept treating HIV/AIDS as an issue confined to homosexuals in the country's largest urban centers, posing no threat to the "general population." According to pioneer AIDS activist Herbert Daniel, since its beginnings, Brazil regarded HIV/AIDS as "something foreign and strange," as well as "something inevitable, almost a kind of price to be paid for the modernity of our cities" (1991, p. 542).

The Health Ministry established a national AIDS control program in 1986, but even the health minister hedged its significance by pointing out that "we are talking about a serious disease, but which is not our priority" (Parker and Daniel 1991, p. 77). The government's initial refusal to address the epidemic seriously would allow HIV/AIDS to spread through the most vulnerable populations (Parker 1997; Scheper-Hughes 1994). However, the government's sluggishness followed international patterns of slow anti-AIDS policy development: only in 1986, for example, did the United Nations (UN) recognize AIDS as a problem that required attention (Galvão 2000, p. 92).

In those early years of HIV/AIDS—amid fear, stigma, and lack of national and international support—effective responses sprang from grassroots movements, most notably, from gay activist groups that pressured municipal and regional health services for information and treatment and that also carried out their own prevention campaigns. Founded in 1980, the Grupo Gay da Bahia (Gay Group of Bahia, or GGB) was already actively at work during Carnival in 1982, distributing brochures that alerted people to the "gay plague" or "pink cancer." In São Paulo, groups like Outra Coisa and Somos also distributed information on the disease and played a key role in creating a provincewide public health HIV/AIDS program in 1983, the first of its kind in Latin America. Its supervisor, Dr. Paulo Teixeira, would bring his know-how to the national HIV/AIDS program and later also to the World Health Organization (WHO) (see Teixeira 1997). Here, grassroots and local-state interventions were not antithetical to each other. Already, a mutual implication of activism and state had become characteristic of AIDS mobilization. The local activists and governmental actors had a common progressive political commitment; both understood the need to integrate education and care, as well as to establish pragmatic alliances with health professionals and philanthropic/religious institutions—these interventions proved to be quite efficient.

The HIV/AIDS epidemic also occasioned the creation of several new NGOs throughout the country, bringing together AIDS victims, progressive intellectuals, and activist migrants from other social movements on the decline. In 1985, the first GAPA (Grupo de Apoio à Prevenção da AIDS, AIDS Support and Prevention Group) was created in São Paulo, and it would soon set up affiliates in Porto Alegre and Salvador. The GAPAs worked on prevention and also mediated the treatment and legal demands of AIDS victims. That same year, Herbert de Souza created ABIA

(Associação Brasileira Interdisciplinar de AIDS, the Brazilian Interdisciplinary AIDS Association), which played a key role in the production and dissemination of HIV/AIDS knowledge. ARCA (Apoio Religioso Contra à AIDS, Religious Support Against AIDS) was created in 1987 to mobilize response in religious institutions. In 1989, Herbert Daniel founded Grupo Pela Vidda (Respect, Integration, and Dignity of AIDS Patients). Operating in Rio de Janeiro and São Paulo, Pela Vidda specifically addressed the rights and medical and treatment concerns of HIV-positive persons. A language of solidarity and citizenship punctuated the various initiatives of these NGOs (Aggleton and Pedrosa 1994).

Based on his work with a Mumbai-based urban activist housing movement, Arjun Appadurai (2002) discusses how activists move outside the legal and bureaucratic frame of the nation-state to work with new global networks in order to change things on the ground. Brazilian AIDS activists, too, successfully integrated themselves in "global geographies," but they also crossed the border into nation-state policy making, becoming activists within the state. Activist organizations played a decisive role in shaping policies for the dissemination of information on AIDS prevention, in promoting the 1986 law that made the registration of HIV/AIDS cases compulsory, and in reforming dangerous blood bank practices (Galvão 2000, p. 73). In 1988, activist mobilization also helped defeat a congressional resolution to restrict the entrance of HIV-positive people into the country.

Representing socially vulnerable groups such as homosexuals and sex workers, AIDS activists also developed a strong public voice in the dispute over access to ever scarcer public and medical resources. In 1988, for example, activists successfully lobbied the congress to extend disability status and pensions to all people with AIDS (Law 7670; Teixeira 1997, p. 61). Along with these activist victories came the idea of formally establishing a national nongovernmental HIV/AIDS network. But ideological differences and disagreements over partnerships with regional and national governments proved to be insurmountable obstacles (Bastos 1999; Galvão 2000, p. 63).

As of 1991, Brazil still did not have an efficient national program for controlling the epidemic. By that time, the accumulated number of notified cases was 21,650, nearly 60 percent (14,138) of which came from the state of São Paulo. In 1991, after a trip to Brazil, anthropologist Nancy Scheper-Hughes reported to the Ford Foundation:

The rate of undernotification of actual AIDS cases is extraordinarily high as HIV testing is random and haphazard and many of the symptoms produced by AIDS (pneumonia and other respiratory ailments, wasting, weakness, diarrheas) were already endemic to the rural and urban poor before the AIDS epidemic. Given the poor quality of medical care available to poor women and children, in particular, many Brazilian women first learn of their infection with HIV virus only after a spouse dies or after the birth of an infected child. (1994, p. 993)

As under-funded and under-staffed public health care services were increasingly paralyzed in their capacity to address the growing complexities of AIDS, grassroots and pastoral spaces of health care began to emerge—still today, the so-called *casas de apoio* (houses of support) such as Caasah in Salvador (see introduction) bear the medical and social burden of the AIDS crisis among the poorest.[9]

In fact, the inept populist administration of Fernando Collor de Mello (1990–92) represented a setback for the AIDS-activist agenda. The administration of the national program was very centralized and lacked direct cooperation with the NGOs; the program's director had a belligerent attitude toward international AIDS programs like that of the WHO. The program also suffered from discontinuous leadership at the Health Ministry, which shuffled through five ministers during the administration's tenure. In spite of HIV/AIDS having spread widely in all regions, public health services were mostly aimed at prevention and were operating at the level of the social imaginary (outdoor advertising, TV campaigns, or events), not infrastructure. A poll published by *Folha de São Paulo* (1991) revealed that only half of the respondents remembered any of the HIV/AIDS campaigns, while less than a third remembered anything about most recent campaigns.

Activist Herbert Daniel argued until his death in May 1992 that "adopting an abstracted and imported model, lacking definitions drawn from actual conditions in Brazil proved to be a way of doing nothing at all" (1991, p. 541). Anthropologist Richard Parker joined Daniel and other activists in denouncing the meager governmental initiatives, especially for not having taken seriously local constructions of sexuality as part of the phenomenon and trajectory of HIV/AIDS. They argued that the education provided by the state had been inappropriate—HIV/AIDS campaigns did not address Brazilians.[10]

In the early days of my research, my interviewees echoed Daniel's and Parker's indictment of national AIDS policy. Consider the attempts of fifteen-year-old Ana Silva to relate HIV/AIDS to her immediate situation and horizons. She had been living in the streets of northeastern Recife for over five years when I talked to her in July 1992. She openly spoke of prostitution as key to her identity and of her fear of HIV/AIDS. In her account, HIV/AIDS is nearby, entangled in everyday violence and the pervasive unaccountability of public services:

> For most prostitutes, the problem is not cholera; it is AIDS, AIDS, AIDS. That's all I hear in the streets. There is a lot of fear. We don't want to let ourselves be killed. I am afraid to be tested. The poorest are the ones who end up dying, the women out there.
>
> A prostitute, a friend of mine, told me that there's no time to use a condom when a man offers good money. The risk is hunger. So she does it without a condom. The guilty one is the government that doesn't give people the means to work. It's not just with a prostitute that you can get AIDS. Housewives aren't careful either. You can have just one man and still get the AIDS virus. A lot of men say, "I don't use condoms because I'm not a homosexual."
>
> The government only speaks of the cholera epidemic; it has lowered the number of projected AIDS cases so that people might think that AIDS is not a problem. Officials want to sterilize women as a form of prevention. Sterilize them so they don't have more kids. Then she runs the risk of getting the virus through the blood. If political candidates can talk so much nonsense on the TV, there should be more education about AIDS.
>
> When the government counts AIDS cases, they don't include prostitutes. Poor people don't die of AIDS, the only ones who die of AIDS are the rich. Poor people don't appear on television.

A Transnational Policy-Space

Amid major political changes (such as the impeachment of President Fernando Collor de Melo), the national AIDS program was restructured in 1992, under the leadership of biologist Lair Guerra. The following year the

World Bank and the Brazilian government signed a $250 million loan agreement to support new prevention and control activities ($160 million came from the World Bank and $90 million from the government). The agreement was called "AIDS Project I" and was effective from 1994 to 1998.

The new AIDS program aimed to reverse what international experts were already calling the "africanization" of AIDS in Brazil.[11] Experts were predicting that by the year 2000, Brazil would have some 1.2 million people infected with HIV. The country's epidemic was neither "nascent" (as in Chile or Morocco) nor "generalized" (as in sub-Saharan Africa), the experts said. Rather, it was "concentrated"—meaning that HIV was found in more than 5 percent of the so-called risk groups and in less than 5 percent of all women undergoing prenatal care—and was thus technically manageable.

In the 1990s, with the IMF and World Bank figuring prominently in policy decisions, fiscal austerity was on the rise and the idea of the social contract was on the decline. The well-known "Washington Consensus"—with all its support for structural readjustment, market deregulation, and trade liberalization—was developed specifically in response to Latin America's problems (see Williamson 1990; Rosário 1996; Stiglitz 2002). According to the international financial institutions, governments had let budgets get out of control, loose monetary policy had led to rampant inflation, and excessive state intervention in the economy had thwarted sustained economic growth.

Alongside these policy shifts, AIDS became increasingly cast as a development problem, prompting social mobilization and demands for public intervention. Various circumstances and actors met in an empty "space of policy" (Hirschman 1995, p. 179). According to Dr. Paulo Teixeira, one of the key articulators of the changes in the national AIDS program, "there was a strategic convergence of interests. Of course, the World Bank was interested in the country's overall economic restructuring, but Brazil was also a concrete site for the Bank to test the financing of such an abstract area: the control of an epidemic through prevention, in the absence of a vaccine" (personal communication, June 2005).

With new national and international funds available, both mobilized citizens and governmental institutions were to infuse this policy-space with specific rationalities, technologies, and claims of human and medical rights. Activists gave up their antagonism toward the state and organized, together with politicians, social scientists, and public health professionals, an impressive apparatus of HIV/AIDS control. The infrastructures and networks previously developed by NGOs and afflicted communities became a key asset in the development of a centralized and efficient AIDS program, dealing with international monitoring and regional demands for intervention. Epidemiologists, demographers, and statisticians working within both the program and local health systems were also beginning to make the human scope of the epidemic legible.

Just as in other policy areas, the World Bank attempted to shape the Brazilian AIDS program. But this time, according to Dr. Teixeira, "the Bank's team included experts that had very progressive views, very similar to those we defended. They supported actions compatible with our national needs—for example, work with injecting drug users. They also agreed that NGOs would have access to the financial resources and would execute the projects. Of course, our view of the NGO was more of a grassroots type, and they had in mind something much more institutionalized."

The main disagreement between the World Bank and the Brazilian AIDS policy makers was over treatment, stated Richard Parker (2001), who also participated in the first meetings with the Bank's experts: "In this negotiation process, the Bank's pressure not to have free dispensation of medication was always in the air. There was pressure for the resources to be used mainly for prevention, because within a neoliberal logic of costs and benefits, it is prevention that would bring more economic benefits. This was the logic that guided the Bank's work to a certain degree, and in spite of some changes, continues to guide the Bank's investments in health and in AIDS in general."

The politically progressive and socially minded activists and health professionals that now run the AIDS program kept open the possibility of medical assistance being part of the government's response to the epidemic and, in many ways, fought for it to happen on a larger scale. AIDS activists were among the first groups to interpret the new constitutional right to health as a right to medicines. In 1988, medications to treat opportunistic diseases were already available, if only on a limited basis, in the public health care system. Then in 1991, the government signed into law the free distribution of AZT and medication for opportunistic infections, but in practice, the supply and dispensation remained irregular (Galvão 2002a, p. 214; Oliveira et al. 2002, p. 1430). As Dr. Teixeira put it, "Brazil had enough mobilization and capacity to continue its treatment policy. The

World Bank did not finance medication, but we were able to direct some of the funds to medical training and improvement of diagnostic and laboratory infrastructure."

The majority of new AIDS funds were allocated to prevention, mostly through NGOs (which grew in number from 120 in 1993 to nearly 500 in 2000), and to the institutional development of regional and municipal AIDS programs that operated like NGOs. Massive, community-mediated prevention projects sought to contain the epidemic's growth, with a particular focus on safe-sex education, condom distribution, HIV testing, behavioral change, and harm reduction (CN 2000; Galvão 2000; World Bank 1999; Levi and Vitória 2002).

In my work in several regions of Brazil, I documented how the local implementation of HIV/AIDS prevention projects corroborated at least three cultural processes: (1) the individualized ingraining of a health-based concept of citizenship mediated by risk and vulnerability assessments; (2) the management of subjectivity in public health sites through testing technologies; and (3) the shaping of an ideal form of communitarian sociality. Social ties were being recast in nongovernmental sites, anonymous epidemiological clinics, and short-lived community initiatives (Biehl 1995; Biehl 2001a; see also Larvie 1997).

At any rate, at this moment in the AIDS policy's life course, NGOs represented afflicted populations within the state, and at a local level, the NGOs themselves were ruled by what anthropologist Jane Galvão calls the "dictatorship of projects" (2000). Also at local levels, religious and philanthropic institutions were triaging AIDS patients' access to welfare and medical goods.

Gerson Winkler, one of the founders of GAPA Porto Alegre, was then one of the most vocal critics of the ways activism "was becoming the state" and of the emergence of a new class of "AIDS professionals": "With AIDS funds coming in from all sorts of international organizations and with an increasing demand for us to sit in governmental committees, activism turned its face from real AIDS and underwent an acute institutional crisis. As we left the street demonstrations behind and sat with governments, we could no longer sustain our juridical activism. In some ways, AIDS NGOs began to justify the state. And in a more perverse move, they were justifying their own existence and economic survival" (personal communication, June 2005).

Meanwhile, AIDS medicine and the individual experience of AIDS were shaped in most unequal ways at both public and private health institutions. As anthropologist Cristiana Bastos (1999, p. 100) said of AIDS treatment in Rio de Janeiro in the mid-1990s:

> AIDS may have brought [development and underdevelopment] together, but access to treatment and medical care separated them again. An upper-class gay man from Leblon and a poor housewife from Baixada Fluminense, for example, might have shared the same AIDS-related illnesses; they might have even shared the same infectious disease specialist, who served at a peripheral public hospital in Baixada in the morning and a private office in Ipanema in the afternoon. Yet their experiences with the disease were quite different. If the hospital's pharmacy ran out of medicine, she could not buy it elsewhere; he, on the other hand, could send an order for the medication by fax to a network of air stewards, who would bring them to New York within twenty-four hours. . . . Their doctor traveled daily between the two worlds. In the morning, she could not give more than a few minutes of her time to the many patients who stood in line for hours. . . . In the afternoon, [patient and doctor] cooperated with each other in discussing and choosing treatments and exchanging information about the latest ones.

The Activist State

After researchers presented the combination antiretroviral therapy at the 11th International AIDS Conference in Vancouver, British Columbia, in 1996, Brazilian AIDS activists and patients—together with politically progressive specialists working within the national AIDS program—were able to mobilize public opinion and garner the support of various political parties in guaranteeing the right to these new medical technologies.[12]

In November 1996, President Fernando Henrique Cardoso signed a law (proposed by Senator and former president José Sarney) that made ARVs available to all registered HIV/AIDS cases. Federal Law 9313 mandated the free dispensation of these drugs through the public health system. Technical specialists, at national and regional levels, generated criteria for identifying AIDS patients and for implementing this intervention through SUS (Sistema Único de Saúde), the country's unified health system. Doctors were required to report cases to the Health Ministry for patients to be able to obtain the medication from their local public health services.

The immediate results of this pharmaceutical policy were striking: an estimated 55,600 individuals received ARVs in 1998 (Galvão 2006b). By the end of the previous year, the national program was already reporting that the therapies were decreasing the number of AIDS deaths and treatment costs (CN 1997f). In São Paulo, the number of reported AIDS deaths during the first three months of 1997 was 35 percent lower than the numbers of deaths in the same three-month period in 1996. The reported death decrease for the same period in Rio de Janeiro was 21 percent (Oliveira et al. 2002).

In considering these shifts, the program emphasized that the decrease in AIDS deaths paralleled a substantial reduction in hospitalization rates among AIDS patients for diseases such as tuberculosis and pneumonia. Use of emergency services and day hospitals was also said to be on the decline: "In São Paulo, the demand for treatment in day-hospitals decreased 40 percent. The reduction of the demand for this kind of service led to the closure of one of the two floors of the AIDS Unit of the Hospital das Clínicas" (CN 1997f). The economic gains were reported to be immense. Although the national HIV/AIDS program and the Health Ministry had spent some $300 million on medication in 1998, the government would save at least $500 million in these new medical transactions.

"This drug policy increased self-reporting and as a result, we have achieved near universal registration," epidemiologist Pedro Chequer told me in an interview at the Health Ministry in August 2000. He had been director of the AIDS program since 1996 and played a key role in the implementation of the drug rollout. Indeed, ARVs were now available, but the claim of universal access and demand sounded like a strategy to bolster the success of the policy, and thus add political value to it, as a way to ensure sustainability. As I show in chapter 3, the supply of AIDS services in public hospitals in poorer regions remained precarious, and many AIDS victims were left without adequate care. Moreover, the country's computerized register of individual patients on ARVs does not include specific social indicators. As a result, it cannot yet give us a detailed profile of this population.

All this technical infrastructure and medication "is not a gift," added Chequer, "it is the governmental response to a very well-organized social demand. The state has to continue to invest in pharmaceutical production, and it will." Dr. Tais Nogueira, one of the national program's chief pharmacists, agrees: "It is social mobilization that gives us the political legitimacy to make the medication available. We are an instrument of social mobilization; we give it rationality and make it work. Politicians give priority to this kind of social pressure. It is time now for AIDS to transfer this experience of both social mobilization and treatments to other pathologies, like TB and Hansen's. We have to revolutionize the health sector" (personal communication, January 2000).

These committed health professionals and activists have established medical survival as a matter of citizenship and politics. They are well aware of how to maximize demands for equity within the reforming state. With an agenda of social inclusion, they defend national autonomy and back a productive state (at least to account for medical needs). At the same time, they articulate an awareness that this lifesaving policy should be cut to fit a neoliberal paradigm that emphasizes economic criteria. "Now we have concrete data on the decline in mortality, showing that the investment has been worthwhile," Chequer told me:

> All this is so important that we have to leave behind the rhetoric of rights and citizenship and work from the point of view of cost-benefit because that is what really matters to people in the economic area of the government. . . . For them, the talk on rights and ethics is nonsense, you must say we spent that much, we saved that much, the policy is valuable because of this. . . . We demonstrated that even though the investment is high, the indirect savings are higher in terms of treatment of opportunistic diseases, less family disruption, and loss of productivity. In order to increase our budget, we included items related to TB control and blood screening, for example, and this minimizes internal disputes in the Health Ministry. The AIDS experience also challenges other disease areas to work from this management perspective and use us as a template.

Given this innovative pharmaceutical management and the apparent containment of HIV/AIDS associated with the first World Bank loan, a second loan agreement, "AIDS Project II" (with funds reaching $300 million), was approved and implemented between 1998 and 2003. By 1999, the World Bank was reporting that its joint project with the Brazilian government, NGOs, and regional and municipal AIDS programs had successfully led to "an estimated 30 percent decline in morbidity levels among the leading risk groups" (World Bank 1999; see also Garrison and Abreu 2000; Levi and Vitória 2002). Estimates now held that 600,000 people were infected out of a population exceeding 170 million. That same year, UNAIDS named the Brazilian program the best in the developing world (CN 2000).

"The Bank's loan is small if compared with what the government has spent on the AIDS program, but the Bank presents it as one of its most important success stories. I would say that they are exaggerating," economist José Serra, Brazil's former health minister, told me in an interview in June 2003 at the Institute for Advanced Study, at Princeton, where he was spending the year. Serra had run for president the previous year, and the AIDS program played an important role in his campaign. "Nonetheless, in spite of its traditional focused approach, the Bank never limited the scope of our action," added Serra. "Overall, the Bank's participation was positive, as it obliged us to do something well organized, to manage things efficiently, and to have a transparent accounting of all projects."

The World Bank, along with the IMF, had been harshly criticized in the mid-1990s for the negative impact of structural readjustment plans, particularly on the ability of local governments to reduce the spread of HIV infection (Lurie et al. 1995). The Brazilian success story came at a time when the Bank was reconsidering its approach to eradicating poverty and the need to involve governments more directly in the design of policies (Stiglitz 2002; see also Parker 1997). As Serra (now the governor of São Paulo) noted, "Informally, the Bank's leading figures told us that we were doing the right thing with medication distribution and challenging the pharmaceutical companies to reduce prices."

Evidently, the state does not completely compromise its regulatory functions as it negotiates loans, adjustment plans, and the transfer of technology with international agencies. Nevertheless, global and local arts of governing are definitively recast in the process, particularly in the context of AIDS, where "life and survival [are] at the heart of political action" (Abeles 2006a, p. 493). Consider the novel forms of regional "horizontal" cooperation crystallized through AIDS, which counter medical agendas established through first world politics. Until the early 1990s, the Pan American Health Organization (PAHO) and the WHO had been playing an important role in providing AIDS-related technical assistance to Latin American countries and supporting activist efforts. But in the mid-1990s, political differences regarding policy design emerged between these organizations and the region's AIDS programs. "Inspired by what was happening in Brazil, other countries refused to limit their interventions to models from the North," Dr. Teixeira told me (personal communication, June 2005).

Twenty-two Latin American countries formed a group of horizontal cooperation for HIV/AIDS in 1996, without the participation of the PAHO and the WHO. The group's executive administration is mobile and has already been hosted by Mexico, Brazil, Cuba, and Argentina. With a website and a newsletter, the group facilitates regional and subregional debates, Dr. Teixeira added. "Our group was extremely important for the diffusion of a progressive agenda and advancement in the area of HIV/AIDS and human rights, the participation of minority groups in the design of policies, and access to ARVs. The more conservative agenda had to retreat, at least for a while. Today, 50 percent of all AIDS patients in treatment in the world live in Latin America and the Caribbean."

According to Lucas Duarte, a public health expert working at the national AIDS program, the group also helps countries to chart the practices of drug companies: "In conversations we learn that Paraguay pays one price for a drug and Uruguay another, for example . . . so we try to identify various strategies for reducing the price of drugs and for making them available. The group also helps national AIDS programs to formalize specific partnerships over the exchange of information and technology."

Until 2002, the WHO did not list ARVs as "essential medicine." In fact, according to Dr. Teixeira, the WHO "was actually telling governments not to invest in AIDS therapies." The political decision to make ARVs available worldwide to all in need, as exemplified by the "3 by 5" initiative, "is a direct result of our activism in the global sphere," he concludes.

Intellectual Property Rights and World Trade

Most social scientific accounts explain the Brazilian ARV revolution in terms of the strength of social mobilization. Minority groups, AIDS activists, and experts on the disease all played critical roles in forcing the federal government to fulfill its constitutionally mandated health obligations. As Galvão writes: "If the decision to distribute medication can be seen from the technical-political angle, the mobilization of civil society has been key to its maintenance" (2002a, p. 16). Galvão cites the public mobilization of 1999 and 2000 that forced the Ministry of Economics to continue importing medication in spite of the devaluation of Brazil's currency. In 2000, at the World AIDS conference in Durban, a manifesto from the AIDS program demanding treatment for all in need and offering aid to other developing countries stirred international debate. Brazil also coordinated efforts that led the UN to pass a resolution in June 2001 that recognized access to medication as a fundamental ingredient for the human right to health. The success of these events, argues Galvão, is due to local activists' alliances with international organizations that have politicized patents as a question of fair global exchange and social justice.

Indeed, much of the inventiveness and success of the AIDS policy is due to the encroachment of social mobilization within the state and its transnational ramifications (MS 2002; Levi and Vitória 2002). What remains largely unconsidered are other political, technological, and market forces that have also been determinant of the AIDS policy's form and course. In what follows, I elaborate on the *pharmaceutical form of governance* that comes out of these new interactions of collective action, a neoliberalizing state, and the pharmaceutical industry.

Let us first consider how the AIDS treatment law fit into former president Cardoso's plan to internationalize Brazil's economy. Not by mere coincidence, just a few months before approving the law in November 1996, the government had given in to industry pressures to strengthen patent protection. Brazil had signed the Trade-Related Aspects of Intellectual Property Rights treaty (TRIPS) in December 1994, and because the government was eager to attract new investments, it allowed a quicker change in legislation.

TRIPS turned over to the WTO what was previously a sovereign right. It obliges all WTO member states to provide at least twenty-year patent protection in all fields of technology. Before TRIPS, every country defined its own regime of patent protection; likewise, prior to the treaty, no supranational power existed to enforce intellectual property rights across borders. Furthermore, in order to obtain WTO membership, all countries had to sign TRIPS, a clause with staunch support from the pharmaceutical industry (Angell 2004; Drahos and Braithwaite 2004; Sell 2003; Cohen and Illingworth 2003; Westerhaus and Castro 2006).

The TRIPS agreement allowed countries to self-identify as developed, developing, or least developed. This permitted least developed countries to postpone implementing TRIPS patent protection requirements until 2016 and gave developing countries (such as India and Argentina) until 2005. To become more viable in the global economy, Brazil and South Africa chose to elect patent protection immediately.[13] Brazil's new intellectual property legislation became effective in May 1996. As a result, the country could no longer produce generic versions of pharmaceuticals that were patented after the legislation was introduced (although it retained certain rights to produce under the use of a compulsory license).

"TRIPS entered in all negotiations with the IMF, the World Bank, and IDB [Inter-American Development Bank], where the United States has a fundamental role, not to mention the direct pressure from the U.S. Treasury," José Serra told me in June 2003. "This pressure led us to radicalize the initial project. In the senate, we even granted protection to products in the pipeline, waiting in line to be approved. The United States always put patents on the table. They had a few hanging things with Brazil, the nuclear thing, human rights, indigenous people, and patents; these were always on the agenda in the early 1990s." Middle-income countries were offered a very strong "developmental" justification for adhering to TRIPS, as is often the case in the call for neoliberal reforms. You provide patent protection in your country, the logic goes, and we, the investors, feel confident in investing there, which translates to more foreign investment and development for you.

I asked Serra whether Brazil had achieved any immediate gains from the patent reform. He was adamant: "No. In general, with the United States, these things end up not weighing in your favor. Later, when we brought it up in Washington, saying, 'We opened the economy,' they

*is marginalization, or exacerbation of inequality,
an inevitable consequence of rapid change?*

replied, 'You did it because it was in your own best interest.' It is always like that; it is not cumulative, things don't add up with the United States, and there is no reciprocity." Meanwhile, parallel to the new legislation, pharmaceutical imports to Brazil have increased substantially. Between 1995 and 1997, the trade deficit in pharmaceutical products jumped from $417 million to $1.277 billion (Bermudez 2000).

"Brazil bet a lot on the WTO and dove into it, body and soul," Serra continued. "We adopted all trade rules that the developed world wanted, like not being able to use measures of economic policy that link investments to exports." According to him, "Neoliberalization developed abruptly; it anticipated events. In one or two years Brazil changed commercial policies in place since the 1930s. From a closed and protected economy we went to the opposite. Today, Brazil is an economy that is much more open and unprotected than the American one. This openness was unilateral. It was not a negotiating process through which the country gained something in return. The developed countries didn't make any concessions with textiles and agriculture, for example."

As a policy maker, Serra was painfully aware of the loss of room for maneuver. "Brazil also dove into the free flow of capital. Dependence on this free flow to ensure the growth of the economy can provoke a generalized instability; and this, combined with external vulnerability, can be volatile indeed. So our government was conditioned by this. And even though there was no direct pressure for privatization, nonetheless the external environment favored it. If you need aid or credit from the World Bank, conditions are always embedded."

Serra also suggested that the early and mid-1990s was a transition period that left little time to reflect critically on the wide-range implications of the terms of economic readjustment—"things were not so clear." The long-term effects of TRIPS did not generate a great deal of public debate, for example, other than recognition that it marked countries' conformity to global trade reforms. In particular, there was a lack of discussion over the impact of pharmaceutical patents on drug prices and accessibility. The president and his team took hasty decisions which were legally binding. And from this new landscape defined by globalization, government was built (see Sassen 2006). "We did not hesitate to abolish all taxes for the import of medication," Serra recalls. "Many in the national industrial sector complained, but we did this to hold down the impact of exchange rates on inflation and to increase competition, to stimulate the production of generics."

Like Cardoso, Serra also denied a causal relationship between globalization and state reform. He was unapologetic about privatization, saying that "shrinking the size of the state does not mean less participation." Both politicians spoke of privatization as a means to make the state more agile so that it might both fulfill its market-regulating role and attend better to society. With the country's economy under siege, these politicians must insist that Brazil is not subservient; clearly, some form of independence and inventiveness is exercised in public policy—and that is what happened with AIDS, as it became technologically manageable.

In 1999, Serra championed the entrance of generics in the Brazilian market and gave incentives for their local production in public and private-sector laboratories. "Reverse engineering and the production of generics was the only way of keeping the lifesaving policy going," he said. For Serra, besides strong social pressure, the AIDS policy "basically worked because it was within the structure of the government and, in fact, because it revitalized part of the governmental structure. You find AIDS programs in every corner of the country, and, generally, they are government funded." A new relationship between government and society was taking form: "In my work, I was also always trying to establish new mechanisms of cooperation between government and society, working with a whole array of philanthropic organizations. To deepen the idea that public is not just what is governmental" (see Serra 2000). One could also argue that local communities increasingly compensated for the state's lack of administrative capacity, particularly as far as public health care infrastructure is concerned.

At any rate, these various practices—a technological reaffirmation of the universal public health care system, state production, and the outsourcing of care—materialized into a new politics of pricing as Brazil threatened to issue compulsory licenses on patented drugs in order to guarantee the sustainability of the policy. "After the strong international support we got at the Durban AIDS Conference," recalls Dr. Teixeira, "the Health Ministry authorized us to begin talking about breaking patents. . . . If it had not been for Serra, this would have been a much more complicated struggle" (personal communication, June 2005).

Serra says that "there was no a priori strategic plan at the national level. It was in practice that things happened the way they did." In spite of the national production of generics, prices of patented drugs were seriously jeopardizing the sustainability of the ARV rollout. Because it had the know-how and capacity to produce some of these drugs, Brazil wanted the pharmaceutical industry to lower prices and was threatening to break patents to do so. The United States, on the other hand, had been threatening to put in place economic sanctions against Brazil at the WTO. In June 2001, the United States and Brazil reached an agreement: Brazil would not export products it manufactured under compulsory licensing, and it would officially notify the U.S. government before it intended to break patents.

With drug prices still not reduced, Serra and the national AIDS coordinators devised the strategy to bring the treatment question to the most unthinkable of places, the WTO, which had a meeting scheduled for November 2001 in Doha, Qatar. By then, there was growing international support for the Brazilian initiative, with endorsements from the likes of AIDS activist networks, the *New York Times*, and UNAIDS. Furthermore, the U.S. government was in a weaker negotiating position after it had itself threatened to break Bayer's patent of Cipro, seeking cheap supplies in the wake of the post-9/11 anthrax episode. The health minister tactically worked his way into the Brazilian delegation at Doha and managed opposition within the government. With the support of key diplomats and NGOs, the Brazilian delegation thus articulated a southern bloc that succeeded in drafting a declaration that, at least symbolically, recaptured the developing country's sovereign right to operate out of the bounds of TRIPS for the purpose of public health (CN 2001c).

In Dr. Teixeira's words: "There were events. We had to focus at every moment on the empirical steps that had to be taken to make things work, and also have a temporal perspective, that is, to see things unfolding over time. The AIDS policy is a process." In practice, Serra stated, "the AIDS policy ended up working as a kind of counterweight to the economic orthodoxy in place internationally."

It is through this specific constellation—the AIDS policy—that globalization assumes concrete form and meaning for some segments of the Brazilian population today. Flows of money, knowledge, and technology—mediated by international financial institutions, NGOs, trade-related treaties, lawmaking, reverse engineering, and a new state capacity—constituted a strategic terrain for novel social and political articulations that, in turn, recast AIDS and its treatment.[14]

I am reminded here of Georges Canguilhem's discussion of the "decline of the idea of progress" and his call for an analytics of "motion": "continuous progress is a conservative epistemological concept. Predictions of progress turn today into tomorrow. But it is only when tomorrow comes that we can speak of yesterday" (1998, p. 318). The cumulative experience of the "unpredictability of the political and social effects of technological inventions," argues the philosopher of science, are also epistemological breaks.[15] The AIDS policy, one can argue, both illuminates what was at stake in past political decisions and economic maneuvers and gives evidence of how these "origins" can be somewhat remediated and medical possibilities developed.

In a talk in October 2003, Serra reaffirmed his vision of an activist state inescapably linked to demands for profit from finance capital. In being activated by civil society and defending public interests, governments can at least make their hurried economic policy decisions more flexible, if not bypass them altogether: "On intellectual property, our administration did not begin from a position based upon a doctrinaire principle or even a political-electoral strategy. . . . No, the reason underlying the measures adopted was our legal obligation to ensure the free-of-charge distribution of medicines. . . . Our position as government was not aimed at proposing the abolition of intellectual property protection, but rather to suggest and defend a position stating that patent rules must make it possible to achieve a balance between the objectives of the private and public interests" (2004, p. 9).

Serra also conveyed that social mobilization is a means for the state to constitute itself as a global player and to pursue visible advantages in the marketplace. That is, after successfully operating within the government, activism is now seen as a political technology of sorts, a counter-poison to the restrictive economic rules now in place: "While pharmaceutical companies may claim the right to charge monopolistic prices, thereby generating resources needed to invest in innovations, the public interest in broad and immediate dissemination of technologies capable of saving lives must also be recognized. To achieve this end, there must be an organized and countervailing power that lies in the countries that consume medicines of essential importance to public health" (2004, p. 10).

A Country's Disease—Public-Private Partnerships

In 2005, global pharmaceutical sales reached $602 billion—a growth of 7 percent from the previous year. According to Murray Aitken, a senior vice president at IMS Health, the market intelligence company that produced this sales report, "As growth in mature markets moderates, industry attention is shifting to smaller, developing markets that are performing exceptionally well."[16] The Brazilian market is of key importance for pharmaceutical companies operating in Latin America. Sales in Latin America grew "an exceptional" 18.5 percent to $24 billion in 2005. Febrafarma (Federação Brasileira da Indústria Farmacêutica), the Brazilian pharmaceutical umbrella organization, estimated the annual cumulative pharmaceutical market at $8 billion as of July 2005. However, according to Espicom, another business intelligence company, the market reached $10 billion in 2005, equal to $54 per capita.

Brazil began its pharmaceutical production in the 1930s, largely dependent on the import of chemical components. National production intensified during World War II, and by the late 1950s several foreign pharmaceutical companies began to invest in the country. Some 600 pharmaceutical companies operated in Brazil in the 1960s and 1970s. Due to the country's economic instability this number fell to 400 in the 1980s (Bermudez 1992, 1995). The international pharmaceutical industry welcomed the country's recent economic reforms and friendly drug-pricing policy. Between 1996 and 1999, the pharmaceutical business environment became more regulated, with the approval of the intellectual property law, the creation of ANVISA (Agência Nacional de Vigilância Sanitária, the National Health Care Monitoring Agency), and the regulation of generics (Cosendey et al. 2000; Luiza 1999). Currently, some 550 pharmaceutical companies (laboratories, importers, and distributors) operate in Brazil and compete for a slice of its lucrative market.

The Brazilian case is much in line with global trends. Consider this recent statement by another market research company: "Positive economic growth, stabilizing political structures, growing patient populations and increasing direct foreign investment in the emerging markets of Brazil, Russia, India and China (BRIC) are creating significant opportunities for pharmaceutical companies to expand into these markets and maximize future revenue potential. Pharmaceutical sales across the BRIC economies grew by 22.3 percent in 2005, compared to single digit growth in the major markets of the U.S., Europe and Japan."[17] By 2010, the developing world is expected to account for approximately 26 percent of the world pharmaceutical market in value, compared with 14.5 percent in 1999.

Dr. Radames, a Brazilian infectious disease specialist and adviser to the WHO, explained to me: "Pharmaceutical companies had already recouped their research investment with the sell-off of AIDS drugs in the United States and Europe, and now with Brazil, they had a new fixed market and even if they had to lower prices they had some unforeseen return. If things worked out in Brazil, new AIDS markets could be opened in Asia and perhaps in Africa" (personal communication, August 2000).[18]

Dr. Jones, an executive of a pharmaceutical multinational that sells ARVs to the Brazilian government, does not put things so explicitly but he asserts that "patents are not the problem. The problem is that there are no markets for these medications in most poor countries. Things worked out in Brazil because of *political will*" (personal communication, May 2003, emphasis added). In his company, Dr. Jones oversees ARV distribution in the developing world. He explained to me what he meant by "no markets," citing an anti-HIV drug his company launched in 1994:

> Demand generation and market creation is the standard thing the industry does, and by 1995, we were already making this antiretroviral drug available to African countries. We would, for example, partner with the French Red Cross in West Africa. We also created education programs for physicians. We sent physicians from Senegal and Burkina Faso to international congresses so that they could absorb new information. But this ended up doing nothing in terms of creating demand. Sales of the drug remained minimal in spite of the fact that, if you look at it from an epidemiological perspective, the patients were there, the disease was starting its rampaging ascent, but there was no demand for treatment. Of course, the issue of price came in, and we as company decided to look at pricing. The price for Western Africa could not be the same as in Europe and I think we launched the drug initially at 50 percent of the European price.

But that combination of high prices with insufficient demand, inadequate health care infrastructure, and a lack of barriers to generic drug imports was topped, argued Dr. Jones, by a "lack of political will." Here, "no markets" in Africa dovetails with local governments' lack of a holistic vision of public health, in which the public and private sectors work in tandem:

AIDS lays bare all of the inadequacies of a country's approach to public health. Unless the government recognizes this and addresses the totality of public health, you don't deal with HIV/AIDS. Step number one: HIV/AIDS is not the Ministry of Health's problem. In most developing nations, the Health Ministry is weak; it is not the place where important decisions are made. Unless all of the government's arms are fully aligned—health, education, transportation, etc.—AIDS programs will not work. This requires massive political will. We see an evolution in countries that have coordinated efforts, a strong national AIDS program, partnership with private sectors, and the country's leader supporting intervention.

Brazil recognized the impact of the disease immediately, Dr. Jones claimed, and "it also approached the problem from a multisectoral perspective." In his recollection (which bypasses the national government's initial disregard of AIDS), campaigns for education and destigmatization led to public dialogue, coupled with a changing vision of public health: "Health is not an area that the Brazilian government allowed to deteriorate anywhere near the degree of what we see in other developing countries. You had an existing structure of STD clinics and World Bank funding helped to strengthen the infrastructure."[19] In this rendering, Brazil's "political will" to treat AIDS coincides with the country's partnership with both international financial institutions and the pharmaceutical industry:

Different than in Africa, in Brazil we had a successful business with our first antiretroviral products. And we will continue to have tremendously successful businesses based on our partnership approach with the government. Brazil continues to be an example of how you can do the right thing in terms of public health, understanding the needs of both the private sector and the government and its population. The government was able to take advantage of existing realities. There was no intellectual property protection for our early products, and given Brazil's industrial capacity, they were able to produce the drugs.

I asked Dr. Jones how the pharmaceutical industry reacted to this strategy. "We were angry," he said. But rather than withdraw from Brazil, the company used the issue of pricing and generics to negotiate broader market access in Brazil. "The down side could have been 'why bother and continue to invest in Brazil?' But anti-HIV products are not the sole bread and butter of most companies. So from a portfolio perspective, any private company balances its specific activities vis-à-vis the entirety of what it is doing. This one sector was being affected but our company had been in Brazil for a long time and we continued to be ranked as a top company there. So we had to look at it in a much broader perspective than an action taken in one product category." He was adamant that sheer market calculation kept the company operating in Brazil.

The industry's capacity to neutralize and redirect any form of counter-reaction to its advantage is indeed remarkable. Just as big pharma has played a key role in setting global trade rules (through TRIPS, for example), it has also helped to shape the international health agenda. The advocates of the neoliberal reforms of the 1990s encouraged the participation of the private sector in resolving social problems. This discourse of corporate social responsibility did not translate into large-scale partnerships to eradicate disease among the global poor, though (Farmer 2003; Pogge 2005; Sterckx 2004). But it definitely enabled the private sector to enter the decision-making process at institutions of global governance, and from there to defend its vision and interests.

The growing participation of the private sector at the UN and the WHO, particularly in the field of AIDS in the 1990s, meant "obliterating decisions to allocate funds for treatment," Dr. Teixeira told me during a workshop on Global Health Governance in São Paulo in June 2005. He recalled that, in 1997, UNAIDS piloted a treatment project where the pharmaceutical industry subsidized anti-HIV drug prices: "There was already a growing activist pressure for AIDS treatments for all in need, and the Brazilian experience with the production of generics was getting international visibility. . . . But these pilot projects, which generally last a few years, were strategically used by governments and international organizations to postpone large-scale ARV rollouts." He mentioned Chile as a "shameful" example: "The government carried out this kind of pilot project as a way of not taking up the responsibility of AIDS treatment until 2002, when it asked for financial assistance from the Global AIDS

Fund . . . a first world-type economy taking funds away from poorer countries."

Moreover, "only in December 2001 did UNAIDS for the first time publicly announce its support for antiretroviral treatment for all in need." In the past few years, following the consolidation of the Brazilian policy and other treatment initiatives (by organizations such as Partners In Health, Doctors Without Borders, and the Clinton Foundation, for example), an international consensus has emerged over the feasibility of delivering ARVs to the neediest in resource-poor settings. The industry is again exercising its flexibility and turning these unexpected fields of medical action into market opportunities.

As I continued my interview with Dr. Jones, I told him that I had recently read a pharmaco-economic report on emergent HIV/AIDS pharmaceutical markets—namely, Brazil, Thailand, India, China, and South Africa—that argued that if these governments were to provide the simplest version of the "AIDS cocktail" to 30 percent of the affected populations at 10 percent of the current U.S. price, the industry would still profit an additional $11.2 billion. He refuted this idea of emerging AIDS markets in the developing world, evoking Africa and corporate philanthropy once again: "We will supply ARVs to Africa at low cost, there will be some demand, there will be increase in volume of products sold, but by definition it is not a market for us. . . . We know that the more we sell the more we lose."

By juxtaposing the arguments of both corporate actors and policy makers one can identify the logic of such a pharmaceutical form of governance. Here, political will means novel public-private cooperation over medical technologies. Once a government designates a disease like AIDS the "country's disease," a therapeutic market takes shape—a captive market. As this government addresses the needs of its population (now unequally refracted through the "country's disease"), the financial operations of pharmaceutical companies are taken in new directions and enlarged, particularly as older lines of treatment (generic ARVs) lose their efficacy, necessitating the introduction of newer and more expensive treatments (still under patent protection) that are demanded by mobilized patients. Patienthood and civic participation thus conflate in an emerging market. Development agencies (such as the WHO, UNAIDS, and the World Bank) assist this process, which has crucial ramifications for the nature and scope of national and local public health interventions. Magic-bullet approaches (i.e., delivery of technology regardless of health care infrastructure) are increasingly the norm, and companies are themselves using the activist discourse that access to ARVs is a matter of human rights. This *pharmaceuticalization of public health* has short- and long-term goals, as Dr. Jones puts it:

> We are seeing changes there where governments try to find out the role they can play in the field of health, health as a fundamental issue they need to deal with. At what point does it get to the government that today citizens put a huge premium on access to health? And it is not just a matter of guaranteeing access to the available medications but of the new ones being developed. If you don't have the capacity to produce this new medication, then you have to find a way to align yourself and partner and trade with those who are doing it. With a global disease like AIDS you must play together and not on your own. So, the question is not so much on what is the role that society or government or industry ought to play but what are the opportunities for further cooperation and partnership as opposed to unilateral actions on either side.

Decentralization and a Magic Bullet Approach

One of the unintended consequences of AIDS treatment scale-up in Brazil has been the consolidation of a model of public health centered on pharmaceutical distribution. The ARV rollout was implemented across the country through an ailing universal health care system. This specific policy was aligned with a pharmaceutically focused form of health delivery that was being articulated by the Cardoso administration. Indeed, Brazil has seen an incremental change in the concept of public health, from prevention and clinical care to community-based care and drugging—that is, public health is increasingly decentralized *and* pharmaceuticalized.

As part of a policy of rationalization and decentralization of assistance, in the mid-1990s the government began to recast the costly and inefficient basic pharmacy program whereby municipalities distributed state-funded essential medication to the general population (this program preexisted the ARV rollout). Provinces and municipalities were urged to develop their own epidemiologically specific treatment strategies and to administer federal and local funds in the acquisition and dispensation of basic medication (MS 1997, 1999; Wilken and Bermudez 1999; Yunes 1999). According to government officials, the policy would contribute to reducing hospitalizations (which tended to dominate state funding) and to making families and communities stronger participants in therapeutic processes. This program took root in key states, which then became models for other regions (Cosendey et al. 2000).

Overall, as I discovered in my fieldwork in the southern and northeastern regions, the universal availability of essential medicines has been subject to changing political winds; treatments are easily stopped, and people have to seek more specialized services in the private health sector or, as many put it, "die waiting" in overcrowded public services (Acurcio et al. 1996; Arrais et al. 1997). Local services can rarely plan alternative treatments because their budgets are as restricted as their pharmaceutical quotas. State plans and medical demand are uncoordinated. The flow of this universal and pharmaceutically mediated health care delivery is discontinuous.

Nonetheless, this is not the case within certain specific disease policies. Even though the responsibility for distributing medicines has become increasingly decentralized, the lobbies of both patients and the pharmaceutical industry kept the federal government partially responsible for the purchase of medication classified as "exceptional," as well as medication for diseased populations that are part of "special national programs," such as the AIDS program. The federal government approved a decree on pharmaceutical dispensation in 1995, in addition to incorporating an official list of drugs into the Health Ministry's budget. The content of the list most likely reflected the demands of interest groups. An increasing number of patients are filing legal suits and forcing regional governments to maintain the inflow of high-cost medicines. According to public health expert Jorge Bermudez (Bermudez et al. 2000), "an individualized rather than collective pharmaceutical care" is being consolidated. A critical understanding of the AIDS policy's success must keep in sight this mobilization over inclusion and exclusion as new markets, regulations, and certain forms of "good government" are being realized.

Brazil's trajectory toward specifically targeted medical interventions, especially those comprising a particular medical technology or treatment, mirrors the development of international health care methods throughout the late twentieth and into the twenty-first centuries. With the release of the Declaration of Alma-Ata in 1978, the WHO sought to refocus international public health efforts to remedy growing inequalities in health care (WHO 1988, p. 3). The principal resolution of the document and of the international conference that produced it was that "primary health care should be widely adopted as the cornerstone to health development" (D. H. Mahler, director-general of the WHO, address at Alma-Ata, September 6, 1978). The philosophy of "comprehensive primary care" (Beaglehole and Bonita 2004) was put forth as the basis for what subsequently became the WHO's central objective: "The attainment by all peoples of the world by the year 2000 of a level of health that will permit them to lead a socially and economically productive life" (Declaration of Alma-Ata 1978; WHO 1988).

But, in practice, international public health efforts in the decades since Alma-Ata have been characterized by targeted responses to specific threats, a type of primary care dubbed "selective" or "vertical" instead of comprehensive (Beaglehole and Bonita 2004; Das 1999; Castro and Merrill 2004). Indeed, Brazil is not alone in having adopted a magic-bullet style of intervention, as illustrated by the push to deliver vaccines to children worldwide, the campaign to fight malaria, and the effort to eradicate poliovirus across the globe. At the same time, basic public health issues are often unaddressed. For example, more than 40 percent of Africa's population (some 300 million people) has no access to safe water, and a tenth of all the diseases suffered by African children are caused by intestinal worms, which can be treated for $0.25 per child (Farlow 2007).

Quite often, public health policy makers prioritize targeted and expensive technological interventions, made possible through public-private partnerships. Such partnerships have proven somewhat effective in combating HIV/AIDS in select African countries (examples include the partnership between Merck and the Gates Foundation in Botswana and also Bristol-Myers Squibb's Secure the Future Initiative). As is the case in Brazil, these efforts have depended on the cooperation of a national governing body, NGOs, and industry. However, critics have complained that the strategies underlying new global health interventions are noncomprehensive and ultimately of poor quality. Many question their sustainability in the absence of more serious involvement by national governments and greater authority for international institutions to hold donors and partners accountable in the long term.

I asked José Serra whether the state had the capacity to address other large-scale diseases pharmaceutically. "Without a doubt. But the problem does not lie in government," he said. "The government ends up responding to society's pressure, and with AIDS, the pressure was very well organized. You must have a huge mobilization. See the case of TB. It is easier to treat than AIDS, and much cheaper. The major difficulty lies in treatment adherence. But you are unable to mobilize NGOs and society for this cause. If TB had a fifth of the kind of social mobilization AIDS has, the problem would be solved. *So it is a problem of society itself*" (emphasis added).

For Cardoso, too, the management of AIDS is clear evidence that politics have moved beyond the control of parties and ideologies: "There is no superior intelligence imposing anything . . . a party, a president, an ideology . . . but there are assemblages, alliances, strategies," he stated in our interview in May 2003. "Today, Brazilian society is much more open than people imagine and very mobilized. In reality, people do not live in a state of illusion as intellectuals and journalists generally think of them; they have learned to mobilize and know how to make pressure and activate those in congress with whom they have affinities."

This is also true for the pharmaceutical industry and its powerful lobby, I added. Cardoso replied, "Indeed, they also mobilize because there is a struggle going on. A bet on democracy leads to this kind of diversity. The government has to navigate amid all these pressures. It must set some specific objectives and develop directives to that end amid this confusion. It cannot just be on this or that side, it must more or less pilot." A national project, he concluded, "is not something that a few intellectuals whisper in the prince's ears. It is rather the fruit of a wide range of actions, discontinuous and even contradictory."

Public-Sector Science and the Production of Generic Drugs

A politicized science has fueled the Brazilian AIDS policy.[21] The strengthening of the country's pharmaceutical infrastructure has been fundamental to the sustainability of the ARV rollout. On several occasions in past years, the Health Ministry has deployed the country's generic ARV know-how to politicize the marketing practices of the pharmaceutical industry and has successfully negotiated drug price reductions (Galvão 2002b; Paraguassú 2001; Serra 2004). What we have here is not simply a wholesale intrusion of "market forces" into the state—the government has actively exercised its power of advocacy by investing in a once-ailing national pharmaceutical industry and by developing new manufacturing capacities.

Dr. Eloan Pinheiro, a chemist and former manager of a British pharmaceutical subsidiary, was until early 2003 the director of Farmanguinhos, the state's main pharmaceutical company that produced many of the generic ARVs for the national program (see Cassier and Correa 2003). In an interview in August 2001, she told me that public laboratories accounted for some 40 percent of Brazilian ARV production, and that her technological development division had already reverse-engineered two drugs under patent protection, which were now "ready to go into production if the government deems it necessary."

For Dr. Pinheiro, Brazil's patent legislation is "just wrong": "It makes the country dependent on imports, and hinders local scientific and technological development." In the years following the country's 1996 industrial property law reform, Brazil has only requested seventeen pharmaceutical and biotechnological patents, representing 1.4 percent of the world's total requests (the United States had 46 percent, England 13 percent, and Germany 10 percent—Bermudez et al. 2000). Dr. Pinheiro is resolute in her support for state industry: "Nobody can negotiate price without challenging a patented product. We don't want to compete with richer nations, but we hope to reach a stage of independence."

Given that the production of medication in Brazil "has been a multinational business" since the 1950s, said Dr. Pinheiro, she was not totally surprised when she learned that the state's top laboratory had reduced production to three basic drugs by the time she took over as coordinator of Farmanguinhos in the late 1980s. In her work with the British multinational, she had learned much about drug development and production, "particularly, how to integrate adequate local materials into the drug's manufacturing." She also developed a keen understanding of the market maneuvers that keep drug prices high: "I saw how much fat was put into the products and that the final prices didn't correspond to research expenses at all, it was huge profit, period." After mentioning her antimilitary activism in the late 1960s, Dr. Pinheiro said that she had always wanted to see Brazil become "a stronger country, incorporating technology."

As Dr. Pinheiro denounced unfair market tactics, she also spoke of the social-mindedness and creativity of local science: "The multinationals must become flexible, and we must all deal with the question of whether new technologies are going to benefit man or exclude him from the possibility of surviving. Justice and equity ought to be defended amid globalization." Dr. Pinheiro dismissed criticisms that her way of doing science was sheer copying: "We had to develop our own methods of analyzing the drugs. I traveled to China and India to learn techniques and to buy salts from them. Sometimes, if we want the species to survive, we have to regress from some advanced logics that are in place." Indeed, Farmanguinhos plays a key role in the acquisition of knowledge on these drugs, as sociologists Maurice Cassier and Marilena Correa put it, "which it can then transfer either to Brazilian public-sector laboratories or to private-sector pharmaceutical laboratories in Brazil and, in the future, to other countries in the South" (2003, p. 91).

Interestingly, Dr. Pinheiro did not speak of AIDS activism as key to the country's ARV rollout. She credited "efficient managers," both in government and in science, and the mobilization of experts. During Cardoso's first presidential term, Dr. Pinheiro had been called to Brasília to discuss strategies for drug development. She immediately noted "seriousness and signs of efficiency." In her negotiation with the government, she ensured that Farmanguinhos would not simply become "a factory," but "a center for technological development": "We wanted to produce, to sell to the state and then reinvest the profit in technological development with an eye toward endemic diseases." In 2001, she had some six hundred people working for her, of which only one-quarter were paid by the federal government.

An astute and pragmatic manager, Dr. Pinheiro knew how to confront or bypass strict public administration rules to make things happen. Soon, critics, envious of her success (as I was told by local AIDS activists), began to accuse her of administrative irregularities (in the purchase of materials and in the outsourcing of services, for example). The directors of other state laboratories also resented the fact that they had to keep producing low-price medication, such as aspirin, for example, and could not improve their overall infrastructure.

The fact is that Dr. Pinheiro revolutionized the laboratory and made production happen, not just efficiently, but with high quality. Under her administration, Farmanguinhos increased its production to sixty-eight drugs, most of them aimed at treating diseases such as TB and Hansen's disease, "which are treatable but are of no economic profitability to the multinationals."

The ARV policy is emblematic of a new kind of "state-market integration," added Dr. Pinheiro during our interview, "the realized vision of Minister Serra," whom she described as "a fearless economist with the ability to make the right decisions." The right decision here meant fostering the local production of generics. As epidemiologist Pedro Chequer told me in an interview at the Health Ministry in January 2000, "When we began to produce drugs the international prices went down. Autonomous production is key to the sustainability of the policy." Chequer, who had been coordinating the AIDS program for over four years, said that "the AIDS policy is evidence that there are concrete ways to fight against this pervasive dismantling of the state that is in process." And he added: "Some people might say that this kind of discourse is a remnant of the past, but I am speaking of a possible future."

•••

These realities are not the outcome of a simple progression, nor are they absolutely new. The Brazilian AIDS policy is reminiscent of the much-celebrated technical control of yellow fever, the bubonic plague, and smallpox in Rio de Janeiro from 1903 to 1906, under the leadership of scientist-administrator Oswaldo Cruz. In 1900, after returning from his microbiological studies at the Pasteur Institute in France, Cruz was given a position in the national Serum Therapy Laboratory, later named after him. The laboratory's growing technological and research base and its rise to international fame with the discovery of Chagas disease were closely connected to Cruz's successful public health campaign, which played a crucial role in revitalizing Rio de Janeiro's economy. As historians Nancy Stepan (1976) and Ilana Lowy (2001) have shown, the need to halt the disease in the capital was pressing not only to improve Brazil's public image abroad but also to guarantee the flow of much-needed immigrants and investments.

Cruz's sanitation program tested the nation's ability to be part of the modern world; it replaced the traditional solutions of fumigation and disinfection with a large-scale and expensive plan that included systematic extermination of *Aedes aegypti* mosquitoes, patient isolation, compulsory vaccination, and treatment with serums. Much of the efficacy of this short-lived intervention lay in strong federal financial support; the integration of laboratory science, experimental medicine, and public health; and its urban targeting.

Success came with recognition that the majority of the population—the rural poor, for whom disease was a constant reality—remained unaccounted for. Strong public pressure mounted by leading intellectuals and health professionals finally led to the creation of a national sanitary movement in 1918, which two years later became the national Department of Public Health (Hochman 1998).

The rational-technical control of AIDS that I have been discussing follows this pattern: it integrates science into public policy and experiments with circumscribed populations, its economic sustainability in question. Like Cruz's program, the AIDS policy brings social exclusion into view and has the potential to redirect the work of state institutions.

•••

At any rate, in 1999, 81 percent of the government's expenses with antiretrovirals went to multinationals and only 19 percent to Brazilian companies; in 2000, 41 percent of expenses were already going to national laboratories, both public and private (Cassier and Correa 2003; CN 2001a; Galvão 2002b). By 2001, Brazil was producing seven of the thirteen antiretroviral drugs used to treat people with HIV/AIDS in the country. And as the Brazilian policy created a market for generic drug components, it also raised international competition that led to an overall decrease in drug prices. In 1999, for example, 2.2 pounds of 3TC, an anti-HIV drug, cost $10,000; in 2001, the same 2.2 pounds would sell for $700. The threat to issue compulsory licenses has proved successful. In 2002, Brazil was able to obtain a 40 to 60 percent cost reduction on purchases of patented components from Merck and Roche that are essential to the production of the AIDS cocktail (Levi and Vitória 2002, p. 2379).

Despite all the differences the government and the pharmaceutical industry have regarding patents and generics, there is "a friendly relationship at work," Pedro Chequer told me in 2000: "We have a professional way of dealing with the industry, so there is mutual respect. We already called

Merck and told them that their market alternative is to pass technology onto Farmanguinhos and produce in association with us. The industry must find other ways of making money. If it wants to keep the market, the industry has to produce with us. It must also learn to think in the medium-term perspective." In the meantime, Merck and Glaxo (now GlaxoSmith-Kline) were helping the national program produce educational materials addressing ARV adherence, said Chequer. "Adherence implies therapeutic success, and this is good for the industry, for us, as well as for the patient."

After carrying out interviews at the national AIDS program in August 2000, I flew to Rio de Janeiro, where I visited Glaxo's regional headquarters. I talked to Peter Rich, the company's director of social marketing programs. He told me that the company was helping AIDS NGOs throughout the country in making adherence to ARVs a central component of their work and was also supporting regional and national events organized by HIV-positive people. According Dr. Rich, "Before 1996, Brazil was a very timid market for us, basically a market of private buyers. Access of ARVs was limited. Universal availability changed the scenario, and this new market makes it possible for us to do something other than simply selling."

While Glaxo's personnel were training doctors and patients to make the best of the medication and improve "quality of living," the company was also intensely engaged in public debates over drug quality and safety: "All these salts coming from India, all labs must be updated to test these drugs; if not, the patients are in danger."

Here, unexpectedly and out of constraints and imagination, global market logics and the politics of science and technology are forced into explicitness, becoming a new and productive field of tension and negotiation. The work of this "peripheral science" to fuel political alternatives and human advancement through reverse engineering and generic production has been relatively successful. The crucial move now is to develop infrastructure to be able to break technological barriers, Serra explained. "The strategy of the drug companies will always be to keep the prices high as long as there is a technological barrier. When the barrier falls they will lower their prices so to make the copied products nonviable and thus keep the market. With some medication, even if you have the formula and you issue the compulsory license you cannot fabricate it because of lack of specific technology" (personal communication, June 2003).

Inside the Brazilian state, this pharmaceutical activism has occasioned the creation of ANVISA, a regulatory agency along the lines of the U.S. Food and Drug Administration (FDA). The agency replaced a department within the Health Ministry that was rife with corruption and subject to unceasing political pressure; it became, according to Serra, "an essential ally in the tug-of-war with the pharmaceutical industry." It remains to be seen how in the long run this and other new regulatory bodies will manage a powerful corporate lobby and protect the interests of citizens.

Meanwhile, this constant negotiation with big pharma over drug price reduction has discouraged national companies from investing in ARV generic production. As Michel Lotrowska, a Doctors Without Borders activist working on treatments for neglected diseases in Rio de Janeiro, told me, "It would have been good if at least one patent had been broken during Serra's term as minister. National companies were always asked to produce and then in the last minute, the government settled with big pharma. These small companies know they are used by the government as a bargaining tool. After this happened four or times, they stopped doing reverse engineering. It is also too risky for these companies to invest all their money in ARV production, and then have big pharma coming back with a lower price without the government being committed to buying from them" (personal communication, June 2005).

Amid these debates over technological barriers and despite all the statistics of therapies delivered and money saved, even Glaxo's social program director was well aware of the ways the country's historical inequality shaped the AIDS policy on the ground: "What we see in the field is that the epidemic keeps growing among the poor. Now you have people living in the streets who have no food, but they have ARVs. It is complicated, especially because some drugs have to be taken with food."

Scaling-Up

Multiple institutions and social actors—both national and international—meet in the Brazilian AIDS policy-space. In weaving together the accounts of people directly involved in the making of the policy, I got a clearer picture of the dynamic political framework that is emerging against equally dynamic global economic developments. At the intersection of a "technological surprise"[22] (combination ARV therapy), social mobilization, and the restructuring of both state and market operations, the following are taking form: a new political economy of pharmaceuticals with international and national particularities, a pilot population through which a reforming state realizes its vision of scientifically based and cost-effective social action, and a contingent of mobilized individuals articulating a novel concept of patient-citizenship. These multiple institutions and actors have distinctive interests, are somewhat permeable, and mutually readjust.

In practice, the AIDS policy is neither a global institution nor a novel state apparatus—it is an *intermediary power formation*. The policy comes into existence in the space between international agencies, global markets, and the reforming state. It is implicated in and meddles with the resources of these institutions as it struggles to intervene effectively. Intermediary power formations are not simply extensions of the macro or the micro—they actually exclude the immanence of both. Their operations do not follow a predetermined strategy of control and do not necessarily have normalizing effects. As evident in the AIDS policy, their sustainability has to be constantly negotiated in the marketplace. Mobilized individuals and groups must continuously maneuver within this particular therapeutic formation to gain medical visibility and have their claims to life addressed. The AIDS policy thus becomes a co-function of political and market institutions, as well as individual lives. In this way, the model policy is a miniaturization of a "global politics of survival" (Abélès 2006a).

Dr. Teixeira, the former national AIDS coordinator, told me that "234,000 hospitalizations for opportunistic diseases have been avoided, saving us more than $700 million in medical assistance." Here, human rights are biomedical rights that the state must protect. Through this commitment the market (in this case, the pharmaceutical market) is moralized. Dr. Teixeira explained: "In the international economic field, there is a prevalence of unjust and restrictive rules, but nationally, we see the universal values that ground public health and also the defense of the individual's rights to life."

The Brazilian AIDS policy thus conveys a sense of political participation and responsibility and has become an efficient vehicle in shaping a perception of the reformed state as open to the public, rational and coherent, efficient and ethical. Moreover, says Dr. Teixeira, "new possibilities for developing countries to intervene in global governance were opened through AIDS" (personal communication, June 2005). Internationally, Brazil is presented as evidence that the direly needed full-scale assault against AIDS is indeed possible. Consider this statement by Peter Piot, executive director of UNAIDS:

> The only way the epidemic can be reversed is through total social mobilization. Leadership from above needs to meet the creativity, energy, and leadership from below, joining together in a coordinated program of sustained social action. . . . AIDS is an emergency, but it is a long-term emergency. We are facing the most devastating epidemic humanity has ever known. Our response must therefore be equally unprecedented: the most concerted, sustained, coordinated, full-scale assault on a disease the world has ever known. (The Global Strategy Framework on HIV/AIDS 2001)

Here the "unprecedented" solution to the world's worst health crisis emerges not from privileged contexts or out of universal principles but from a desperate reality and a "political will" to redirect ostensibly inflexible state and market logic toward equitable outcomes. As Dr. Teixeira put it, "We have a just cause and international public support." A 2002 WHO document entitled "A Commitment to Action for Expanded Access to HIV/AIDS Treatment: International HIV Treatment Access Coalition" states that "Brazil shows the way," as it has the most advanced national treatment program in the developing world. Among other cost-effective benefits, the report notes that survival of patients diagnosed with AIDS increased from eighteen months in 1995 to fifty-eight months after the introduction of ARVs. The viability of the Brazilian policy "owes much to effective social mobilization, including representation of affected communities in government, NGO and other fora."

There is now political and economic evidence that a full-scale ARV rollout in both poor and rich countries is possible—but well-designed and sustained international support is crucial, argues Dr. Teixeira:

> We believe that the past objections to HIV treatment in developing countries are not persuasive anymore and there are strong arguments in favor of a widespread treatment access effort. A considerable amount of evidence suggests that an effective AIDS treatment is now possible

even in low-income countries. Reducing prices of antiretroviral drugs can dramatically alter the economics of HIV/AIDS treatment, and possible obstacles to adequate treatment such as poor infrastructure can be overcome through a well-designed and supported international effort. We believe that on moral, health, social and economic grounds, the international community should provide the scientific and financial leadership for a scaling up of AIDS [treatment] in the rich and poor nations of the world.[23]

The ARV scaling-up program that Dr. Jim Yong Kim has initiated at the WHO (aimed at treating three million living with AIDS in poor countries by 2005) requires a new form of governmentality in which national governments bridge the existing disconnect between affected communities and international (pharmaceutical) lifesaving efforts. Here, civil society and science (broadly speaking) are interacting through and beyond states and older paradigms of international health and development.

In global public health and AIDS activist discourses, national governments are now commonly portrayed as a kind of middleman mediating the work of science-minded international organizations and the interests of local communities. The reality of AIDS, in addition to related political-economic and technological developments, confronts governments with a responsibility to develop AIDS policy and to act pharmaceutically in order to avoid accusations of human rights violations, much as the complicated South African management of AIDS reveals (see Fassin 2007; Fassin and Schneider 2003; Epstein 2003, 2005b; Gauri and Lieberman 2006). Filling this policy-space adequately is seen as a triumph of the interaction of democracy and science in contexts of stark inequality and foretold death.

For Dr. Teixeira, who has actively participated in the design of the "3 by 5" initiative, the UN still has a key role to play if this "new era of global public health" is to become a reality for the poorest. "We are now seeing a discursive shift toward making antiretroviral therapies available to all in need, but we are lagging behind in terms of implementation. Where is the money for medication? At the UN, like in other global institutions, rich countries and the private sector always decide things in their favor or withdraw their donations. . . . Only through ceaseless activism and political pressure from the southern bloc can a thorough revision of financial and decision-making mechanisms take place." These developments need to be followed carefully as they redefine the new regime of pharmaceutical accessibility, the terms of public health today, and, more broadly, the shape and scope of neoliberal political action.

The Pharmaceuticalization of Public Health

The transformations of the state and of the concept of public health that my political, scientific, and activist informants emphasized look rather different at the margins.[24] In my ethnographic work in the city of Salvador, Bahia, I observed many poor AIDS patients extending their lives with free access to ARVs. These patients worked hard to keep philanthropic, nongovernmental, and medical support in place to guarantee the effectiveness of the therapies. Fighting for food and housing is concurrent with learning new scientific knowledge and navigating through laboratories and treatment regimes that now coexist with scarcity.

However, although some marginalized AIDS victims exercise their will to live and acquire a form of patient-citizenship, many others remain epidemiologically and medically unaccounted for and die in abandonment. As I discuss in chapter 3, I performed an analysis of AIDS death certificates in the AIDS unit of the state hospital in Salvador and found that only 26 percent of these cases were actually registered by the epidemiological surveillance service. More than half of the total of cases analyzed died during the first hospitalization in the unit. For the poor and homeless sufferers in this hidden epidemic, there are no specific programs other than sporadic charity.

After I told all this to former president Cardoso, he responded that "in the social arena, the main problem is not money" and emphasized that during his administration, spending in social programs went up, from 11 to 14 percent. "The difficulty lies in reaching the neediest, and this is a political question."

Cardoso's remarks immediately reminded me of philosopher Renato Janine Ribeiro's acute observation that in Brazil's dominant political culture, the categories *social* and *society* do not pertain to the same people and worlds of rights: "social refers to the needy, and society refers to the efficient ones" (2000, p. 21). Neoliberal political discourses, Ribeiro argues, transmit the conviction that society is active as economy and passive as social life: "The objects of social action are assumed not to be able to become an integral and efficient member of society" (p. 22). Such political imagination "aims at the end of the social in order to emancipate society" (p. 21).

At any rate, Cardoso then mentioned that "the middle class appropriates most of the government's social programs" and that there is still a generalized lack of local managerial capacity, "a competence to carry out the programs." The real political problem, however, as he sees it, lies within regional governments that are "not yet successfully modernized." He explained that local states "hinder" the alternative social policies that his government developed in tandem with organized civil society, particularly in the areas of health and education: "Where is the highest degree of governmental corruption these days? At local levels. Local infrastructures of governance are not yet tuned with the new structure of the Brazilian state. Things are improving, and they vary from place to place. In the south and southeast regions, for example, you already have reformed administrations, active mayors, a dialogue with society, and less rigid party politics."

Cardoso trusted that it was "a matter of time" before local "corrupt and charity-oriented" governments caught up with the country's transition: "In twenty, thirty, or fifty years Brazil will be like the United States." This sounded to me like another variation of the "country of the future" story (Zweig 1941). For whatever reason, Cardoso was unwilling to take local governmental infrastructures and actions as contemporaneous and contiguous with the forms of governmentality created during his administration. He was also unwilling to question the inexorability and desirability of "becoming like the United States."

In a slip of the tongue he said, "When a mayor comes to talk to the president . . . No, in reality the president does not see mayors . . ." and did not finish the sentence. He was trying to say that party affiliation no longer determines the allocation of state funds. I was intrigued by the paradox he raised: the federal government was locally present through medication, as in the AIDS policy, but there was also a formalized distance between the state and local worlds. And an implicit assumption, I thought, that in the new political game, it is, after all, up to the market or to interest groups to level out poverty and inequality.

Over time, the AIDS policy has indeed acquired important political currency inside Brazil. As the policy reaffirmed a universal public health care system and even the politics of pricing, it promoted a new "state voice," as former president Cardoso put it. In the meantime, however, a new kind of state distance from the local battlefields over inclusion has been formalized as well. Unfortunately, as far as AIDS is concerned, this has not changed in the current government, I was told by disenchanted activists in June 2005. One wonders about the sustainability of the AIDS policy. In spite of local generic production, more than two-thirds of the country's ARV budget goes to foreign manufacturers, and AIDS expenses take up much of the state's budget for medical treatments. Recently, the dispensation of some ARV drugs has been rationed in the state of São Paulo, which concentrates most of the country's AIDS cases (Folha online, February 24, 2005). The Health Ministry has also opened up a medical discussion of cases in which treatments could be discontinued in case of medication shortage (*Folha de São Paulo*, April 14, 2005, p. C1).

Moreover, regional governments have been forced to alter their public health budgets drastically to accommodate the growing judicial demands for high-cost medicines by patient groups formed around chronic diseases and rare genetic diseases, for instance. Patients follow the path opened up by AIDS mobilization, with the exception that many groups are now supported by the pharmaceutical industry.

In August 2004, anthropologist Adriana Petryna and I spoke to medical researchers at the Clínicas Hospital in Porto Alegre.[25] Led by Dr. Paulo Picon, these researchers critically review the available pharmaceutical and medical literature and show that in many instances the efficacy of new and extremely expensive treatments is not much better than previous and cheaper ones. They are also proving that lower doses of some of the new drugs are equally efficacious and are writing alternative clinical protocols. Dr. Guilherme Sander spoke of what the research team was up against:

We doctors are being asked by mobilized patient groups to provide them with the best treatment. And this means what is being tested in clinical trials. Yet, what is being tested is not and will never be what the majority of our population needs. So the government ends up giving out lots of drugs, but the neediest areas remain unaddressed. The pharmaceutical companies decide what will be medically tested in the developing world, and their interests end up defining public health priorities and budgets. This is a central problem in public health today. A great many people are being further marginalized, without access to vaccines and prenatal care, for example. And addressing these questions would, in my view, solve more problems than we are currently solving by using the last line of drugs.

Dr. Picon joined the conversation and spoke of how "deeply ingrained" the pharmaceutical industry is in the Brazilian public health system: "Even when a drug has not yet been approved by ANVISA, judges side with patients and force regional governments to provide the drug. In many case, judges state that drug approval is too slow and base their decisions on a medical report saying that without the drug the patient will die. But that report is based on industry-produced evidence. That's why we think it is so important for us to produce alternative clinical evidence."

The Clínicas research team was working closely with local public officers to ensure that their alternative evidence might have some public health value. We also learned that state prosecutors had recently created a national association to address this problem explicitly. One of the prosecutors we talked to said that, in order to find some minimum guidelines to assess fairly the growing judicialization of health, she and her colleagues must "critically engage commercial science" and "reinterpret what universality and equity constitutionally means."

In sum, as I have argued throughout this chapter, the Brazilian AIDS policy is an intermediary power formation, representing a shift from a crumbling welfare state to an activist state; from public health understood as prevention and clinical care to access to pharmaceuticals; and from political to biological-based rights. Novel scientific and regulatory practices emerged to counter the effects of an ever-expanding pharmaceuticalization. "To disentangle business from medical science and to discuss publicly drug development and pricing are important first steps as we engineer negotiation power," added Dr. Picon. "If we don't find intelligent ways to counter this profit extraction of public health we will be left with an insurmountable indebtedness, a wound that won't heal."

In what follows, I explore the institutionalization of AIDS in urban poor settings. I chart the concrete ways in which politics and personhood are being reconfigured through the rational-technical management of AIDS. As AIDS activism migrated into government, local AIDS care networks emerged and helped consolidate a new form of patient-citizenship, even as social inequality became further entrenched.

Chapter Two

Circuits of Care

How Has AIDS Activism Changed?

"The success of the AIDS program is a consequence of activism, of both affected communities and health professionals . . . an activism of civil society and of government," stated Dr. Paulo Teixeira, former national AIDS coordinator and World Health Organization (WHO) adviser, in a workshop on "Global Health Governance" in São Paulo, June 2005. In the past two decades, Dr. Teixeira has championed a comprehensive response to HIV prevention and care grounded in human rights principles and in the right to health care guaranteed by the Brazilian constitution.

The Brazilian AIDS policy-space, as I have shown in chapter 1, came into being against the background of neoliberalization, within a government eager to rationalize and minimize public services. In the early 1990s, leading AIDS activists turned to government and, together with public health professionals, constituted a central command that mediated international monitoring efforts and local demands for intervention. In partnership with nongovernmental organizations (NGOs) and regional AIDS programs and informed by the basic principles of solidarity and the right to life, this central command defined normative areas of policy making and ways of efficiently allocating the new resources made available by the World Bank and the Brazilian government. Beyond confrontation, interest groups and the state cooperated and reciprocally adjusted. In marrying medical and economic concerns, this AIDS mobilization acquired transformative power.[1]

It does not escape Dr. Teixeira that as "activism became professionalized and perfected itself internationally," it has also enhanced the administrative capacity of the reforming state. That is, "what has maintained the Brazilian state over the past ten years in most areas of government, but mainly public health, has been the outsourcing of services. Did bad things happen in the process? Yes. But without outsourcing there would not have been advancements either. Evolution is never unidirectional—it is forward and backward. We hope that it is two steps forward and one backward."

During our conversation, Dr. Teixeira dismissed the usual critique that dependence on public money necessarily leads to the political co-opting of social movements: "We need to reframe the discussion from social movements being dependent on the government to their *right* to use a certain percentage of public funds . . . and this kind of thinking can take us farther ahead."

A 2002 report by the national AIDS program also celebrated these "joint actions between civil society and state" that have unleashed novel public health measures (MS 2002). By then, the national AIDS program had partnerships with local governmental AIDS programs in all 27 states as well as 150 municipalities, and it was collaborating with 24 major industries. Between 1998 and 2001, the program financially supported the work of 686 civil society organizations, distributing some $30 million to finance 1,618 projects, which were mostly focused on prevention. These partnerships, says the report, "greatly value the citizenry's responsibility and commitment to the public sphere," giving them problem-solving worth. They are said to be more creative and efficacious than governmental interventions, as well as more capable of promoting social justice, that is, of "reaching disenfranchised groups and individuals at higher risk of HIV infection who search for inclusion in public life" (p. 21).

But how, in practice, do these public formations locate the marginalized, situate them in a common forum with others, and encourage them to speak and care for themselves and others?

During my field trip to Brazil in June 2005, I talked to several AIDS activists who were frustrated not with the national AIDS program—which audaciously continues to work toward sustainable antiretroviral (ARV) rollout—but with the "industrialization of nongovernmental work," as Gerson Winkler, former director of an AIDS NGO and of the AIDS program of the city of Porto Alegre, put it. "The social movement has been swallowed by the government. Street and juridical militancy is now replaced by computer-based chats and petition signing—it is truly virtual. Moreover, the control of AIDS became a big market. Thousands of people are employed by AIDS NGOs and work as consultants for the government. How many AIDS NGOs have closed in the last decade?

Very few. . . . Something feeds them, and don't tell me that it is solidarity" (personal communication, June 2005).

Conflicts of interest abound, says Winkler, whose critique has been discredited by other activists as anti-governmental and retrograde. "There is a true promiscuity in this relationship between social movements and the state. You don't know who is who and what is what. You have a researcher paid by the AIDS program working in an NGO studying the efficacy of a harm reduction program . . . for whom are the results written?"

Michel Lotrowska, an economist working for Doctors Without Borders in Rio de Janeiro, agrees that AIDS NGOs have become increasingly run by professional advocates: "You see fewer and fewer HIV-positive poor people in these spaces. There is now a huge gap between the kind of mobilization taking place and the AIDS the poor experience" (personal communication, June 2005). There is little doubt that social mobilization facilitated a change in the treatment of AIDS and the discourse surrounding it. But social mobilization around AIDS is not as coherent and steady as policy makers and reports portray it—it cuts across class lines.

AIDS activists have long deliberated over the question of who could represent poor AIDS victims and how that representative could do so. Sociologist Amélia Cohn, for example, says that contrary to other epidemics and endemic diseases, AIDS did not initially configure itself as a "disease of the poor" in Brazil, and this, she argues, "might explain the privileged space that AIDS occupies in public health and in the state's social policy at large. This might be one of the factors, besides its lethality, that explains the social impact of AIDS among us and the place it occupies inside the health sector, in spite of the serious crisis it is undergoing, a crisis of fiscal nature, but also of governmentality. AIDS programs are particularly exposed to accountability and are objects of specific interventions and normativity, in spite of the inhospitable state of the social domain" (1997, p. 51).

Former health minister José Serra is more explicit in naming what, according to him, made AIDS something other than a disease of the poor: "The homosexual community has not only been an active social force, but has contributed decisively to formulating, implementing, and ensuring the success of the anti-AIDS strategy." When asked to elaborate further on this, Serra suggested that the gay community was able to build on what he calls "failure visibility effect": "In no other disease does the failure of treatment galvanize so much attention, debate, and moral indictment for governments and societies as in AIDS" (personal communication, June 2003).

Both Winkler and Lotrowska are well aware of the real difficulties that AIDS activism faces in retaining its connection to problem solving in a world filled with poverty, inequality, and violence, a world in which institutions of care are disintegrating. Emergent realities that accompany the ARV rollout, particularly those that affect the poor, remain unaddressed, says Lotrowska: "Why, for example, has there never been a mobilization in Brazil for a 3 in 1 pill [to combine the first-line drugs AZT, 3TC, and Nevirapine in one pill and thus to facilitate adherence]? I have a simple answer: the well-educated urbanites, [they] still mobilize, but for the fifth line of treatment . . . who cares about better medication that would benefit the rural poor who are now getting infected?"

For Winkler, too, AIDS mobilization has now become a kind of "parallel polity" that is largely concerned with financial flows and the sustainability of the technological gains in place, as well as the new ones on the horizon. Meanwhile, the everyday battlefields over survival remain unconsidered. "It is indeed very difficult for the poorest with AIDS to enter the hermetically closed world of AIDS NGOs," states Winkler.

I heard a similar critique from poor AIDS patients in Salvador. As Luis Cardoso, Caasah's office assistant, put it, "There are many NGOs that speak in the name of the neediest but who don't help with what is actually needed . . . some food, a job. We don't know where their money goes." AIDS NGOs had not, for example, been able to help poor patients organize to demand free transportation tickets, added Luis, a demand met in a few other cities.

"How," asks Winkler, "have we let the activism of poor AIDS patients be reduced to begging?" Yes, he says, it is a matter of addressing people's real needs, but it is also a matter of being willing to incorporate the practical and difficult knowledge of disenfranchised AIDS subjects into policy and politics at large:

All these prevention materials that have been produced . . . look at the content and format, from 1988 to 2005. . . . It looks pretty much the same. The discourse of solidarity is the same, the norm is the same: use condoms, don't share needles—the imperative tone is the same. There is such a vast and tense history of all people who had some form of engagement with these institutions, goods, and services, people who lived and died trying to access them. . . . The principles guiding NGO action remained the same . . . but where is the plurality of local perspectives and experiences?

In this chapter, I try to understand what is happening to the social mobilization around AIDS in Brazil. In particular, I explore how civil society groups came to set the AIDS policy agenda, to expand zones of accountability, and to act as the primary executors of services. I also consider how this mobilization weakened certain inequalities and created new ones. In contrast to the old form of corporativist social mobilization and politics, this new variety places mobilization and politics in the marketplace. Various nongovernmental, pastoral, and medical networks now link, through biotechnology, the world of marginality and the state. Larger interventions are experienced locally through landscapes of triage and care.

There are many ways to assemble and edit these complicated, nonlinear developments into a story. I have chosen to focus on how poor and marginalized AIDS patients in urban settings have been addressed by the AIDS policy, how they have engaged it, and how their lives have transformed over time. By concomitantly analyzing the ways experts define publics and the lives forged by subjects fighting social extinction, one sees a different institutionality of AIDS, a more nuanced picture of emergent political cultures and value systems. On the ground, one finds discontinuities and awkward connections among institutions, people, money, and knowledge, comprising a general non-intelligibility that this study considers as both a reality and a variable (see Comaroff and Comaroff 2003; Tsing 2005).

The fusion of AIDS mobilization with governmental, private, and international initiatives into a new public health policy has redefined the form and scope of governance. Through the AIDS policy, past victims of social exclusion have an unprecedented opportunity to claim their human and medical rights. In the poverty-stricken settings where I worked, I saw AIDS sufferers engaging in a range of social and medical exchanges in order to be seen by AIDS NGOs and state institutions and to guarantee their own existence. For many, the AIDS policy was becoming the state, so to speak, with people participating in all kinds of administrative and medical exchanges in order to become "AIDS citizens"—to be linked to the national and local branches of the AIDS policy. The processes at work here are not easily reduced to mere access to AIDS disability benefits, condoms, therapies, or food baskets. They also involve personal transformations and new relations. Circuits of care are emerging amid the reality of AIDS, everyday violence, and rational-technical interventions.

By approaching the circuitous paths through which these socially abandoned AIDS patients negotiate their condition, one begins to illuminate present-day fields in which life chances are forged—the crucial economic and moral significance of care. Furthermore, by following this analytical route, one raises an array of micro-political questions concerning the implementation and redirection of Brazil's AIDS policy. The alternative approach—a strict institutional analysis of the policy—would leave invisible these important elements and life pathways.

From Passion to Politics

In 1992, ARCA (Apoio Religioso Contra à AIDS, Religious Support Against AIDS), one of Brazil's first AIDS NGOs, asked me to travel through several regions of Brazil and produce an account of the new social initiatives undertaken to control AIDS. Financial, administrative, and medical matters were changing with respect to AIDS in the country. My task was to chronicle the various trajectories through which the AIDS policy was locally operationalized, the integration of grassroots organizations, the possibilities opened, and the forms of civic imagination and human agency emerging through this new political machine.[2]

I began my journey in May 1992 in Porto Alegre, a city of approximately 2 million people in the southern state of Rio Grande do Sul, a well-known stronghold of the Workers' Party (PT, Partido dos Trabalhadores). Local contacts referred me to Gerson Winkler, who was then president of the city's main AIDS NGO, called GAPA (Grupo de Apoio à Prevenção da AIDS, the AIDS Support and Prevention Group). The first GAPA was founded in 1985 in São Paulo, and by 1992 it had eighteen affiliates across the country. Upon my arrival in Porto Alegre, I stumbled across a local newspaper's headline: "Giant condom broke over

Gerson, 1995

the chimney of the old Gas Power Plant [now a municipal arts center]."
Asked about the apparent failure of that latest AIDS-related cultural
event that GAPA had produced, thirty-three-year-old Winkler remarked:
"This gave us an opportunity to rethink this sort of cultural activism à
la first world. So much time and energy invested in safe-sex education
turns into this."

Winkler interpreted my very presence at GAPA as a sign of changing
times. "To have an anthropologist here is already part of the bureaucra-
tization of our activism." In our first conversation, he emphasized that
an understanding of marginality and spontaneity was key to the activism
he produced. Jargonized representativeness and "depoliticized missions
of NGOs" would be one of his major preoccupations. "As far as I'm
concerned, knowledge of AIDS is produced by transvestites, prisoners,
whores, hustlers." Winkler was blatant and unapologetic at every turn,
speaking against "academic power" or the power "to co-opt fucked-up
experiences" and fix them into social scientific wholes.

It was not a question of being anti-intellectual, he said, but of being
cautious with the ways people from research institutions and other social
movements, "the new professionals of AIDS," were taking up the epidemic as their subject and cause and framing all that comes with AIDS as
a "technical problem." Moreover, said Winkler, "these people are not living with HIV/AIDS and are starting to speak 'in the name of' . . . but one
can never speak from the same place of the Other, and there are always
personal interests to one's politics."

Winkler refused to be victimized by HIV/AIDS and to become a representative of AIDS victims. "I live aggregated to the present," he often told
me. A social worker by training, divorced and with two daughters, he had
come out as gay in the early 1980s. He had grown up in a working-class
district of Porto Alegre "in a family of alcoholics, never well-structured
but not evil either." Winkler learned that he was HIV-infected in 1987,
and that is when he "began to stall the future":

> With HIV, I was no longer the same. It was inside the family that I
> first got the sense that I was now a different being. My coming out
> as gay did not make such a difference. There was more stigma now.
> People began to see and treat me as if I were dying. . . . But as strange
> as this might sound, to know that I would die of AIDS made me feel

A gay bar in Porto Alegre

powerful. . . . The family is where you don't want to be rejected, but it is also from this rejection that you can reconstruct yourself. Facing my death as imminent made me redo life, redo the plans I had for life. For me, the main outcome of AIDS was the sense of the moment, the here and now.

Winkler was embroiled in—and, to some extent, representative of—the tension-filled debate by local NGOs over their participation in the new national AIDS program. The fear of becoming government, the singularity of his own experience with HIV, and the actual need to change the AIDS care infrastructure preoccupied Winkler. He and his colleagues identified their place in concrete power relations, enabling them to force the local government to hear their claims and proposals. They took painful aspects of everyday life with HIV/AIDS as a source of knowledge, translating human rights discourses and claims for citizenship into concrete demands for juridical and medical accountability. "If we truly want to prevent HIV/AIDS we must go there where risk is." They agonized over the fact that participation in governmental committees was replacing "street activism" and over the normalization of lifestyles that came with the control of HIV/AIDS. The figures of the AIDS activist as policy developer and of the patient-citizen were taking form.

Over the years, I returned many times to Porto Alegre to work with Winkler and his colleagues. I documented both the shifting configurations of AIDS activism and the surprising ways in which Winkler remade his life. Through his activism, Winkler also directed me to a number of ethnographic sites that shaped my field of vision and my concerns, such as the asylum Vita (Biehl 2005) and a community of HIV-infected inmates at Porto Alegre's central prison (which I discuss later in this chapter). In the process, I acquired an understanding of ethnography as the art of recording the dynamic trajectories people make in reality, without which, as Gilles Deleuze puts it, "they would not become"—"these internalized trajectories are inseparable from becomings" (1997, p. 66, 67). The ethnographic challenge is to find the means to make visible the mutual presence of trajectories and becomings and to address people's efforts to be at once part of a public and "singularized out of a population rather than determined by a form" (Deleuze 1997, p. 1).

The AIDS Industry

Gerson Winkler first learned of AIDS in the media in 1983, he said, "but it was something very distant, very United States. My second daughter had just been born, and I was then more concerned with bottles and diapers." His coming out as gay was very much connected with "discovering" marginality, as well as mingling this marginality with self-transformation. "After divorcing, I promised myself that I would do whatever I wanted to, that I would experiment with all kinds of things, drugs, sex, everything . . . that I would experience the underworld. I had so many people in my life. My only fear was of being killed by a homophobe. So every time I went cruising, I left a note in my mailbox saying where I was headed. . . . I was afraid of being killed and never found."

In 1985, Winkler met Ricardo, a physician working in a local public health post. "He was the great love of my life. The values I cherish today I learned from living with him . . . of adventure and happiness. . . . I couldn't imagine the need to use a condom because I was fucking a doctor. I put myself completely in his hands, I knew he would take care of me. But he didn't know he was infected. He fell sick in 1987 and we both tested positive. He denied it was AIDS, and liked to say that he had TB, a disease he had contracted from the poor he worked with. I was also filled with romantic illusions, and thought of HIV as a virus of love . . . there was something of him in me now. So, at first AIDS did not frighten me."

As Ricardo started to fall ill, it was "crisis after crisis," recollected Winkler, "and it was in the hospitals that I started to face the ugliness of AIDS, the physical degradation, the pain, and the loss of the human condition." Ricardo died at the end of 1987. "The thought that we would all die, that we were so equal in this regard, comforted me. The human world as I had known it was no longer possible for me."

In 1988, Winkler entered the emergent world of AIDS activism in Porto Alegre. "I was part of the first self-help group of people living with HIV/AIDS. We met with a psychologist and a social worker. Why did we meet? To talk about the hypocrisy of institutions. We felt powerful to speak the truth. . . . But we also met to date. The group became an orgy. We did not want to be rejected, so we dated the ones who were like us. Those early years gave me the clear sense that we create groups to protect ourselves and to have secondary gains as individuals. By the early 1990s seronegative people started to want that too. . . . With a discourse of solidarity, they started to tell us, 'I am like you, but I don't have the virus.' And soon we needed the professionals who spoke English to write grant proposals for us."

During my first visit in May 1992, factions within GAPA were struggling over its control. People bitterly criticized each other, gossiping about sexual intrigue among coordinators and volunteers and alleging corruption (never confirmed). Winkler described the disputes in terms of the ongoing depoliticization of activist efforts: "I don't want to be an AIDS bureaucrat. As an NGO, we have not yet achieved maturity. The AIDS-business trend is gaining strength in here, and people want to produce services either parallel to the state or for the state. For them, everything should be efficiently thought out. They want a time clock and a bilingual secretary to write project proposals." Winkler employed Julio, a gay cross-dresser without such skills but with "knowledge of what happens on the street, with the people we have to reach." He continued to take advantage of all possible ways to counter what he saw as the "encroachment of bureaucratic desire." At stake, he insisted, was keeping in view the immanence of AIDS.

"There are times one has to sit at the same table with the government, but there are times that one does not. The crucial point is the present moment of AIDS," stated Winkler. And as he tried to "not be governed too much," he also optimized resources made available by governmental and international agencies, using them for ends distinct from those prescribed. His activism focused on both pointing out gaps and making concrete, manifold changes within the public health care system: "We ought to be demanding that the University Dental School be closed until it changes its policy about people 'suspected of having AIDS.' We have to demand that more rooms be made available for AIDS patients in the University Hospital. We must be with the patients, fighting for their rights once they are admitted to the hospital. The prisoners don't need to be taught more prevention; they need condoms and access to bleach for their needles."

•••

In 1991, eleven prisoners of Porto Alegre's overcrowded central prison were identified as *aidéticos* (i.e., as having AIDS) and were subsequently

Porto Alegre's central prison

isolated from the other inmates. Soon thereafter, Winkler and social worker Rosana Gouveia began addressing their immediate needs and also mediated requests for reduced sentences. During June and August 1992, I visited this community several times.

The *isolados* (isolated) were quarantined in a chilly cement room, separated into living spaces by clusters of precarious wooden bunk beds. The four transvestites, with their handmade images of saints, accessories, and small furnaces, lived in a far corner. Other inmates who dated the "girls" occupied the central spaces. And others who said they had sex only with women lived close to the door in a cave made of blankets draped over beds, near a television and with photos of naked women hung on the makeshift walls. Windows were few and broken.

The *isolados* knew that their AIDS condition was ordinary. "They stopped testing inmates because they were finding too many *aidéticos* to deal with," thirty-year-old Luis told me. Besides, for a variety of reasons ranging from death threats to relationships with cell mates to drug trafficking, many prisoners with AIDS preferred to stay in the main prison hall. Winkler told me that an estimated one-third of the roughly 1,500 prisoners there were HIV-positive. The *isolados* were cynical but content about the special treatment they were receiving. With the support of

GAPA, they had succeeded in obtaining higher rations. "We have more milk than the others." Their status as "AIDS-disabled" (in 1988, the congress approved disability status and pensions for all AIDS cases) also gave them some hope they might leave the prison sooner. I frequently heard them saying, "I don't want to die here."[3]

Jorge, one of the inmates, recounted that, before prison, he lived for some ten years in the streets of Porto Alegre, his wife and child both gone, victims of AIDS. The thirty-three-year-old man made icons using the image of Our Lady of Prague, coins, wool, and Disney stickers. At the time of the interview, he had just returned from the infirmary. Recovering from pneumonia, with skin lesions covering his body, he complained that care was discontinuous. "We need AIDS specialists." Some of the *isolados* mentioned their hope to live to have access to the AIDS vaccine, even though they were already in advanced stages of the disease and most of them were suffering from untreated TB. Many insisted: "Instead of rats and monkeys, they should use me as a guinea pig for the cure." Their accounts—like Winkler's—mobilized present events. Their desire to become experimental subjects reflected international and national developments around AIDS science. In 1991, the former national AIDS coordinator had ended a partnership with the WHO to have vaccines tested in Brazil. Prisoners constantly referred to basic medical research and treatment, along with milk, care, work, and the longing for family ties.

•••

During May and August of that year, I participated in GAPA's self-reflexive weekly meetings, where I found heated discussions over the fate of AIDS NGOs that did not plan to participate in the new national AIDS program. One overriding concern surrounded the neutralization of the political character of activism, a by-product of involving activism in a system that favored the creation of local services to absorb the burden of AIDS rather than a wholesale overhaul of the AIDS public health care infrastructure. Also apparently at stake at GAPA was the creation of a public space for the acceptance and revaluation of homosexual identity in this very discriminatory "macho land," as Winkler put it (see Costa 1992). Foreign safe-sex posters hung on the walls to shock and reeducate visitors and volunteers. Women complained of being "second-class citizens" in that environment and of being "sick with all those cocks everywhere."

Winkler and his group also seemed aware that "promiscuity" and other "hidden practices" that were formerly condemned could now in this way be publicly cleansed and conventionalized.

In a letter Winkler wrote that June, before embarking for his first meeting with the new national AIDS coordinators in Brasília, he reminded the volunteers of their initial ties to marginal groups and cultural events that brought the image of AIDS death to the public. He was presenting them, I thought, with a map of sorts, a convulsive map of where they had been and now were. He acted not out of nostalgia, though, but to help himself and his peers reflect on the personal and political stakes of their growing connection to the country's new AIDS program:

> Who remembers the encounters of prostitutes or ourselves buying the coffin for the demonstration? Who remembers us being mad? Nobody does this work of memory any longer. I take the liberty of doing it because I think that that is our compass. The possibility of survival will rest above all on exorcizing the institutionalizing process and breaking with this new order, of normalization. Maybe if it is necessary we will have to give back the beautiful computers or the fax machines . . . to gain the time to find ourselves and to do activism. . . . In this search we can perceive lies and truths. . . . Who knows? Is the desire for resources killing our political interventions while we want luxury instead of orgies? I am not talking about sex, but of this new way of relating to each other. . . . Of this fear we have of saying that we gave ass in the park last night and that it was delicious. . . . Don't be shocked, I am reconstituting our past, only that. . . . We should let the carriers of the virus pass through here breaking the pristine crystal glasses . . . we should keep this space open for people to sleep and fuck in here if they want to, or cry about the ugliness of AIDS. . . . For my part, even if it is contradictory, I am disposed to keep constructing a space of struggle, not institutional, not normalizing, in which all these plural beings can discuss new ways of thinking.

Winkler lamented the displacement of an AIDS modus vivendi by an AIDS modus operandi and was concerned as much with the subjectivity of the activists as with the subjectivity of the place itself. While recollecting events the group had experienced together, he addressed thresholds crossed as well as present enclosures and uncertainties. Winkler wanted to keep open the possibility of flight, not from AIDS, I thought, but from the "AIDS industry," which, in his view, limited freedom of movement, of expression, of desire, and "in a sordid way, was making it practically impossible for the deviants who actually live with HIV/AIDS to participate." Yet for him, flight also now meant flying to Brasília and engaging not an abstract bureaucracy but a changed political reality.

Micro-Politics of Patienthood

"We are on the front line. Perhaps because of the very urgency of AIDS we don't have much time for self-reflexivity," Wildney Freres Contrera, co-founder and president of Brazil's first AIDS NGO, told me during my visit to GAPA São Paulo in early July 1992. I wanted to know what actions they carried out, what principles underlay their work, and how they were thinking about their partnerships with the government at local and national levels. "There comes a point when we get sick of our own discourse and have to start all over again." GAPA now represented the poorer sector of the population, said Contrera. "The urgency of AIDS forces you to rethink the political roles you have chosen for yourself. We didn't want to fall into an assistance mindset. But you can't ignore this aspect of the work. . . . Then there is the question of dying, and this is unique to each individual. . . . What does it represent for society? Everything we don't desire."

For activists like Winkler and Contrera, the struggle against AIDS was making it possible to rethink public and private involvements: "We have passed through a period of great repression [during the military regime], and politically speaking, we are still way behind. People have gotten used to being imposed upon, and the question of citizens' rights is now on the agenda. AIDS has taught people that you have rights, that you have conquered spaces, and that you also have to work with individual subjectivity." Entering the AIDS world does not "change people's character," stressed Contrera. Rather, she argued, the possibility of becoming a welfare recipient via AIDS was generating a new kind of politicized sociability: "AIDS brings certain benefits to some sectors of the population. If you get someone who was a marginal, lived in the periphery, was underemployed . . . AIDS brings him certain gains, and he bargains to get things from us."

As AIDS became the disease of the poor, a micro-politics of patient-hood subsumed questions of identity and sexuality, reasoned Contrera. Within those vast urban landscapes of inequality, AIDS gave people the possibility of surviving economically by being "diseased citizens."[4] Contrera recognized that this new category and mode of claiming citizenship had evolved as a by-product of the country's overall economic paradox: Brazil is the fourteenth largest economy in the world, but the gap between the rich and the poor is second only to that of Sierra Leone.[5]

But there was more to the work of these new patient-citizens. They were forcing a democratization of medical sovereignty and enabling alternative health care practices: "In Brazil, it seems to me that doctors learn about AIDS as they face it in practice. AIDS has brought about a reexamination of the patient-doctor relationship. These days, the patient wants to participate in the protocol. Doctors weren't used to that. They've always had absolute power to manage their patients' lives. Now, it's the patient who is going after new treatment information, and so on. We've also been talking with health technicians about their rights to biosafety."

GAPA São Paulo worked in tandem with Grupo Pela Vidda, a medically oriented activist organization, explained Contrera. "We help them in gathering medical information. We contact doctors and get updated. Everything that comes out of the other AIDS NGOs we get and recycle, and we help Pela Vidda to produce their bulletin." In addition to its therapy-related activities, Pela Vidda focused its activism on fostering a systematic dialogue with the scientific production of AIDS knowledge. After studying the international medical journals and local medical practices and findings, the group published a bulletin that critically translated this material, presenting it in an accessible format.

Pedro Bittencourt, Pela Vidda's director, whom I also met during that trip, emphasized that the group wanted to intervene educationally in a way that would enable activists and people with AIDS "to negotiate treatment effectively." Pela Vidda's actions were focused on the adequate use of the goods of global AIDS science. At that time, there was not much talk about a local AIDS science. Indeed, without state incentives and money, and without the technical know-how to develop original protocols, Brazil's complex AIDS clinical practice "could not be converted into scientific knowledge that would be accepted by the international system," anthropologist Cristiana Bastos has noted (1999, p. 150).

GAPA had a series of partnerships with civic organizations and the private sector. "The AIDS education program is practically self-sufficient; it supports itself and even generates resources. We have a course named 'AIDS and the Community' that is aimed at equipping people to pass on information to different sectors of society. We assemble health professionals, community leaders, teachers, and women from the periphery and discuss specific problems." According to Contrera, GAPA's entrance into the corporate world was important, as corporations "can greatly help with prevention campaigns and . . . [can also help us] disseminate information on biosafety and the rights of AIDS patients." A few years earlier, AIDS work was done basically on a volunteer basis. But now, in order to sustain this broad range of activities and networks, "we have to professionalize," added Contrera. "We now have a fee for the provision of services. We charge the rich; for the poor, it's free."

"Information on its own doesn't change behaviors," said Contrera. "It is a political fight. We have to get already existing services to include prevention." GAPA had strategically placed activists in the municipal and regional AIDS committees, helping it keep a close eye on ministerial policies. One of the key areas of conflict, as I saw in Porto Alegre, was the fight over hospital beds and medicines. "We've been pressuring for these things to be made available. Sometimes governmental services do something positive, and then we applaud it. But when they mess up, we'll certainly be right there saying, 'Excuse me, but you're wrong.' Recently, we censured a hospital for a case of misappropriated public funds. We called the press and denounced it. Our patients have started to suffer retaliation from that hospital." For Contrera, GAPA was neither the enforcer of the state nor a representative of civil society; its force lay in its place at the *nexus* of the public and the private.

In the 1990s, Brazilian social movements found themselves in shifting positions between state, market, and culture, argues Ana Maria Doimo (1995). Movements focusing on the rights of such disparate groups as indigenous peoples, women, street children, Afro-Brazilians, and consumers oscillated between defense of statism and claims for advantages from the market. These new political subjects were part of a changing sociality. To defend their rights, they created a syntax that combined "modern values" such as autonomy and citizenship with "traditional values" such as community, religiosity, and solidarity, writes Doimo. And they learned to establish productive partnerships with neoliberalizing governments (Krischke 1990; Mainwaring 1987). The notion of "organized civil society" has replaced the notion of "popular movements"—"that constant stance of negativity *vis-à-vis* the institutional sphere . . . gave way to selective and positive relations with the political and administrative sphere" (Doimo 1995, p. 223).

AIDS mobilization followed this modus operandi of institutional partnership and punctual response to specific demands. The trope of victimization was not at its core. AIDS and gay activists framed their needs and claims as universal human rights, their actions as those of civil society at large. Their claims and practices of solidarity were aimed at halting "civic death" and maximizing the "right to life." No other patient group had claimed state protection on those terms before; AIDS activists strategically activated the new legal mechanisms that the democratic constitution had put into place and made AIDS "a socially recognizable need" (see Serra 2004), thus inventing a new form of biosocial mobilization in the country.

In such a neoliberalizing context, warns Doimo, institutional partnership with a reforming state could easily turn social movements into lobbying groups, and this could in turn reduce their larger public force (1995, p. 224). But AIDS mobilization contradicted this forecast. Activism became integral to the state, both in reengineering decision-making processes, particularly regarding AIDS treatment, and in demonstrating a new state-society partnership. AIDS activists like Winkler, however, remained doubtful about the ability of "governmental activism" to accommodate demands anchored in the lived world of the poorest and the marginalized.

•••

"There is no return for us," explained Brenda Lee, founder of Brazil's first *casa de apoio* (house of support, a controversial type of grassroots AIDS service) in downtown São Paulo, as she shifted her state ambulance into gear. Lee, dressed in a white coat, was referring to the ongoing stream of reporters, students, and social scientists that visited her house, where the majority of patients were transvestites. She told me that the hospice was originally intended to be a "center for São Paulo transvestite history," if not for the immediacy of AIDS in that group.

As Contrera remarked earlier that week, "Places like Brenda Lee's are double-edged swords. Even if we don't necessarily think it's the best service, you end up having to use them." The *casa de apoio*, far less repressive than people had described to me, struck me as an earnest, albeit fragile, attempt by transvestites to make a medical institution on their own—one that was already effectively negotiating with the state of São Paulo to meet some of the immediate needs of the people living there. The ambulance symbolized some level of the group's success.

"The Health Ministry is pushing hard for these hospices," Contrera told me in our conversation. "This needs to be discussed. The question of where to put the ill isn't the community's responsibility; it should be the state's. Emphasizing the role of these community initiatives ends up transferring this responsibility to civil society. The Ministry sees this as an alternative that will make treatment cheaper by keeping patients out of the hospital."

At any rate, for the afflicted urban poor, these were basically the only sites that actually offered them some form of palliative care or the means to obtain medical aid. *Casas de apoio* also challenged neighborhoods, the medical establishment, and mainstream AIDS NGOs by taking in hand the daily risk and the morbid realities of AIDS work, not in the least paralyzed by them.

Consider the activities of a group of male and female prostitutes I met a few weeks later in São Luis, the capital of the northern state of Maranhão. They had created their own NGO, InterAIDS, to counter the lack of local governmental action against HIV/AIDS, with the financial support of a French agency. As forty-year-old Maria da Luz put it, "We don't require schooling to join us. Here you don't need to be a doctor to be a coordinator. We don't even have a psychologist around. We have to struggle each year for financing to continue the project. We don't have telephone or fax. We're forbidden to think about money here, just at the end of the month, when we do the accounting. This forces us to be more creative."

Amid scarcity, these poor activists trained prostitutes to carry out prevention among their peers, publicized their work using free radio time, and did "blitz" prevention campaigns in clubs and discotheques, aside from their more conventional work distributing condoms (from the national AIDS program) in public squares. "People come to us asking for condoms, and we visit brothels. There has been some behavioral change," Maria da Luz reasoned, "but it is not easy to diminish people's resistance to condom use."

To satisfy the demands from possible U.S.-funding organizations for quantitative results, InterAIDS required people to register before giving them condoms. "You have to send a proposal with numbers because, for the North Americans, everything gets transformed into numbers. So then they say 'excellent' and might consider funding us." The bureaucratic technology woven into the registration cards also allowed InterAIDS's customarily sidelined clientele to inscribe itself officially in an organization by which it felt represented and attended. At InterAIDS, people were creating a public space of care. According to thirty-two-year-old Ribamar dos Santos:

> Our clientele doesn't want only the condoms, they want our presence, they want to know that someone is going to hear their problem, that someone is going to show a solution, wipe away their tears, do some activity or game, some diversion where they can forget the life they have to live. . . . It may not be a formal meeting, but we're passing along information, and also receiving it, because we understand what the person is going through. We do our work on the basis of what our clientele is thinking about. In the beginning, they said that AIDS didn't exist. Today, they admit that AIDS exists, a lot of them take precautions and verbalize their fear of dying.

Performing Citizenship

That same summer of 1992, I also traveled to the northeastern city of Recife, one of Latin America's poorest metropolises. There, several well-established organizations like SOS Corpo (SOS Body) and Casa de Passagem (House of Passage) were integrating AIDS into their work with working-class women and homeless girls, with funding from German, Swedish, and Danish agencies. An emergent AIDS NGO, the AIDS Forum, run by sociologist Acioly Neto, was also trying, as he put it, "to place AIDS on the face of the city," that is, to destigmatize AIDS and to counter discrimination through massive information dissemination campaigns, which included public service announcements on television. Neto was working with local artists to use theater for prevention: "We are taking our AIDS play to four districts that have a good level of popular organization and people there can be reproducers of information for neighboring areas."

Neto introduced me to Pernalonga, an HIV-positive performing artist who kept an informal affiliation with a local popular education NGO (where his lover Paulinho had worked) in the nearby colonial town of Olinda. Pernalonga said that he took care of Paulinho until death. "I am poor by birth, but not in spirit," he described himself. "AIDS gave me more glamour. . . . I had the courage to speak about AIDS, to smile, to be happy." In exchange for a minimum wage and a monthly food basket from Olinda's mayor, he mediated claims for assistance by poor people with AIDS:

> People think that I'm crazy, think that I'm in a good financial position, that I'm an opportunist, but I live on the margin without being a delinquent. I make detours with my intelligence and my interior beauty. I also do some things wrong, obviously. I got a job that the mayor's office would never have given me. But now, because I have AIDS, it's politically interesting for him. He's proud to be the first mayor to admit a worker with AIDS rather than expelling him. I help people, I go after an AIDS patient, I go after a basic ration for him, get hold of medicines. I don't take anything for myself. My work is social, it's political. What some people do for gain, I do for free, out of love.

AIDS public health services are extremely precarious there, said Pernalonga: "People are afraid to treat AIDS patients. We don't have basic conditions for it. Here, it is all need, all a ruin."

Pernalonga said that he had many opportunities in life, that he had been involved with people from other social classes, "not from my origin." His parents are divorced. His mother is illiterate, a Pentecostal convert, and his father is an auto mechanic. "My father started another family with a prostitute," he said and then recollected his life with Paulinho:

> When he died, I got desperate because I thought that I was going to die in a month. We had a good life, the two of us. I never felt so loved in

Pernalonga, 1995

my life. So when he died, I sold the shack we had and I went traveling. I was thinking that I was going to die, I went to Brasília and São Paulo. I've been all over Brazil. I've never been out of Brazil though. I went by bus, with friends, I worked in Brasília. I traveled with theater people and occupational therapists. I was living the transvestite life. . . . I like doing theater.

More aware of the possibilities of living with AIDS "and capitalizing on it rather than denying it," as he put it, Pernalonga returned to Olinda and fell in love with a divorced woman, the mother of two teenagers. "I say that I'm homosexual, but I live with a woman. So then the guys get confused. My practice is bisexual. I'm not the sort to go where there are only homosexuals. So there it is, sexual promiscuity continues."

That Pernalonga got his glamour back as he became an AIDS activist is not a coincidence. The roots of his attempts to transform marginality into art and politics go back to the Vivencial (Living) project, a transvestite dinner theater, at the end of the 1970s. An ex-seminarian and an ex-functionary of Febem (State Foundation for Youth Welfare) assembled a group of poor youth and transvestites (among them Pernalonga), who, improvising with all kinds of leftover materials and texts, put up irreverent shows that became a local sensation. Since then, Vivencial has been acknowledged as one of the precursors of the gay movement in the country (Trevisan 1986; MacRae 1990).

AIDS permits Pernalonga to come back to the public scene, now also as activist, functionary of the mayor's office, married heterosexual, and candidate for city council. His glamour seemed to translate into popular acceptance in Olinda's street life, but it probably would not translate into votes, as some of his acquaintances told me. Nor would it translate into employment by a local NGO, something he aspired to. Pernalonga's activism was theatrical: to say what people cannot say by themselves, to make people "have fun." The effect was both personalizing and exotic.

Pernalonga said that it is taboo to talk about AIDS in the state of Pernambuco and that people kept exposing themselves to risky situations. "Many people abdicate the condom, especially homosexuals who are in the closet. Because of prejudice, because of lack of information, because they think that it's a different sensation of pleasure, that it's not pure, it's not the real scream of the flesh. When I fall for a man, I go off with him, using a condom or not. Everyone knows me here." His narrative incorporated the male sexual fluidity of a highly hierarchical society:[6]

> It's the proletariat class that takes more sexual pleasure, homosexual pleasure, even the married ones. Married men love penetration, anal sex with other men. They're usually repressed. Most of them are passive, but they take the active role too. Active and passive, passive and active, it's a trade. . . . If you get out of bed and talk, you're dead. Pure prejudice. It's not prejudice about AIDS, it's social prejudice, the discrimination that you're not macho. Because the fact that I'm talking like this, or that I go to bed with men, doesn't change the fact that I can sleep with a woman and be a man. It's a different thing.

This socially unaddressed bisexual culture was having tragic consequences for women, as official epidemiological indicators on AIDS were beginning to show (Parker et al. 1994; Scheper-Hughes 1994). Facing great difficulties in negotiating the use of condoms, many women were getting infected by their partners.

Pernalonga's theatrical activism made me uneasy, particularly as it concerned prevention. That exercise of citizenship takes on a countermodern, romantic air when viewed in light of its failure to deal responsibly with the available medical knowledge about AIDS. Pernalonga's view of condom use seemed akin to what psychoanalyst Jurandir Freire Costa has called "moral privatization in the evaluation of risk of infection" (1992, p. 192). In his study of HIV prevention among Brazilian men who have sex with men, Freire Costa found that many of his informants oriented their decisions by factors that were exclusive to themselves. These individuals, argued Freire Costa, had by and large internalized dominant sexual codes and simultaneously exhibited "an anarchic rebelliousness" against these codes. They did not seem to make use of "models of identity that are sufficiently coercive to be integrated into a predictable system of behaviors when faced with the risk of AIDS. These rules lacking, decisions . . . don't always correspond to a caution that benefits the collectivity."

Ambiguity remained. As Pernalonga put it, "There aren't many who do what I do. A lot of people don't do the HIV test because they're afraid. What kind of progress is that? What sociology are you going to give me if you are afraid of doing a test? I talk about AIDS because I was there, my lover died in my arms, I saw the whole progression of it. I'm not pretending about the disease. I give interviews on the radio, in the newspaper, on TV. I'm public and notorious." It was quite disturbing to hear Lurdes, Pernalonga's wife, say, "We agree not to use condoms. We love each other. I accept him totally, as he is." Her "unconditional" love and his "free" libido were intrinsic to local social dramas, I thought, and needed to be understood in terms of cultural trajectories rather than risk assessment models formulated by outsiders.

Individual particularities of affect, cognition, moral responsibility, and action are entangled in everyday violence and scarcity (Scheper-Hughes 1992). Here, gendered domination and changing moral apparatuses coalesce in the emergence of a different kind of subjectivity and citizenship, or at least the idea of it. "It is not enough to have the plague of the century," Pernalonga said with an anguished voice. Things in the world of AIDS activism were changing, and to do something, anything, "now you need a diploma." Pernalonga, however, could only present "a theater resume; what I do is perform." What the performatic activist lacked, then, was "a friend to translate my project proposal into English, the help of a psychiatrist, an epidemiologist." Pernalonga hoped to replicate the *casa de apoio* of Brenda Lee, herself an émigré from Pernambuco:

If I had my own organization we could give family support, dentistry, psychological help. It would make my life better, too. I think that Brenda Lee's work is really beautiful. I've never been to her house in São Paulo, but I've heard many good things about it. She revolutionized things, in a way. But she had capital to invest. That's why I want to be on the city council. I admire the courage it took for her to face the biggest prejudice of all in that stone jungle, São Paulo. And she won. She got everything. As a transvestite she made it to the top. She made a hospital for desperate people who try to kill themselves. That's what I think is beautiful.

Pernalonga's performance—which took place on the street, among NGO people, before political party members—featured him as an adult cross-dressed as court jester. He mimicked famous *marginais*. He had neither a diploma nor "capital to invest." After the performance was over, he went home to his shack in the *favela*: "It's horrible."

Grassroots Health Systems

During that initial field trip, I learned of Caasah, a grassroots care facility for marginalized AIDS patients in the northeastern city of Salvador. Caasah had been founded in 1991 by the philanthropist Maria Luiza and the transvestite Romildo, Bahia's first AIDS case. When I first visited in early June 1992, a man without teeth looked through the semi-open door and greeted me with the statement "Eu sou portador" (I am an HIV carrier)—or what almost sounded like "Eu sou a porta da dor" (I am the door of pain).

Caasah was a two-story old house in the poor central district of Tororó, where twenty-two AIDS patients clustered together in a foul space, some on precarious beds, others on the floor. Apart from these patients, sixty-six others regularly visited Caasah from the abandoned buildings of downtown Salvador to receive food baskets. At Maria Luiza's small office, the telephone rang all the time, donations passed through the window from the street (from people who wanted to avoid contact but also to see people with AIDS), and patients came in and out, endearingly calling her "Mainha" (Mommy). Each day, three or four people—including men, women, and children—asked to join Caasah, she told me. Maria Luiza had just adopted an HIV-positive orphan.

Caasah's very existence was of a sentinel nature, I thought, a public acknowledgment that AIDS had long ago gone beyond the so-called risk groups; AIDS in Bahia was, since its beginnings, a disease of the poorest. Individual charity and help from pastoral organizations such as the Catholic mission made Caasah's maintenance possible: "Here in Bahia the government is not doing anything to fight the epidemic. The other day I threatened the mayor that we would camp inside the city hall if he would not release the promised money to pay for our rent. That's the least they can do," said Maria Luiza. Caasah's trademark was to engage in a realpolitik of public visibility. Its founding members demonstrated in front of theaters and in buses to remind the public of their presence and to maintain the flow of resources so that the institution could function, albeit minimally.

These marginalized subjects had bitter words for local AIDS activism and government bureaucracy. Romildo's story crystallized the group's antagonism; it was to become a founding narrative of Caasah. "Where did it all begin? I was the first case of AIDS in Bahia, in 1984. . . . But now, NGOs want to be the owner of AIDS; they don't want to deal with people with AIDS." He said that in 1990, while hospitalized in the state's AIDS unit, the local and independent branch of GAPA (founded in 1988) had offered him support, but that offer was never to materialize. After being discharged, he was left on the streets. "Finally, Dona Conceição, who is a nurse and God-fearing woman, helped me rent a room. . . . GAPA's president went traveling through Europe, participating in conferences, and I stayed in the gutter, in reality."

Adventurous, remorseless self-making and street violence punctuated Romildo's account of how he became an *aidético*. Born in 1964, Romildo left his parents' house at the age of ten, after which he started "running in the world," which means prostituting himself: "I adored being a transvestite . . . so much so that today here I am, I still have these little breasts. I liked to feel like a woman. I was a good professional. I let them all satisfy themselves. But to satisfy myself I needed someone indifferent to money, not looking for anything—purely physical and interested in me, then I felt pleasure." In the beginning of the 1980s, Romildo traveled a lot between his native state of Minas Gerais and Rio de Janeiro. In 1984, he settled in

Bahia permanently: "I got into a bit of trouble with the police. I'm not so easy. If they beat me, I went right back after them. I injected drugs, drank, smoked pot, did downers—all sorts of drugs in order to have the courage to face not the people, but the police. The police are barbaric here. A group of us went to the secretary of the civil police to protest. They took our money, our clothes, forced us to have sex, left us there naked as the day we were born, and we had to go home that way."

Romildo got to the point of cutting himself with razor blades so that the police would stop beating him: "I'd have ten blades in my mouth. When the police approached us and told us to take out the razor, we'd spit one out and still have nine in our mouths. When they came to beat me, I cut myself. They'd be afraid of the blood and say if this fag does this to himself, imagine what he's going to do to me."

In mid-1991, the police caught Romildo selling drugs and sent him back to the AIDS unit of the state hospital, "a human deposit," as he put it. There, he met Maria Luiza, who was taking care of her brother. After her brother's death, Maria Luiza decided to adopt Romildo, hiring a lawyer to arrange house detention for him. "The hospital would have thrown him back into the street where he would have died alone. . . . There are many like him; families and friends don't take them in," she told me.

Caasah emerged from the AIDS public health system as a response to its basic inadequacies and to the vanished family ties of AIDS victims. "Whoever it was, HIV-positive, suffering from AIDS, homosexual, homeless, addicted to drugs, crazy, she brought them in and took care of them. . . . Here in Bahia, the government has money to restore the historical center, to have carnival music played all over the place, but it doesn't have money to help a house that provides this kind of service," added Romildo.

For over a decade, I have closely followed the developments of this outcast AIDS collective. Through hard work and a surprising turn of events, Caasah became Bahia's most important institution of care for the poorest facing AIDS. With the advent of AIDS therapies, an unexpected future came into being for some of these AIDS sufferers as well. My ethnography of the unfolding of life in Caasah illuminates alternative visions of existence that dominant medical, political, and economic institutions do not routinely address or even foreclose. Here, "minor voices" reveal the limits of governmental and nongovernmental interventions and make relative the "truths" that the institutions of AIDS disseminate.[7] In words and deeds, these AIDS subjects complement institutions and invent what is actually needed to survive AIDS. With almost no money and learning to care for themselves and each other, they "set free . . . this creation of a health or this invention of a people, that is, a possibility of life" (Deleuze 1997, p. 4).

Minutes from Caasah's founding meeting refer to it as "a civil society, without profit motives," and its statutes are based on "the natural right that rules human life." "AIDS is not an offense," read Caasah's slogan at the time. "AIDS is a disease." The only requirement to be admitted to Caasah was the "desire to be assisted." On admission, people had to show laboratory or medical proof of being HIV-infected. Duties of inmates included respecting and defending the rules of Caasah, accepting and facilitating treatment provided by the state's AIDS unit, cleaning and maintaining the house, and preserving transparency by reporting to Caasah's coordinators any personal, financial, or social problems. Caasah's president had the duty "to imbue life in the institution" with a "necessary spiritual sense, so that faith in God might strengthen its daily practices and might contribute decisively to raising the morale of the patients." The house's creed was expressed in the following principles: "To love God and to do acts of kindness and understanding towards persons suffering from discrimination, social contempt, and a definitive death sentence."

Maria Luiza constantly confronted health professionals and religious leaders in the name of these principles. The week before I arrived, she had fought with a pastor from the wildly popular Pentecostal Universal Church of God's Kindgdom. "After he said that someone with AIDS doesn't deserve the cure of God's grace, I yelled at him. He told me that I was possessed. So I kicked him in the balls and said, 'Well, I am possessed now.'" Maria Luiza also mentioned fighting directly with the prejudices of medics: "I smacked a radiologist who refused to take some X-rays of a patient." She was angry about the slow tempo of international AIDS science: "I can't get into my head that they can build the bomb, that machines work for human beings, and that still the intelligent human animal sees others die and has so little interest for disenfranchised human life."

Mainha and her patients were turning AIDS and social pathogenicity into "dramas of inclusion" (Appadurai 2002, p. 40). As I had seen in Brenda Lee's work and in InterAIDS in São Luis, these unwanted and otherwise invisible subjects were bringing their demands for respect and care to local institutions, not just as sufferers, but as experts in human life. Demanding a proper balance between the moral, the political, and the medical—at both individual and institutional levels—the actions of these mobilized patients and their caretakers were redefining the "problem of AIDS" and the ways to solve it. As they made themselves into objects of public attention, they demanded exams and medications, criticized slow and corrupt bureaucracies, administered scarce public moneys, ran alternative and family-like health services, lectured in buses and schools, and shouted in the streets that "prevention is fundamental," trying by all means to become "socially useful" so as to guarantee accountability and resources.

Maria Luiza told me that I was just witnessing "the start." In a few weeks, they would invade and squat in a public building. She showed the banners they were painting, with the acronyms of all the major parties in Bahia, to ensure that they would have a chance to get support from all of them. "Since it is election year, all parties will support us. This invasion will give us true public visibility."

From its beginnings, Caasah claimed to be a kind of surrogate state—a site from which these marginalized Brazilians could pursue recognition of their "natural rights." These rights were understood in terms of a long overdue fulfillment of claims of "legitimate citizenship in all legal senses (federal, state and municipal, work-related, and legal and juridical doctrines)," as stated in Caasah's statutes. But these subjects were caught in a paradox. In claiming their rights, they also needed strong public institutions, something that had never before existed for them, something that was in fact disappearing at all levels of government, particularly in the realm of social assistance.

The "family" care providers of this public house had to rely on patronage. It was up to Maria Luiza and her patients to activate the available institutions and put them to work. Not only did they demand supplies and adequate treatment but they were also required to invent the local means to provide it. They became involved with activist and philanthropic organizations that actually redistributed the goods associated with those rights (basic food baskets, for example). In exchange, from a political perspective, Caasah played a symbolic role of sanitizing the streets and ghettos of the capital, removing the penurious reality of AIDS and making it look at least partially contained. This dynamic reflected some of Contrera's concerns with the political exploitation of AIDS-related pastoral programs and community initiatives.

Caasah was also becoming entangled in shifting forms of urban governance in other ways. By caring for patients and overseeing their medical treatment, as I would discover, Caasah became a venue for an incipient AIDS public health triage system. It mediated the relationship between AIDS patients and public AIDS services, which were haphazard and extremely limited. More specifically, it selected the patients who could benefit the most from the scarce resources available (the state's AIDS unit had only sixteen beds).

In short, Caasah provided a means through which these individuals could accede to a distinct (and tentative) *patient-citizenship*—something unavailable to them in the past and something that an increasing number of groups were articulating in Brazil. As I will show in chapter 5, this late-born democratic practice of citizenship via patienthood (or at least a claim to it) would transform in subsequent years (at an impressive speed) into a sophisticated pharmaceutical form of care. These marginalized subjects and their "AIDS society" would become less confrontational in politics, less absorptive of street life, and more integrated with the lifesaving mechanisms and technologies associated with the national AIDS policy.

A New National AIDS Program

By July 1992, Brazilian health policy circles had already become aware of Caasah's efforts. Dr. Paulo Teixeira, then one of the key articulators of the changes in the national AIDS program, mentioned Caasah to me in an interview as "the sort of program to which we would like to be directing money." I met with Dr. Teixeira in Brasília, following my first visit to Caasah, while it was still functioning at the house in the Tororó District. A new combination of government and activism was emerging to serve the neglected, poor AIDS patients.

Dr. Teixeira, who had pioneered public health responses to HIV/AIDS in São Paulo, made clear that he wanted to form partnerships with state-critical NGOs and suggested that community initiatives such as Caasah could be vital extensions of national assistance programs. As he put it,

"I propose to establish systematic interlocution with local initiatives, always respecting differences, and enabling them to intervene as much in prevention as in alternative assistance to patients, in order to be sure that our initiatives will actually arrive in the communities . . . a consensual, democratic project, with guaranteed resources."

International financial agencies have strongly influenced the setting of national political agendas and interventions, substantively altering the nature of economic discourses and redirecting policies toward specific problems and circumscribed populations. The World Bank loan to revamp the Brazilian AIDS program was directed to the technical reorganization of the state's public health policies. Themes of fighting poverty and promoting focused and efficient social protection programs permeated much of the Bank's agenda and the experts' arguments about state reform in the 1990s. The Bank depicted NGOs' interventions and the outsourcing of care to community-based initiatives as the "true alternative" to the bureaucratized, impersonal, and inefficient structures of Brazil's doomed (and never successful) welfare system. These substitutes, the thinking went, could ensure that benefits would reach individuals and groups in need. At the new national AIDS program, small initiatives like Caasah were viewed as important linkages between the government and the local and urgent realities of AIDS.

For Dr. Teixeira, the "big social victory" related to the World Bank loan came in meeting the demand for political inclusion by AIDS representatives and groups. But one could also argue that, in Brazil's restructuring context, AIDS provided the Bank with a ready-made situation in which it could operationalize its notions and strategies of "adequate" public investments: that such investments should be focused on the most vulnerable—prioritizing the control of infectious diseases that predominantly affect children and the poor—and that they should be informed by prevention rather than treatment. According to the Bank's experts, public health programs should be based on the criteria of cost-effectiveness and rationalization of supply. Public financing should be directed to only a restricted package of basic health services, whose execution should be decentralized. High-cost medical procedures should be bought in the health market.

This health agenda grew out of the various restrictive cost-containment and managed care programs that were put into effect in the 1980s, particularly in the United States, and that altered almost beyond recognition

Programa Nacional de DST e AIDS, Brasília

traditional patient-practitioner relationships. As a template, the "managed care revolution" has informed transnational technocratic alliances aimed at rationalizing expenditure, even in societies with low per capita investment in health care (Chen, Kleinman, and Ware 1994).

Consider a 1988 World Bank report denouncing Brazil's poor social welfare performance as exemplifying mismanagement of resources by the state. This inefficiency was due to "overly centralized systems of financing and providing services, a complex and unwieldy system of governmental transfers, and inadequate use of the private sector in service delivery" (World Bank 1988, p. 3). By 1993, the World Bank critique had changed in tone. Specialists endorsed as a "significant achievement" the 1991 decision to implement the constitutionally mandated universal access to health care through the Sistema Único de Saúde (the Brazilian unified health care system known as SUS). They especially supported the establishment of local administrative, managerial, and regulatory infrastructures that would "take over the oversight and delivery functions previously under purview of the federal government" (World Bank 1993, p. 41). The international agency praised the country's amalgamation of different financing and delivery mechanisms: "The range of experimentation and activity is impressive, and the innovation across these many initiatives reflects the creativity of both the Brazilian medical establishment and the business community. The government has been at the forefront in attempting reforms as well. As a result, Brazil has a unique health care system with one foot in developed countries and the other in the developing world" (1993, p. 149).

Many AIDS activists and health professionals who helped shape the new AIDS program felt they did not have to relinquish their own conceptions of the proper approach to AIDS as they engaged with these new directives and resources. "The World Bank loan does not include medication," stated Dr. Teixeira. "The discourse of these international organizations does not deal with the already-ill. They are taken as lost causes. It is as if you had to invest in prevention because nothing else merits investment. This analysis is wrong. It doesn't hold up to the most minimal epidemiological analysis, never mind social or ethical considerations."[8] The activism and work of people with this alternative vision in the government would be fundamental to the pharmaceutical course of the policy.

By 1992, AIDS drugs such as AZT were unevenly available through SUS, and the new HIV/AIDS funds were directed to prevention through NGOs and newly created NGO-like regional and municipal AIDS programs. Until then, most AIDS activism opposed the federal government, which refused to make AIDS a public health priority, and local discriminatory public health practices. With a new national AIDS administration and the international funding available, the situation changed. As Dr. Teixeira saw it, "The biggest modification that occurred in the national program was the systematization of the relationship between the state and NGOs, which are now an essential part of the program."

According to Dr. Teixeira, NGOs had to learn to translate community voice better, to make generic and vague claims for participation into something tangible and workable. He specifically mentioned the resistance of GAPA Porto Alegre to working with the government (as voiced by Gerson Winkler). "It seems that a lot of what happens there is mostly the product of a kind of recalcitrance, which is a priori immature, absolutely unproductive, if at times somewhat cathartic." For Dr. Teixeira, AIDS mobilization was still, at times, too glued to identity-making dynamics and political ideology, and the accumulated experience showed that it was possible to get things done and to be "absolutely straightforward and critical" at the same time.

Dr. Teixeira also noted that AIDS NGOs had both constituency and institutionality, and that this social space, if rightly managed, could help make the new investments somewhat measurable. "There are many things being done, but they are scattered. It is indisputably necessary to invest in strengthening and training NGO personnel—that is, to create instruments that enable the process of reflection, that help with even the more theoretically prosaic aspects of things, with the management of projects, the systematization of work, the choice of targets, the establishment of goals." Carrying out such a large financial and social investment required the professionalization of NGOs.

As I have shown in chapter 1, this cooperation among AIDS activists, the government, and the World Bank indeed led to the constitution of an efficient and transparent national program. Grounded in human rights principles, the program's activities were marked by scientific standards, the treatment of culture as a major technical variable, nongovernmental partnerships (which had grown from 120 in 1993 to 480 in 1999), and cost-benefit analyses of interventions.

According to epidemiologist Pedro Chequer, coordinator of the national program after 1996, much of the novelty of this approach to AIDS lay in the elaboration of a "realistic perception of the social and cultural aspects of the epidemic." The program funded social research in an effort to identify local variables related to risk, sexuality, and social vulnerability to HIV. These studies criticized the epidemiological and behavioral assumptions underlying international prevention models, and the results, according to Chequer, were given "the same relevance as clinical efforts" (CN 1997c; see also Bastos 1999, p. 103; Galvão 2000). Local experiences with prevention among marginalized groups such as homosexuals, prostitutes, and transvestites were also seriously considered.

However, as AIDS kept spreading among the poor, and as public health interventions moved from prevention to treatment amid a weak if not chaotic SUS, grassroots institutions such as Caasah ("social instruments of substitution" in the experts' language) would become overburdened with both care and triage. The fact is that Brazilian response to HIV/AIDS crystallized amid the paradoxical coexistence of novel democratic claims to health and a structural readjustment agenda, which allocated no specific resources to health care and considered prevention to be a private rather than public matter. In the words of anthropologist Richard

Parker, "Even if we try to avoid seeing it, we also helped to materialize some aspects of economic neoliberalism" (2001).

On the Street: Violence, Charity, and Pleasure

In August 1992, Caasah's members squatted in and managed to secure an abandoned maternity ward in the outskirts of the city. The building belonged to the Red Cross but had been built on municipal property. When I returned to Caasah in 1993, I learned that immediately following the takeover, *Mainha* Maria Luiza had been expelled. Romildo, who led the effort to drive her out, accused her of authoritarianism, ill-treatment, and embezzlement. The patients chose Celeste Gomes, a trained nurse, to preside over them. Aided by her niece, Naiara Santos, the new president implemented strict rules of internal governance and set up mechanisms to sustain the flow of funding for infrastructure and maintenance.

"Caasah had no government," recalled Celeste. "There were fights with knives and broken bottles. The volunteers had to hide in the bathroom.

Caasah

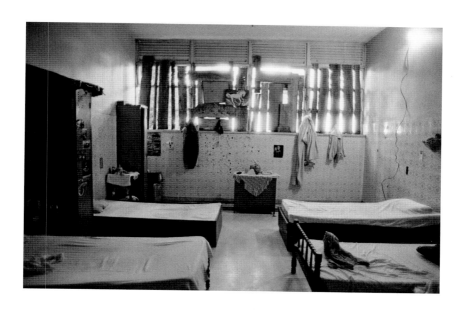

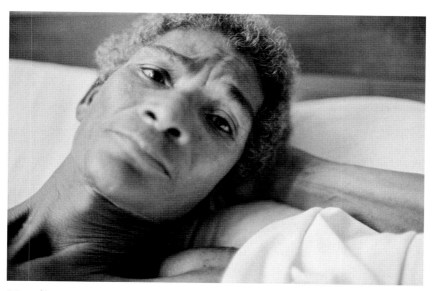

Marcelino, a patient at Caasah

Police officials came here all the time, threatening to kick us out, saying that all these people were *marginais*. We had to show them Romildo in the wheelchair and the children to get them to leave us in peace for some time."

Soon Caasah became implicated in local politics, used as a foil for the heated disputes between the traditional provincial government and the progressive city administration, which was led by a coalition of the Socialist Party and the Workers' Party. According to Celeste, "Bahian state officials said that they could only help us after we had become an institution of public utility. But there is so much bureaucracy and so many documents involved in the process that we always fall short of approval. They play tricks to postpone real action—it is a perverse process. In fact, they do not want to improve this facility."

Left-wing city officials and AIDS activists helped Caasah gain legal status, and in 1993 it became an NGO. The national AIDS program also approved funding for two of Caasah's projects. From then on, the core maintenance of the institution, its technical upgrading, and its normalization have been closely tied to funds channeled from the World Bank loan. I returned to Salvador in February 1995, this time accompanied by photographer Torben Eskerod. Caasah's direct link with the national AIDS program had strengthened its local political influence. "Relations began to improve with the local government," Celeste told me. "They are acknowledging us not only because we actually help the hospital's AIDS unit but also because of the power we now have with the Health Ministry. The projects gave us some credibility."

In spite of new funds and institutional improvements, Caasah was still circumscribing its services to thirty AIDS patients and providing home care (in the form of basic food baskets) to seventy others. Strict discipline and demanding medical regimes had led many patients to leave the house. While inquiring about their fate, I was led to Dona Conceição, who became key to my understanding of the other side of Caasah's incipient patient-citizenship: economic and social death.

•••

Every Wednesday at noon, Dona Conceição, a fifty-year-old nurse, cooked large pans of food and, with the help of her religious friends, distributed them to dozens of poor people and families who lived with AIDS and

Dona Conceição and her homeless AIDS patients

very little else in the abandoned corners of historical downtown Salvador. A procession of young men, women, children, and transvestites—of the ugly, scarred, and wounded—followed Dona Conceição and the food through a side alley. She provided free meals and some care (medication, clothing, and rent aid) to a total of 110 adults, most of them involved in prostitution and drug dealing, and to their children.

The first day Torben and I met Dona Conceição and her patients, she referred to them as "my street family." "These people have no protection, they are at the margins of law and life. They are nomads. When the police discover where they are, they have to leave. Today here, tomorrow elsewhere. They then contact me and say where they are and I go there. . . . I am tied to them in spirit."

Dona Conceição maintained her household by working three nights a week in a private hospital. Her older children helped her with the household expenses and with the pastoral work during the weekends. She had some support from her friends, but she basically had to generate money for the AIDS work on her own (mostly through handicrafts sales and donation campaigns). She began her work in 1990, first helping Romildo, Caasah's founder, after finding him asleep outside of the state hospital, where she then worked as a nurse. "He was semi-paralyzed in his legs. I rented a room for him in the Baixa dos Sapateiros, and started to take pre-made food and clothing there. I also bathed him. Then soon there was another case like that, abandoned by the family; his name was Manoel, he was a drug user and was sleeping outside the hospital. I also rented him a little room. Manoel died soon, after six months."

"Civil society has abandoned them," Dona Conceição says of her patients, so she has made them part of a charitable family. "They need to know that at least someone cares, that they are not just left to themselves. I give them a little comfort and help alleviate things a bit." She saw her work as a direct consequence of generalized moral indifference and inadequate public health services: "You know, the services never meet the demands; there is always shortage." Most of her patients presented AIDS symptoms, coming to her after being treated at an emergency room, discharged without having recovered or having received no assistance at all:

> They come to me with a medical history. They had bad diarrhea, high fever, TB. Since I go every week to the hospital to visit my patients, I meet the new AIDS patients. The old ones tell them about me: "Today my mother comes." Every time I get there, there is already somebody else who wants to talk to me, "Are you Dona Conceição? For today is my turn to get to know you." They also tell me of the ones who are living on the streets and who cannot go to the hospital. I visit some and try to help.

One day, as we were observing the procession of people waiting for Dona Conceição's aid, policemen arrived at the scene, guns in hand, and forced all men in the group against a wall—not the anthropologist or the photographer, though. Over our protests, the policemen searched them, finding no drugs or guns.

After this common, insensitive show of force and intimidation, we reached the place of that day's dinner: a cement platform, designed by one of Bahia's most famous architects in the 1970s as an exclusive restaurant. Now a modernist ruin, it overlooks the bay where for centuries millions of African slaves arrived mostly from Angola, Mozambique, and central Africa. Syringes were scattered across the ground, as was garbage.

Ten people, including children, lived in huts sized for single bodies, made of cardboard, under the stairways. "I know them, but I don't have

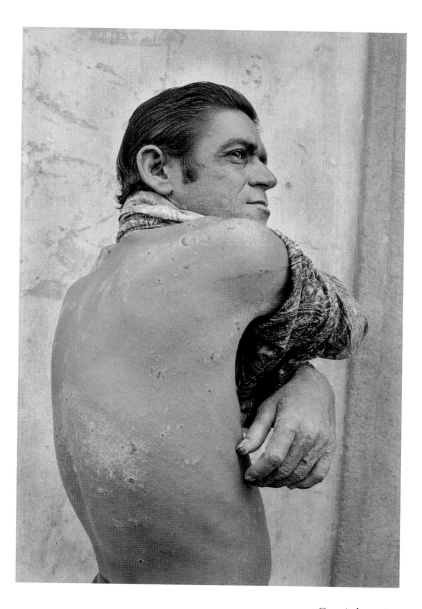

Caminhoneiro

time to do a thorough inventory of who they are," Dona Conceição remarked. "Besides, there is such a flux. Many die, new ones ask for help. Every day I get to know three or four more poor people with AIDS. . . . And I am only one."

Yet, injury and suffering was not all these urban figures displayed. While some helped Dona Conceição and friends to set up the lunch, two young men performed *capoeira* and a third paraded as a model. In a corner, a young man and a woman whispered to each other. Several posed for Torben and joked about seducing him.

During lunch, I overheard people reviewing recent events in their lives, asking each other about acquaintances and planning joint ventures for later in the day. As they moved in and out of social AIDS, they navigated through all kinds of networks that, ultimately, I thought, made up the city and the life they looked for (see Bourgois and Schonberg in press; Le Marcis 2004).

●●●

On several occasions during that field trip, I talked to the group. Soft-spoken Jorge Antonio Santos Araújo said that he was born on January 1, 1963. "I will not lie to you. I injected drugs, and I have AIDS," he told me without hesitation. "I abused drugs and myself. I had to amputate my left leg. When I got to the hospital it was too late. And on top of losing the leg they told me I had AIDS."

Jorge had lived by himself and on the streets since the age of fourteen. "I left home because of my stepfather; we didn't get along. I did little jobs, here and there, sold drugs. I think it is a thing of destiny, right?" At some point, he lived with a woman and had a child, but he eventually left them. "I think they already take me as dead, and I don't want them to know that I am like this. . . . Of course I think of them, but what can I do?"

He seemed resigned to what had become of his life; the only question he asked was rhetorical, implying the answer "I can do nothing else." He had tried, though, to become a patient and to fight "the thing," meaning AIDS. "I am part of Caasah's first generation," Jorge said.

Jorge had lived in the old Tororó house and had helped in the takeover of the abandoned maternity ward:

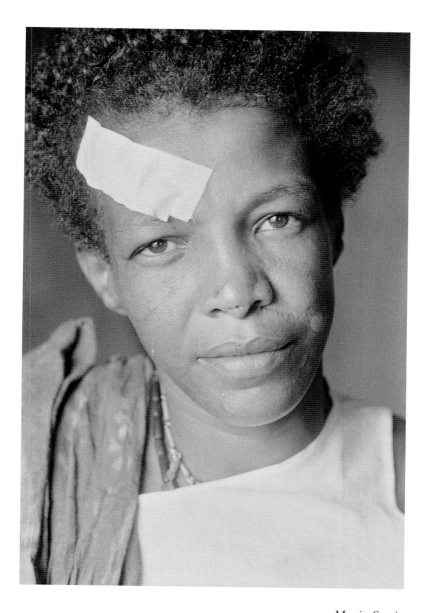

Maria Sonia

That was something we did. But I couldn't stand being locked in. So I left. I like to play around. Here I drink my *cachaça* [a popular sugar-cane liquor] and smoke my pot in peace, that's my medication, that's my woman. If I had taken all those pills I would already be dead. If I kept thinking about AIDS, I would already be dead. I don't live for the disease. That's what I do, I make as if nothing were happening. I am not taking any special medication. When the time comes and if necessary I will take it though. Meanwhile, I roll my life as God wants it. One must forget. One cannot put in one's mind that one is the disease. If we dwell on the disease, then one starts to say, "Maybe I should not do this or eat that for it will harm me," and then one is left with even less. . . . To be a patient one needs things. What is there here to have?

Nothing but some form of enjoyment, I heard him mumbling. The enjoyment and dependency that came with drugs had replaced sexual desire, suggested Jorge. For him, health had no intrinsic value. And addiction had become entangled with both "choosing" not to live a life of exclusive suffering and countering the life that comes with medical technology and patienthood. The scarcity and violence these street patients navigate day after day are overwhelming. I have always been disturbed and puzzled by the sense of impossibility they voiced, the sense that they had no real references to guide another kind of lifework.

Jorge praised Dona Conceição, saying that she played a crucial role in the larger network that he and his friends had to engage to guarantee daily food and drugs. "Besides her help, we do all kinds of things to-gether—we have a few houses and businesses that we visit periodically in the afternoon and get the leftover food. Sometimes we have to sweep the floor or carry things. I will not lie to you, sometimes when the need is just too great we rob things and sell them." Jorge had bitter words for the limited help he was getting from Caasah. "As a founding member, I deserved more respect. If it were left only to Caasah's help, I would al-ready be dead. But I still go there to get my food basket. Two pounds of rice, one pound of black beans, six eggs . . . it is very little, and I still have to pay the bus ticket and split half with the colleague who carries it for me, for I am on crutches. . . . [The] overall situation of AIDS care is pretty bad, that's what I have to say."

Police search

However, through this circuitry—begging, work-for-food, petty robberies, and AIDS charity—a sense of belonging also takes shape, eliciting a new constellation in which inner life is reframed and social death endured. "People here are all my friends; we have little or nothing, but we give each other strength. Of course, nobody will do anything for the other for free. But they talk to me. They make me laugh. This gives me a bit of extra life. That's how I think: I live a little bit longer day by day. That's how I take my situation, until the day God wants it to be. One was not born to be forever, and that is how my life unfolds."

●●●

One should not expect these patients to adhere to medical treatments, says Dona Conceição, because "they just use medication until they recover." And she did not blame them: "How can they comply if they live on the streets? Until they have a home, no therapy will work." The simplest reason for their disinclination to adhere to treatments is lack of money, she reasoned. "I know many people with AIDS who are psychologists and astrologers . . . they have money to pay for private medicine, to guarantee that the AIDS drugs are there for them." Illegal substances were in fact the only connection most of the street people had to the market and to the social order of consumption. Dona Conceição suggested that they actually needed these substances to participate in the economy: "It gives them courage to do something, like robbing. . . . It is wrong, but it keeps them circulating."

Twenty-eight-year-old Luzeide said that she was depressed, that she had not seen her two children for a long time, that her sisters did not want to hear from her, and that drugs helped her "forget." Others there told me that they had stopped taking illicit substances. However, Dona Conceição later said that she knew that this was not true. Most of them exchanged the medication and food they received from hospitals and Caasah for drugs: "I love them. But one cannot take what they say at face value. They fabricate realities. If I give a piece of clothing to someone he later tells me that someone else robbed it. But I know that he has sold it and bought drugs."

Dona Conceição did not judge her street patients and their actions in terms of right or wrong, in terms of normality or pathology; she understood that structural violence compounded substance- and self-abuse. In

The place of that day's dinner

Capoeira and Adair

Children

doing so, she implicitly made their condition a public affair, a Brazilian social symptom, I thought. But to complicate things further, she refused to treat them as a collective, and that's what drew them to her. She helped them singularize, and she literally struggled in their place: "Each one has a history, a life left behind. Jorge suffers emotionally—all the discrimination he goes through, and he is unable to overcome his personal failures. *He does not struggle for health; I struggle for him.*"

Quite often her patients are caught by the police and Dona Conceição is then called to intervene. "They know me at the police station. Sometimes they are stripped there and I have to bring them clothing. When I mention that they have AIDS, they are freed and put under my custody." She is also well-known at the hospitals: "The social workers call me and ask me to prepare the funeral when they die there." Assembled by the plight of AIDS, most of these poor people had no medical or family assistance other than Dona Conceição's, no public health agency or NGO assisting them. For the most part, they were not counted as AIDS cases by the health and surveillance services they passed through en route to their deaths.

After that initial contact with Dona Conceição and her nomad patients, I began to visualize an AIDS epidemic that existed in an advanced post-clinical stage among the marginalized, an epidemic that remained publicly unaccounted for. This "hidden epidemic" was part of a larger political economy of AIDS, which included family abandonment, the vagaries of AIDS public health services, the remedial work of Caasah, and the drug trade. The street patients were not included in population counts, and they did not think of themselves in such a bounded way. As Jorge had alluded to, his dying, although inevitable, was an intricate and open-ended process, mediated by an intense search for continuous enjoyment and the tweaking of existing infrastructures, all amid escalating violence.[9]

Against a background of brutal inequality and institutional neglect, the national AIDS program had emerged as a hard-fought island of accountability. Grassroots initiatives like Caasah and the charitable work of Dona Conceição made it possible for a few of the poorest and most marginalized AIDS patients to have some form of engagement, albeit minimally, with this emergent and responsive governmental landscape. Here again, the infamous impenetrability of regional politics and the restrictive terms of medical engagement come vividly into view. Amid the continuous search for goods, another domain, both tender and cruel,

Jorge, a few years later, still living in the streets

comes into existence in the human aggregate around Dona Conceição: at least for a few hours, one day of the week, these abandoned subjects have a place to deal with the imminence of death and where they are asked to consider another form of self-administration.

In the Mainstream

After our March 1995 visit to Caasah and our work with Dona Conceição's group, Torben Eskerod and I stopped in Porto Alegre and met with Gerson Winkler. Much had happened to him since 1992, when I had last seen him. As planned, he had campaigned to become a city representative, but he had not been elected. A few of his collaborators and opponents from GAPA, as well as several of his friends, had died. Winkler himself was in good health, he had moved in with his new boyfriend, and he had left GAPA to coordinate the city's AIDS program.

Most local NGOs had partnered with the governmental sphere, and major AIDS activists had joined national and regional governments. Winkler had brought some of his GAPA colleagues to work with him in city hall, which was now administered by the Workers' Party. "Given the way AIDS now works," he reasoned, "I can make things happen by being an administrator. Moreover, NGOs are too far from the people, from what is at stake for them."[10] Winkler was referring to the monopoly NGOs held in prevention work (which he deemed "not audacious enough") and the diminution of AIDS assistance to a local and municipal affair. He thought he could make a difference, as far as AIDS and poverty were concerned.

In May 1998, the national AIDS program reported that NGOs were carrying out 593 (50 percent) of the total 1,175 national AIDS projects established through the World Bank loan. In contrast, federal organizations were carrying out just 158 projects (19 percent), with regional and municipal organizations implementing 224 (24 percent) and 200 (17 percent), respectively. The NGOs were administering $23.2 million in services (45 percent of the $51.2 million available for projects). The problem with this joint governmental-nongovernmental venture, said Winkler, was not just that NGOs had been politically co-opted but that a new power machine emerged in the process:

On the ground, a clear division between those who have access to the material provisions of the AIDS world and those who do not is evolving. Poor people are lost in the fight for access. The machine does not absorb the demand. My overall sense is that the AIDS mobilization remains elitist, most AIDS patients have no care at all, and a few faggots, and I am not excluded, chat via e-mail about our medical recipes. We see AIDS turned into misery, and NGOs don't deal with that; they are concerned with keeping their space and funds to themselves. For me, this is a fictional government and an abstraction of AIDS.

I told Winkler what I had seen following Dona Conceição and her street patients. "These are precisely the places," he replied, "that neither NGOs, government, solidarity, nor order and progress get to. This is another form of life, with its own security systems, and people rarely open their doors; disease is considered differently in these places."

But a new relation is emerging, I said, between these people and AIDS, and once they acknowledge AIDS, it works as an institution for them.[11] Community representatives and philanthropists connect the HIV-infected with medicine and state institutions in the form of drugs and food aid. Here, we see not social conversions (as in Caasah) or struggles against the limits of social inclusion, but an agonizing postponement of when dying begins. This complicated paternalism and the bleak social life that accompanies it are not the signs of old, "pre-modern" times (as former president Cardoso would refer to them); rather, they are actually fueled by the emergent local structures of assistance to poor people with AIDS.

At any rate, given his influential new position and the funds available, Winkler was devising ways to address the immediate concerns of poor AIDS patients, he said. "We had no other choice but to invest in direct social assistance from bus vouchers to food baskets. The ideal would be to connect these actions to systematic visitation programs and so on . . . but at least we are bringing them into the program." The logic underlying his work was that of "partnerships not just with civil society organizations but with those vulnerable groups that our actions are meant to reach": "We are bringing them into the program to work with us. We have a group of intravenous drug users who help us with research and with peer prevention. They have practical knowledge and it is fundamental to put them in conversation with the technical team. Of course, the experts feel threatened; their knowledge is being questioned. . . . Yes, it gives me great joy to find institutional ways to counter technocracy."

Winkler was also politicizing prevention among the central prison's inmates. During that 1995 visit, Torben and I were only admitted into the central prison after much negotiation. Stricter security measures had been established after a July 1994 rebellion that left the general director wounded and wheelchair bound and seven prisoners dead.

Sandra Perin, a psychologist whom I had met in 1992, told us, "The AIDS community has been dismantled. No more preferential treatment." Six of the eleven *isolados* I had met back then were already dead, she said. She called an HIV-positive inmate to her office; upon arriving, Eliane, a transvestite, asked to be photographed and summed up the terror in the cells: "They take my plate of food and only give it back after they fuck me. Nobody cares. They are animals. For me, the virus is theirs now."

●●●

By August 1996, the civilian police had replaced the military forces in the central prison's administration. As the new commander-in-chief put it, "We are looking for a more democratic and humane form of administration." With the support of the state's human rights commission and funding from the national AIDS program, Winkler was able to circulate a monthly newsletter produced by the prisoners themselves. The newsletter was called *Arpão* (drug needle or harpoon).

The project was immediately useful on all sides. The prisoners used it as a vehicle to fight for better medical treatment and also to expedite their claims for retrial or pardon. The prison's new administration used it as a proof of its "cooperation" with human rights issues, despite its ironic lack of concern for the AIDS that was rampant among the prisoners. As for activists like Winkler and his team, the editorial meetings with the prisoners were a way for them to smuggle in syringes and bleach water: "We are messing with the bureaucracy from inside. We are putting the World Bank money to good use; we are exposing what the government is not doing."[12]

An anonymous prisoner described his predicament:

The government and AIDS organizations forgot the prison system. Here there are too many people with AIDS and too few thinking that it is worth fighting for the rights to better treatment. I know that there is a cry for help in each one of us. . . . I must say that the doctor of this place is not at all interested in our health. Why, then, is this doctor here? Only at the last moment, they take us to the infirmary and leave us there to die on our own. . . . My struggle is psychological, for I don't have access to AZT, DDI, DDC, those drugs that combat the virus. Thus, the only way to fight the disease is by engaging the administrative and legal authorities. They are the ones who can send us into the world so that we might find ways of surviving there.

Eliane

Measures of Success, Undesirable Realities

Winkler's prevention work with the prisoners was not the norm. As I mentioned earlier, AIDS prevention was mostly carried out by NGOs, as they were supposedly close to local communities and able to reach specific vulnerable groups. All over the country, NGOs were training community leaders to become "HIV/AIDS prevention disseminators." Consider the language of one of the community prevention manuals issued by the national program in 1996 (Larvie 1997, p. 101):

> The only medicine [against HIV] is the substitution of risky situations by behaviors which present little or no risk. Behavioral change is always possible, since we, as human beings, are not controlled by instinct as in the case of irrational animals. All of our behaviors, including those related to sexual and erotic attraction, may be controlled or modified. This intervention fundamentally seeks to transmit information and techniques about how to avoid STD's and HIV infection. It is not an attempt to change behaviors because of moral convictions or out of ethical condemnation.

Anthropologist Sean Patrick Larvie (1997) analyzed community-oriented prevention programs fostered by the national AIDS program in Rio de Janeiro in the mid-1990s. The model that informs these safe-sex interventions, argues Larvie, "suggests that health—defined as the absence of illness—is a universal and invariant value as well as a powerful motivator of individual and collective action" (p. 102). This emphasis is part of a change in public health that locates "the nature of the problem as well as the possibilities for its solution, within the minds and instincts of individuals" (p. 99). The key idea is that risky behavior can be brought to a public forum by community representatives, and it can then be rationally and technically manipulated through "'correct' and neutral information based on scientific principles," says Larvie (p. 100). Central to these prevention strategies is an articulation of medical knowledge with non-medical theories and techniques (including social anthropology, behavioral psychology, pragmatic philosophy, and community activism) and the social reorganization of selective segments of a larger population.

In my work with AIDS NGOs in the southern and northeastern regions, I saw how prevention projects were being idiosyncratically tooled by organizations, professionals, and target groups. Some of the immediate spin-offs of these projects were the improvement of the particular NGO's infrastructure and the opening of new staff positions. Through these projects, underemployed social scientists were technically upgraded and had additional earnings. The community leaders involved had the opportunity to consolidate their "representativeness" and widen their networks of contacts, in addition to reaping personal financial benefits. However, something unexpected was also emerging during prevention workshops and activities. Participants usually referred to the proximity of AIDS in their family and neighborhood circles. This local AIDS commonly gained a character of public secret—it was unrecognized as AIDS in the nearby health services and denied by people who refused to identify themselves as *aidéticos* and be stigmatized in the "community."

In the fall of 1996, I was hired by the Health Division of the city of Salvador to evaluate local HIV prevention projects financed by the national AIDS program (see Biehl 1996). At that time, I examined a community prevention project—AIDS & Women—developed by GAPA Bahia. Founded in 1988 by young social scientists and gay activists, GAPA now had seventeen employees (four institutional coordinators, five project coordinators, eight assistants and administrative personnel). The organization had three lines of action: (1) psychological and legal support for HIV-positive individuals; (2) prevention and education (projects included AIDS & Prostitution; AIDS & Periphery; AIDS & Teen-agers in Public Schools; AIDS, Gender & Ethnicity; and AIDS & Homo-/Bi-sexuality); (3) public policy making (e.g., public debates, political demonstrations, petitions). The organization also had an AIDS hotline service, a library, and a video service. A group of registered volunteers helped carry out training programs and conferences.

The project AIDS & Women sought "to train some 25 women living in the periphery, through a participatory methodology, as STD/AIDS prevention educators" (GAPA Bahia 1996). These women leaders were identified in local health associations operating in ten peripheral districts. The project emphasized that these community associations served as privileged spaces for AIDS prevention because women gathered there both as clients and as volunteer health providers.

I interviewed two of GAPA's coordinators, and they acknowledged that this intervention was loosely modeled on prevention work done in gay communities in the United States. It was founded upon the presumption that scientific and technical AIDS information should be translated into a popular form by representatives of local communities and that AIDS was "a means of advancing civil rights causes and demands," as sociologist Maria Miranda told me. Psychologist Cristina Vasconcelos added: "It is a way of making information available for the masses . . . more people can be protected once they have access to information."

It seemed to me that GAPA's project was framed within the social dilemmas that confronted a large number of AIDS NGOs in the mid-1990s. They searched for their own activist identities in the lexicons of international communities, all the while having to address people living in extreme poverty. How did this specific project account for the multiple variables that lead people to assume that they are not at risk? How did it relate to local public health care services? How did GAPA and the community workers measure success?

I found out that the women's project was part of a larger project called AIDS & Periphery, which began in 1993 with funding from a German agency. The first part of the project was aimed at producing a detailed profile of these poor communities. It also sought to assess the availability of information about HIV/AIDS in each community, as well as the meanings and representations associated with the epidemic. Sociology and psychology students administered structured questionnaires to a broadly conceived general population. After three years of study, not surprisingly, the study identified a generalized intransigence with respect to adopting safe-sex practices.

What hinders safe sex? "A thing of the collective imaginary that the well-behaved, the normal ones, will not get infected," Cristina Vasconcelos, one of the principal investigators, told me. Another finding of the study was that people "greatly reject individuals living with AIDS in the community." The investigators took this as a sign of a discriminatory ethos that prevents people from seeing themselves as subjects at risk. "Even the female leaders show fear and awe in the face of the proximity of AIDS, of people dying with AIDS. They discriminate against their own neighbors," added Maria Miranda.

These social scientific findings did not lead GAPA to address the difficult reality of an AIDS already there. Rather, the study was followed up by a workshop focusing on "women's self-esteem" (funded by the partnership with the city and the national AIDS program): "These women resist accepting the fact that they are a target of AIDS, they still see the Other as the target. We chose to work on a gender perspective, because we realize the social vulnerability of women at this moment of the epidemic," Miranda told me. The AIDS prevention workshop was basically aimed at helping these community leaders understand themselves as potential targets of AIDS. A thorough vision of the social, economic, and intersubjective dynamics surrounding AIDS remained unaddressed.

The training workshop questioned the place of poor women in the epidemic from the perspective of women's difficulties in negotiating the use of condoms: "AIDS presupposes a dialogue between partners—that you talk about these things, that you make joint decisions. There must be an affirmation of the legitimacy of woman in these negotiations," said Vasconcelos. There were many doubts, even among GAPA's coordinators, about the efficacy of a prevention model centered on the dissemination of information. This was volunteer work, and the majority of people working in this field had to fight for their own economic survival. There was not much time for in-depth technical training. Thus, the immediate positive result of the workshop was to have helped these women to ask themselves, "How do I see myself in relation to AIDS?" Miranda explained: "I can only begin disseminating AIDS prevention information when I internalize this for myself—at the moment, when I discuss this with my husband, my partner . . . it depends on how I handle things at home."

The project allowed the community representatives to discuss publicly their own sexuality, or, according to Vasconcelos, "to break down some barriers. . . . At first there is the barrier of speech. It is very difficult to expose and speak what you feel about sexuality." GAPA's project of "gluing AIDS to other community problems" seemed reduced to this educational process in which a few women were taken out of their communities, epidemiologically trained, empowered with a "reflexive and dialogical sexuality," and upgraded to be representatives of "community initiatives." For the most part, the workshops did not reflect on the experiences of these poor women as community health volunteers, nor did they account for these women in planning future lines of action. Why?

GAPA framed AIDS as a "social problem"—the social being understood in quite fragmented terms, through identity categories. Alongside this fragmented notion of the social, demands for medical assistance were not prioritized or given political legitimacy. Simply put, rhetorically, the NGOs channeled the government's prevention efforts to vulnerable populations. In NGO practice, however, impoverished realities were contained in a selective group of "prevention citizens." These "prevention citizens" came to represent (and thus symbolically generate) "communities" that were brought into the NGOs' headquarters and trained as alternative health providers.

What I saw at the micro-level in GAPA also reflected what was happening with prevention programs all over the country: as these technically construed and culturally sensitive prevention programs helped to create a governmental credibility of sorts and to partly reshape the public discourse on sexuality and risk, they failed to integrate themselves into the country's precarious public health care system. In the meantime, nobody addressed these women's "unexpected" recognition of a local and nearby AIDS reality. This "undesirable reality," using the words of Wildney Contrera (GAPA São Paulo), opens up new questions of how to relate prevention practices to immediate health care and how to consider the complex interplay of AIDS publicity and secrecy in the formulation of prevention strategies.

The Undetectable Virus

"Today Roche, Merck, and Glaxo are the ones who control me. They are in my body, I am them." Gerson Winkler told me this in June 1997, as I had concluded my long-term fieldwork in Bahia and was participating in a symposium he had organized on the new tendencies of the AIDS epidemic in the south. Winkler was taking the "AIDS cocktail" that was now allegedly available to all HIV/AIDS cases in the country. "My HIV is now almost undetectable. Can you imagine this?"

Winkler was alive and well, trying to make sense of his new condition: "I am aging again. I might no longer die of AIDS. My oldest daughter will turn fifteen this year. This is fantastic, to know that more than eleven years of infection and the fear of death being near have passed, and that now there is a future. Is AIDS over?"

"At least one form of AIDS, and for some," he added. As enthusiastic as Winkler was about the possibility of extending his lifetime, he was ambivalent about the lifestyle and values that accompanied the combined antiretroviral therapies:

I have friends who count how many pills they take a month. It is a different calendar and sense of time. I am still kind of lost in this new reality. People congratulate me for being "undetectable," as if I were responsible for this. It is the chemicals, I say. They should congratulate Merck. People create these myths of AIDS patients now being somehow "superhuman." Truly, I feel that I am in a vacuum. I am no longer part of the "I will die of AIDS world," but I also don't feel that I am back to the world of the humans, the ordinary that has been robbed from me. My life is regulated by a complicated math of pills, exams, private doctor, the bureaucracy that keeps the flow of drugs from Brasília to Porto Alegre. Before I had an overpowering enemy in the HIV virus; now I have to develop this capacity to control it at all times.

Winkler said that he was privileged to be under the care of one of the city's best immunologists, who helped him bridge the incalculable distance between the disciplines with which he was once familiar and those required to adhere to the ARVs. "As a public officer I have private health insurance. My doctor is thorough, I cannot play the victim. He keeps me on track."

In our conversation, Winkler also pointed to the inequalities embedded in the emergent pharmaceutical management of AIDS: "When people with AIDS meet, you can right away identify those who take antiretrovirals and those who don't. It was not the virus that created this difference. There are new kinds of criteria at work here; all criteria are sordid in one way or the other. But this one is the worst, for now you can have the sign of Roche in your body but you don't have the trace of rice and beans."

Winkler had opened the symposium by warning people not to draw quick conclusions from epidemiological data alone: "Many times, the data that comes from scientific production is late; it does not correspond to what happens over time. We must place the data in the map of real life." And to do this, "we must develop visitation programs to households, we must identify local variables related to AIDS incidence and the ways people deal with the disease."

This local knowledge was ever more important now that AIDS patients could actually extend their lives, added Winkler. He told us of a homeless man who had recently come to his office asking for help. "He was ragged, dirty, and slept on the public square. He said that he was undergoing treatment. He didn't mention AIDS. He pulled a bottle of Crixivan from his pocket. The bottle was big, white, hermetically sealed, exhibitionist. That is inequality too. That bottle will not save his life, but he can now have the illusion that it will, or better, use it as a way of begging."

Yes, the fact that Brazil now had an AIDS treatment policy had to be praised. But this was no reason to dampen critical political discourse. "Should we not critique or point to gaps because we are getting the drugs? Which compromise is this?" AIDS was now treatable in the country, Winkler pondered, "but it continues to spread and who is concerned with the new forms of the epidemic, with the real difficulties poor people face to enter this medical regime?"

•••

Against all odds, AIDS activists and health professionals working within the government had invented a public way of treating AIDS. Epidemiological accounts of HIV/AIDS in Brazil in the late 1990s conveyed a sense that, on the one hand, there was a decline in AIDS mortality and a relative stabilization of the number of new HIV infections. On the other hand, HIV/AIDS was said to continue to spread beyond "risk groups" and among the poor and most vulnerable (CN 1997a). Both politically and medically, the Brazilian AIDS program was a rare, if partial, success story, and the World Bank approved a second loan (without funding for treatment). The "AIDS Project II" began to be implemented in 1998. One of the key concerns now revolved around decentralizing actions, as well as reaching women and low-income groups that kept defying prevention and assistance efforts. The integration of poor communities in the planning and implementation of alternative actions was a priority (CN 1998b). Policy makers were now mentioning civil society organizations alongside NGOs.

As I read the "AIDS Project II," I was reminded of Nilton Rosário da Costa's comments that, in times of structural readjustment, the defense of "communities of poor people" can easily work as an ethical imperative through which old state institutions of social protection are identified and judged as corrupt or useless (1996, p. 27). Such discourses, with their participatory and ethical underpinnings, thus legitimate administrative reforms. In reality, no specific analyses and few policies were being developed to address the increased misery and social fragmentation linked to market-oriented reforms.

In mid-1998, Gerson Winkler sent me an e-mail to inform me of his change of address. He had left the AIDS program of the city of Porto Alegre and had begun working with the state's Human Rights Commission, with a focus on violations in the penal system. "AIDS has become a chronic condition to me, to the extent that I now can live without it," he wrote. "I leave the status of death to that of reconstruction."

In the e-mail exchange that followed, Winkler expanded on what he meant by reconstruction: "Initially, many people living with AIDS constructed their identities around disability; for some, AIDS was also a way to go further into marginality. The therapies have radically altered this. The social and cultural formations that made AIDS real have reached an endpoint, at least for me. There was something very religious about our activism: we surrendered to the idea of death. My body is now reacting differently. I feel life . . . and then this sense emerges that I actually have a say of what will be next."

Winkler said that the sum of the AIDS cocktail with the testosterone he was taking to get back his "sexual appetite," and the prospect of working with people who had all kinds of emotional, legal, and social concerns, "has recombined the mirror of my life. It is a moment from which I don't want to return. I got tired of being a whistleblower. I now see other life forms than HIV alone. I want to do normal things, like renovating my house." A pragmatist at heart, Winkler credited his new take on life to his marriage to Junior Batista, an Afro-Brazilian journalist, "a person who is from the side of the good, who opens the door for you."

Winkler also told me that that in spite of all the important advancements in terms of free access to ARVs, the poorest and most marginalized people living with HIV/AIDS were still a long way from being adequately addressed and that this had a lot to do with local politics and the overall deteriorating public health care system. Out of the world of AIDS activism, he remained painfully aware of the ordinary contexts in which the technical extension of life takes form and kept asking fundamental questions: "Are assistance services being remodeled? Are medical professionals being trained to address the diversified epidemiological, social, and biological demands of all AIDS patients that are now being found? Are the vulnerable populations actually getting special treatment, and if so, what are they doing with the therapies, how are they adhering to treatment, in which ways are their lives changing?"

•••

The next time I met Winkler, in December 2000, he spoke happily about aging and "the wisdom that comes with recognizing other possible identities and life pathways" outside AIDS activism. He had just been awarded a prize from the city hall for his struggle against AIDS. "I was kind of confused," Winkler said with a tone of disdain. "I was not sure whether the prize was to honor the work I had done for over a decade and therefore it was a proof that I was from the past, or because I was still alive and I was supposed to symbolize the hope of a world without AIDS, which is not what one sees on the ground." At any rate, "the prize reaffirmed my distance from the ways government and NGOs are reducing AIDS to a technical matter."

Winkler also had bitter words against an emergent public health discourse that blamed "ill-informed AIDS patients" for low adherence to treatments, rather than understanding the many social variables at work: "Adherence programs are being clinically framed. They are by and large imaginary and too authoritarian." Winkler knew from his own experience how difficult it was "to live by numbers, a pill at this hour, then the next in two hours, after eating such and such food . . . then the expectation of the next CD4 and viral load count. . . . How is the treatment adherence of a mother who works and has children? Of a young man who hides his AIDS from the family, or of a factory worker? Not to mention how difficult it is to get the medication to prisoners, or how difficult it is for poor AIDS patients to get adequate information from doctors in public services. We need to get a notion of the walls to which treatment can adhere, and this understanding must be carefully composed."

"It is all about medicines now"

I visited the headquarters of the national AIDS program in Brasília twice in 2000. At the time, the program was undergoing a transition, with Pedro Chequer leaving and Dr. Paulo Teixeira taking over as coordinator. Several AIDS activists I had met over the years were now working at the program. People were quite busy, but they found the time to talk to me. Their comments revealed an impressive capacity for strategy and critical reflexivity, even amid the nationalistic pride that came with the success of the ARV rollout.

Several of my informants there had no qualms about saying that both the World Bank and the government were capitalizing politically on the AIDS policy. "They don't want to lose this," anthropologist Jane Galvão told me in January 2000. AIDS was indeed a central topic in both the Health Ministry and the Foreign Affairs Ministry. After presiding over ABIA (Associação Brasileira Interdisciplinar de AIDS, Brazil's most influential AIDS NGO) for over six years, she was now coordinating the program's interface with civil society: "The partnership between government and civil society is always praised by politicians and donors, but in reality, what we have are people from NGOs and AIDS researchers in all kinds of governmental committees . . . from this perspective, civil society has been reduced to very few people."

Yes, there was "strong presence of AIDS within the state, and the program has great autonomy," but Galvão was concerned with the sustainability of activities carried out by NGOs and civic groups, as there was uncertainty over the likelihood of a third World Bank loan. "Bank funds were used for infrastructure building, and without these funds, it will be hard for NGOs to survive," Galvão explained. "They will have to look for partnerships with the private sector. GAPA Bahia, for example, opened a store in a local mall." Donor agencies were also shifting their focus from Brazil to Africa and Asia. "They conclude their participation here with a success story," said Galvão. "Organizations such as ABIA, which produced more than ten volumes on the social history of the epidemic in the country, now have no audience for this production."

Money from international agencies "is less flexible now. It must be used to execute a project with a beginning, a middle, and an end. And any citizen can do that, it does not require an NGO." Now the concept of "civil society" was less linked to the idea of "political participation" and ever more to the "execution of services." And Galvão thought that assistance institutions, such as Caasah, would have more chances of surviving because, "for better or worse, they do something that nobody else wants to do."

NGOs were no longer the sole means through which people participated in the program. Galvão told me that the Health Ministry had now extended the "Dial Health" initiative—a hotline service that had been developed within the AIDS program—to all areas of the Ministry, and that the initiative was fielding more than 10,000 phone calls per day: "Anyone can use the service . . . to ask for information, to place a complaint that a medicine is not available in the health post, that he has been mishandled by health professionals. . . . Many also call to denounce corruption in NGOs." Dial Health was channeling a new culture of democratic citizenship, suggested Galvão: "Today, people use whichever channel is available to denounce. Here, the person can remain anonymous; he has only to fill out a form. . . . Local services are immediately contacted and asked to fix the problem. When it is a question of corruption, for example, an investigation is undertaken, and more information is sought."

"For various political reasons," an immunologist working at the program told me in August 2000, "the government is finding it strategically advantageous to keep the AIDS policy going, and our day-to-day here has become quite tumultuous. Things have to be done from one day to the next . . . it's a bit complicated, there is some confusion on our priorities . . . but things have to get done." Another expert, a psychologist specializing in HIV prevention, said, "It is all about medicines now. All our actions, national and international, are basically aimed at maintaining the flow of medicines. There is little discussion of anything else." There was no way back on this pharmaceutical course, he added: "The distribution of medicines is a technical thing, right? After the rollout began, it's like food, you must guarantee that it gets to people's body every day."

In all, I heard these experts saying that powerful institutions like the World Bank, political parties, the government, philanthropic foundations, and now the pharmaceutical industry were continuously altering the landscape from which the sustainability of AIDS actions had to be construed. It required a lot of thought and maneuvering on the part of these governmental AIDS activists to keep their place in a moving reality that these powerful institutions were attempting to monopolize, and to wrest this monopoly from them, action by action, day by day.

Political will, a strong national pharmaceutical industry, and international visibility and exchange were key tools to ensure the continuity of the policy, Pedro Chequer had told me earlier in January, as he provided me with graphics and statistics showing how cost-effective the AIDS policy was. "We can prove that there is a decline in the number of AIDS deaths. In 1998, we spent $300 million in medication, and actually saved $500 million due to reductions in hospitalizations and costs related to treating oportunistic diseases." A national laboratory system now offered free CD4 and antiviral dose measures in every state of the federation. The program was now taking steps to implement a nationwide network of free access to genotyping for AIDS patients. In spite of the generic production by public- and private-sector companies, 70 percent of the spending on anti-HIV drugs went to foreign companies. "This figure will be reversed," added Chequer. "The state must produce, and it will."

Brazil was sharing its ARV rollout know-how with other poor countries and developing all kinds of governmental and academic partnerships. The immunologist I interviewed was coordinating several of these international efforts: "Right now we are developing a cooperation program with Mozambique's Ministry of Health, working with the German government on an STD program, and renewing partnerships with the Johns Hopkins University. We are also helping to organize a meeting in Rio in November that will gather 3,000 people working on AIDS in twenty-two Latin American countries." The Latin American AIDS Network had become a kind of "proxy-PAHO" (Pan American Health Organization) that facilitated the exchange of information on drug pricing, the production of generics, and the "nuts and bolts" of managing a large drug distribution program. Internationally and nationally, the AIDS program stood, in many ways, for the state and for the new right of individual citizens to intervene effectively in the sphere of policy and strategy.

In Search of a Comprehensive Approach

"The AIDS policy is the only thing that is actually working well in a government that is clearly uncommitted to social causes. Fernando Henrique Cardoso is a prince. He lives in a palace that does not correspond to the reality of the population and sells this policy as if it were a model of his administration," said psychologist Monica Vieira, director of the AIDS NGO Pela Vidda, in Niterói, a city in the state of Rio de Janeiro, in early August 2001. "We are on the eve of presidential elections, and the health minister will be the government's candidate. The AIDS policy is part of this political game; it is being used to mobilize public opinion." Like Gerson Winkler's, her position, in principle, was neither for nor against the government. Rather, both incorporated in their critique and work the processes unfolding through the policy and at its edges.

Vieira has been working with Pela Vidda since 1991, and she was proud to say that her organization focused on the "real world" of people living with HIV/AIDS and that it blended juridical activism and assistance. "We had to provide people with basics such as food so that they could begin thinking in terms of citizenship." She had just returned from New York, where she had been the civil society representative in Brazil's delegation to the United Nations (UN) Conference on AIDS and Human Rights. "We fought for a declaration that considered access to ARVs a human right. Of course, this will not be put into practice immediately, but such a declaration is an important tool for treatment activism," stated Vieira. This pharmaceutical approach to AIDS, however, coexisted with a moral politics that kept "sanitizing AIDS discourse," she said. "Our biggest loss was that the declaration did not mention specific vulnerable groups such as homosexuals and intravenous drug users. . . . There is a strong pressure from the United States and from Arab countries to keep them out. And this has all kinds of negative implications in terms of projects that are funded, etc."

"But there is no doubt," Vieira continued, "that the Brazilian ARV rollout led to a change of paradigm in AIDS activism here and elsewhere. And this is great. If, in the late 1980s and through the mid-1990s, we struggled against discrimination so that the person living with HIV/AIDS remained a citizen and had access to disability pensions, now the struggle is for the quality of their lives, and this means continuous access to therapies and lab exams, and to employment." As with most things in Brazil, she reasoned, "the AIDS policy is perfect as a proposal, but when you put it into practice, it becomes something different. . . . We know the day-to-day difficulties involved in this 'blessed model.' On the ground, it is not an island of fantasy; there are gaps, many things that are not working and should be improved." Here the ordinary plight of poor men and women living with HIV/AIDS was treated not simply as a residue of policy but as grounds for improvements that were both urgent and entirely possible.

That same week, I also met psychologist Carlos Passarelli, who had just left his position as coordinator of HIV testing at the national AIDS program and was "back to activism," as he put it, at ABIA, in Rio de Janeiro. "The truth is that we advanced a lot as far as HIV/AIDS is concerned in Brazil, but with the basic focus on therapies and foreign relations, prevention efforts have stalled, and that is troubling." Like Vieira, Passarelli also preferred not to use the word *model* to describe the AIDS policy: "The image the policy became is entangled in politics, and [this characterization] is not so true." He added: "Yes, it is important to show to national and international publics that this policy is a good thing, and this helps sustain the policy, but not at the cost of losing sight of everyday problems. . . . We already proved that the World Bank was wrong: it is actually possible to treat AIDS among the poor. We invented this possibility. Now we need to know whether we have many other infected poor who are not taking ARVs."

Vieira and Passarelli agreed that the ARV policy's biggest obstacle was the deteriorating public health infrastructure and the lack of specific training of health professionals. "Patients might have to wait for a long time, but they leave the service with the medication. Many, however, don't get adequate information and follow-up, which hinders their adherence," said Vieira. "And when they get pneumonia, there is no Bactrin at the local health post. It takes an average of three months for a patient to get an appointment with a medical specialist. Unfortunately, people get used to this and do all kinds of thing to move through the system. Seen from this angle, the celebrated response is not so perfect."

"By and large, from my own experience," Passarelli told me, "health professionals still convey a sense that they have been forced to treat us, that we are privileged deviant faggots, giving them lots of additional work. And I am speaking of São Paulo, the best place in the country to get your therapies." Moreover, said Vieira, "there are large numbers of poor people still dying with AIDS and without assistance. They don't have access to information, to medication, to decent care. They lack, period. These people need *casas de apoio*, but in Rio these initiatives never took off."

Passarelli also mentioned that much more had to be done in terms of HIV control: "We cannot let the success of ARV rollout suggest that prevention efforts are actually working. HIV/AIDS is part of the public imaginary, but to turn this knowledge into a personal risk assessment is a big jump . . . and there are many social and economic variables that make this happen or not, and they must be addressed by specific prevention strategies. The antiretroviral policy must also integrate prevention. Informally, one hears that some homosexuals on ARVs are no longer using condoms . . . this must be seriously addressed." There was also concern over the accuracy of representations of the course of the epidemic and the design of alternative research: "The rates of HIV vertical transmission are shameful. The epidemic keeps growing among the poor, and we cannot simply say that it is because our surveillance services are now more refined and can capture more cases. Moreover, there are no clear criteria guiding the discussion of AIDS and poverty at the program. Who specifically are this poor? A person does not get infected just because he is poor; there are practices and connections in time and space."

As for the government–civil society partnership, both activists also had a view similar to that of Gerson Winkler: financial mechanisms and participation in committee work had created a confusion of roles in this partnership, which had curtailed the critical capacity of civic groups. "Social movement is strong when it acts at the level of jurisprudence. But right now there is a tacit conception that civil society is useful inasmuch as it says to the government what it already knows it has to or wants to do," Vieira said.

To make his point, Passarelli used the debates over the shortage of kits for viral load tests in the Rio de Janeiro health care system: "The problem was brought up at the regional AIDS program. I was there. The officers said, 'Let's mobilize the NGOs to do something about it.' Things are indeed out of place. What kind of circuit is this, with the state pressuring

NGOs to pressure the state?" For him, this incident was "symptomatic of a confused situation, in which there is no clear sense of the basic role of an NGO, of government, and of health professionals who see themselves more as social movement than as government. At the same time that we have to sustain the policy, for we are now, supposedly, the best in the world, we are also caught in this confused state that, after all, contributes to the postponement of action."

Their task as activists, I thought, was to look a bit behind politics. In their hard work and stubbornness, it was a question of being close to reality, without a doubt. Having identified cracks and dysfunctions in institutions and practices, they laid out a context from which a different kind of work could be done, not forgetting that this very work was also part of their biographies.

"There is not just one death"

At the end of that 2001 trip, I stopped in Porto Alegre and met one more time with Gerson Winkler, who now had an important and well-paid position at the witness protection program of the state of Rio Grande do Sul. He was indeed the reconstruction of AIDS: healthy, in a steady marital relationship, a public officer. Our conversation centered on four recent events. Early that year, while repairing a shack on his farm, he had been infected by a rat and almost died of the bubonic plague. "I was beating AIDS and now had to overcome this ancient plague," he said. "There is not just one death."

Winkler had also become a grandfather. His youngest daughter, at the age of fifteen, had given birth to a girl, and both Winkler and his former wife had to provide emotional and financial support to the single mother and her child. He asked whether I had not noticed anything in his face. I said that he looked great. "I had a face lift." He went on to elaborate that the cocktail had returned a sense of aesthetics to AIDS patients. "You see new bodies, the sequelae of surviving AIDS . . . antiretrovirals leave profound marks in your face, the fat is dislodged all over the body. Before the physical compromise was related to death, but now it is to life . . . and this raises all kinds of anxieties over appearance and self-esteem."

Winkler had also briefly reentered the world of AIDS:

I just came back from [northeastern] Recife. The Workers' Party administration asked me to evaluate the city's AIDS services. It is horrible. I never imagined that the epidemic was in such chaotic situation. Nothing but drug dispensation works. Surveillance does not work, social assistance and prevention do not work. You see figures of the epidemic from long ago, those skinny and cadaveric bodies—not the AIDS bodies of today with lipodystrophy—coming to the medication counter. They passed through a *via crucis* to get there, their health completely deteriorated in a totally compromised health care system. . . . How irresponsible of the state to have left this epidemic abandoned to its own course in a place of extreme misery, tourism, prostitution. All very somber. The total opposite of this picture of NGOs, big money, representations of this and that. AIDS there is desperate. I suffered doing this consultancy. I was a consultant in a desert. The wind that comes every night takes the sand to another place, and you cannot change the force of the wind.

•••

Recently, I learned of the tragic death of Pernalonga in the outskirts of Recife in late 2001. One night, on his way home he was attacked and left bleeding. As he was crying for help, neighbors approached. Afraid of getting contaminated, they did not provide immediate care. They called the police though, who came and had the same attitude. Pernalonga agonized for hours until morning broke and a special medical unit removed him from the scene and took him to the hospital, where he arrived dead.

Chapter Three

A Hidden Epidemic

The Limits of Surveillance

In December 1996, a consortium of epidemiologists and social scientists met to prepare an official evaluation of the state of the AIDS epidemic in Brazil. I participated in the final workshop in Salvador, Bahia. Discussions ranged from the validity of risk categories and biostatistical models to the failures of the epidemiological surveillance system. After several days of these discussions, a representative of the national AIDS program entered the final plenary session, asking the participants, "Which version of the epidemic will we present to the journalists?" All knew the tone had to be positive—a success story was in the making, and a second World Bank loan was on the horizon.

The final report acknowledged that the epidemic had spread to diverse populations and that it was now characterized by impoverishment and feminization, but it also pointed out that these trends were accompanied by a "decreased speed of growth of AIDS in the country" (CN 1996, p. 113). In March 1997, the Health Ministry's *Boletim Epidemiológico de AIDS* reported a general decrease in the incidence of AIDS for the first time. Furthering the success story, an impressive drop in AIDS mortality followed, arguably a direct result of the now freely available antiretrovirals (ARVs).

The AIDS reality I saw in the streets of Salvador and in grassroots sites of care like Caasah contradicted these optimistic trends (which also pervaded local epidemiological reports). In reality, surveillance efforts and medical services overlooked a large number of poor and marginalized AIDS victims. Those whom I met were young (ranging from eighteen to forty-five years old) and had no stable family ties. Having lost their formal identities, they lived in the streets or abandoned buildings and survived through informal market activity or begging. In their troubled existence with AIDS, they had only sporadic contact with the government's testing and medical services or with nongovernmental forms of support;

however, no specific programs of prevention, treatment, or support were aimed at them. Their experience of dying was ordinary, met by political and moral indifference.

In what follows, I locate their dying—which occurs in *apparent invisibility*—within a local public health pattern of *nonintervention*. I then explore the paradoxical coexistence of this hidden epidemic with the national AIDS policy and the country's overall reform. I found it necessary to employ both qualitative and quantitative methods. And I teamed up with epidemiologists who were concerned as much with reflecting on the politics of AIDS knowledge as with developing analytical frameworks to give us a more realistic assessment of AIDS in Bahia. A caveat: AIDS politics in Brazil is extremely dynamic, and the local control of the epidemic discussed in this chapter is also fluid and changing.

AIDS in Bahia

In 1996, Bahia's Health Division issued a general report on the state's public health situation. Citing evidence of mortality decline, the report began by positing that population health had improved in Bahia, although this statement is somewhat misleading. Infant mortality, for example, had indeed fallen in Bahia, but at a slower pace than it had in the richer regions of the country. The report also suggested that the region was undergoing epidemiological transition, with traditional infectious and parasitical diseases on the decline and chronic and degenerative diseases on the rise. Along these lines, it stated that, from 1984 to 1993, the number of hospitalizations due to infectious diseases stabilized, "in spite of a discrete increase in the end of the period"—probably because of the cholera and dengue epidemics, described as "natural" exceptions (Secretaria da Saúde do Estado da Bahia 1996, p. 16).

The report pointed to a continuous tension in recent decades between the changing nature of public health and the enduring presence of poverty and inequality. "Problems of misery," the report posited, have intensified through uncontrolled social processes—such as migration, urbanization, and industrialization—that are themselves linked to the economic development of the region (p. 18).

As I read the report, my first concern was with how such epidemiological accounts rhetorically "modernize" the population profile in question. The great majority of poor people, public health researchers told me, have continued to be the target of largely imaginary state plans, and they have increasingly been left to devise their own community health alternatives. A physician who had worked on the Bahian epidemiological surveillance system during the 1994 cholera epidemic went further in his criticism: "it is immoral, it is a crime." He confided that he terminated his collaboration with the surveillance system upon realizing that it was withholding mortality data from the general public "in order not to generate panic." For him, this "containment" amounted to a "death decree."[1]

What did the Bahian AIDS numbers say?

National risk data placed the state of Bahia at an intermediate level of risk, with lower incidence rates than those of São Paulo and Rio de Janeiro but higher rates than those of other states in the northeast, center-west, and northern regions. The scarce AIDS data available for the northeast region confirmed that there was no temporal lapse in the progression of the epidemic there (Dourado et al. 1997a). The first Bahian AIDS case was registered in 1984 (Romildo, who was first cared for by Dona Conceição and then became the co-founder of Caasah—see chapter 2).

A study carried out at the AIDS ambulatory service of Salvador's main public hospital during July 1989 to January 1991 identified 111 AIDS patients (Moreira et al. 1993). These patients represented 65 percent of the 170 AIDS cases that the epidemiological surveillance service reported to the Ministry of Health during the same period. The 102 men and 9 women reported homosexual/bisexual activities (60 percent), intravenous drug use (19 percent), or both (6 percent); heterosexual activities (11 percent); and blood transfusions (2 percent). Two percent belonged to an undetermined category (p. 687).

As shown in Figure 3.1, the period of 1985 to 1993 was marked by a rapid increase in AIDS incidence rates. After 1993, incidence started to drop, apparently stabilizing by 1997. The "Health Plan 1996–1999" of the Secretaria da Saúde do Estado da Bahia stated that "the proportion of AIDS patients in Bahia who stay alive increases every year, and this is a positive effect of the therapeutic assistance directed to these patients by public health care services" (1996, p. 71). Note, however, that healthier people with AIDS might have been unregistered before, such that the

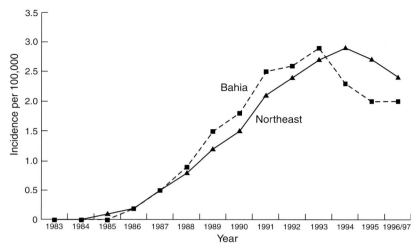

FIGURE 3.1 Official AIDS incidence in Bahia and the rest of Brazil's northeast

proportion formerly included mostly severe cases. Now that registration had improved, the healthier patients might have been better represented in the survival data, thus inflating changes in the proportion that stayed alive.

According to the February 1997 *Boletim Epidemiológico de AIDS*, Bahia had 2,308 reported AIDS cases, with a cumulative incidence rate of 20.9 cases per 100,000 inhabitants since 1984 (CN 1997e). Salvador accounted for 67 percent of these reported cases (2,308 cases), with a cumulative incidence rate of 81.7 per 100,000 inhabitants since 1984. As Dourado and colleagues (1997b) put it, "The majority of these AIDS cases are young men with homosexual and/or bisexual behavior, and intravenous drug-users."

Data collected at the Bahian epidemiological surveillance service in 1995 (Dourado et al. 1997b) showed that the majority of Bahian cases were young (ages twenty to thirty-nine) and that the male/female ratio was 4:1, which was consistent with the national profile of the epidemic. Of this official Bahian AIDS population, 66.1 percent had been infected through sexual transmission and 23.2 percent through intravenous drug use (perinatal transmission amounted to 1.5 percent; hemophiliacs to 1.4 percent; and blood transfusion to 1.7 percent).

Figure 3.1 shows a similar pattern of increase and subsequent decrease in the annual AIDS incidence rates of both the state and the whole northeastern region—with the Bahian epidemic reaching a plateau between 1995 and 1996–97 (Dourado et al. 1997a).

Interestingly, around the same time that this plateau was reached in 1997, the incidence of AIDS in the northeast (Brazil's poorest region) finally coincided with the national statistics. This happened at a point when national AIDS incidence was decreasing. And as of 1998, data from the Bahian Health Division pointed to sharp decreases in both new registered AIDS cases and mortality among those registered (Teixeira et al. 2002).

I found a much bleaker reality in both the field and the media, a reality that called into question what was actually being epidemiologically described and what was by default not recognized. Consider the story of Sebastião Santos, age thirty-two, as reported by Bahia's main newspaper, *A Tarde* (October 12, 1996). Without a family, he spent his days lying in a ruined cubicle at the end of an unpaved street in the Cosme Farias District. The shack had no windows, the ceiling was only partial, and the faucet was a hole in the sandy ground. The arms and legs of the dark-skinned man were semi-paralyzed; he was just skin and bones. Sebastião was quoted saying, "A vida é só pra penar." Life is just for affliction. Some of the neighbors lamented Sebastião's fate and told the reporter that he had "mental problems."

Before falling ill, Sebastião used to wander through the Pelourinho (Pillory) District in search of work. Most of his little money was spent on *cachaça* (liquor). Sebastião was treated for TB at the Nossa Senhora Hospital. After being diagnosed with AIDS, he was sent to the AIDS unit of the state hospital to be tested for HIV. Sebastião said that he had never received the result of the test. He refused to have AIDS, the unspeakable "it": "I don't have it. This is one thing that I know for sure; I know that I don't have it."

Sebastião did not understand why the doctors discharged him from the hospital in such bad health. For some time, Dona Carmô, a neighbor, cared for him, arranging his cubicle and administering the medication with which he had left the hospital. But Dona Carmô had a stroke. "She used to feed me; she cleaned me and bathed me. She was my world. Without a mother, there is nothing. Now, all I want is to be hospitalized."

Next to the report, there was a photograph of Sebastião covering his mouth. What else was there to be said?

In what follows, I describe this complicated web through which an individual's life chances take shape, specific AIDS populations are revealed and concealed, and a local triage-state is constituted.[2]

Economic Death

I returned to work with Dona Conceição and her street patients in 1996 and 1997. In our first encounter, I showed Dona Conceição the photographs Torben Eskerod had taken almost two years before:

This one I never saw again.

This one died. This one is in prison for robbery.

The boy and the girl are in Candeias, an orphanage, it's very far from here. Both have AIDS.

He is alive, a wasted body. He was from Caasah but now is in the streets.

Look at them, both died. He had another woman and had two more kids before dying.

This one died too. People called him Caminhoneiro [Truck Driver] because he ambushed and robbed trucks on the highway. He got AIDS, went to Caasah, had leprosy, and had to leave. I rented a room for him and cared for him until death.

She sold *acarajés* [a local shrimp cake], then people discovered she had AIDS. Nobody wanted to buy the food she handled, so she began to sell drugs. She slept in a homeless shelter but got very ill and I rented a room for her. She kept selling drugs. I have the newspaper with a report of her being caught by police. I took care of her until she died too.

This was her husband. Besides losing their *acarajé* stand, they both got ill, got TB, herpes, diarrhea . . . then couldn't find jobs anymore . . . coughing, spitting, who will employ them? He also sold drugs with her.

This one is still in the streets . . . here she is much younger than she now looks. Those who died, died so ugly, only bones.

At one point I had ten rooms rented there at the Pelourinho for them. They get food from me that a priest provides, some one hundred people, and some also get a basket from Caasah.

Maria de Fátima is alive, she recovered. She works in a government-run needle exchange program and gets a monthly salary. Her doctor made this happen for her. She is no longer taking drugs and is taking AIDS therapies. She lives in Cajazeiras. When given the opportunity, things change for them.

This one died, this one and that one too.

This one is alive. Lia is a transvestite. His mother takes care of his daughter. The real mother is in Italy, she went in one of those operations that smuggles prostitutes. He does medical treatment at the University Hospital.

This was Maria Sonia, she died.

Joel . . . He is so beautiful here and died so ugly.

Andarina died in the emergency room, she stayed two weeks in a gurney for there was no bed in the AIDS infirmary. The services did not change, the cocktail is available, but there are still only sixteen beds. And the ones who take the medication have to take it with water for there is no milk. We give out a package every week, but it is not enough.

Maria Madalena . . . I heard that she ended up at Caasah.

Gil also died. He was a homosexual, sold drugs, and was shot in the head while sleeping.

There are no limits here.

Yes, they are capable of taking other lives. It comes from ignorance, survival, the discrimination and exclusion they experience. They are excluded from family and society, and for them to kill or to die is the same thing. They extend violence to each other, one kills the other, and then you might be caught in the middle, blood flows. I am there and alleviate things a little.

•••

Long before it made the epidemiological headlines in the mid-1990s, Dona Conceição knew that impoverishment, drug use, sex labor, and inadequate health care marked the course of the AIDS epidemic in Bahia. In her account, Dona Conceição suggests that the very way these lives are handled—discharged from public hospitals and left in the streets on their own—in some way participates in disease outcome. Her actions supplement a health care system that is predisposed not to intervene.

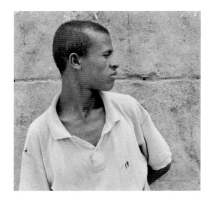

"This one I never saw again"

"This one is in prison"

"He is alive, a wasted body"

Andarina

"This one died too"

Maria Madalena

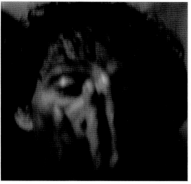

Gil

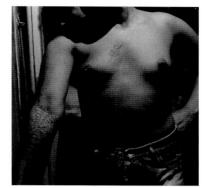

Lia

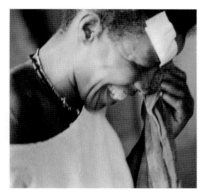

"This was Maria Sonia"

Joel

Maria de Fátima

"This one is dead"

Through Dona Conceição, these AIDS victims acknowledge their disease (as a means to access goods and care), but they remain engaged in injurious marginal economies. They know of AIDS and deny it at the same time. As Dona Conceição puts it, "People from Caasah identify themselves as *aidéticos* [persons with AIDS]. Everybody knows that that is the place of AIDS in Salvador anyway, and they do not have anywhere else to go. But the ones from the street do not want to be known as *aidéticos*. The parents do not want the children to be tested. They are afraid of further discrimination."

Mariza Silva comes to the communal meals with her three kids. She sums up her position, without uttering the word *AIDS*: "If I am with *it*, I would rather die without knowing *it*." Anthropologist Don Kulick found a similar denial of AIDS among the transvestites he worked with in Salvador in the mid-1990s:

> Asking *travestis* to estimate the number of their friends and colleagues who have died of the disease is pointless. Whenever journalists approach them on the street at night and ask them about AIDS, individual *travestis* will come up with a number, usually a large one. But in conversations with one another, those same *travestis* will later be quick to point out that *travestis* die of many things, and AIDS—which is most commonly only referred to euphemistically as a *menina*, "the girl," or a *tia*, "the aunt"—is only one of them. And besides, they ask, "how does anyone know that a particular *travesti*'s death was caused by AIDS?"

The majority of *travestis*, like most Brazilians, argues Kulick (1998, p. 26), "spend their lives self-diagnosing their infirmities and curing them by using pharmaceutical products recommended by friends of pharmacists, who dispense a wide array of powerful drugs over the counter with no medical prescriptions. In a context like this, AIDS is more a matter of opinion than of medical test results." But as I saw in the field in 1996 and 1997, this was beginning to change with the boom in harm reduction programs, the opening of HIV testing centers, and the growing availability of AIDS therapies. Marginalized individuals and groups were learning to integrate whatever resources they could draw from these specific and sporadic interventions in their circuits of survival.

In other poor areas of Salvador, people demonstrated a general knowledge of HIV/AIDS and prevention measures (Biehl 1996). Many knew others (relatives, acquaintances, or neighbors) who were either living with AIDS or had died from it without having been tested or treated. Without having undergone services that would make them detectable in the Bahian epidemiological surveillance service, these AIDS cases were left unregistered. In this local AIDS epidemic, individuals chose to remain anonymous, that is, to live with "public AIDS" for as short a time as possible. Facing poorly qualified and discriminatory assistance in the public health sector, people opted "to die at home," so to speak, with only relatives and neighbors, if anyone, to care for them. In the absence of prenatal care, many women learned of their HIV status only after an abortion or a complicated delivery. People knew of the difficulties involved in qualifying for ongoing good treatment and in accessing resources that might help treatment adherence.

In acknowledging AIDS, people also had to confront the normative practices and values of their most immediate family and neighbors. By denying it, people opted to remain integrated with partners, family, and community until the very last moment: the clinical manifestation of AIDS. Their refusal to affirm AIDS thus also functioned as a register for their self-defense and for reiterating libidinal economies and the normality of their risky practices. These practices undermined the dominant view that equated information with behavioral change and expected people with AIDS to seek assistance immediately.

At any rate, when I saw Dona Conceição in October 1996, she was exhausted and overwhelmed but not giving up. She was critical of her patients' reckless behavior: "It's crazy, I have to find money, go to the supermarket, prepare food, distribute it, then there is one who got hospitalized, the other is in the police station." She was starting to ration her help: "It would be good if there were ten more people doing this kind of work, for alone I cannot take care of so many. I can only help the ones who are mostly in need. So I must choose. How can I give hospital-like assistance to so many, when I still have to find ways to make money?"

Alongside her poor patients, there was another hidden AIDS population in Salvador. Dona Conceição also found time to provide home nursing care for rich AIDS patients: "Many families from high society call me to care for their relatives who are dying with AIDS. The patients don't want to be taken to hospitals, and the families prefer to keep things in secret anyway. They trust my secrecy." Dona Conceição does not accept money for this work with the diseased rich: "I only ask them to donate something for my street patients. Many times, the families of the diseased call me and donate medicine and clothing. Thus, there is an exchange."

Pelourinho

Most of Dona Conceição's patients lived in abandoned buildings in the Centro Histórico (Historic Center) of the old city of Salvador (also known as Pelourinho, or Pillory). Once a place where African captives were auctioned and rebellious slaves punished (until abolition in 1888), Pelourinho is today a lively site of Afro-Brazilian activism as well as tourism. The poorest wander around, selling things, picking up cans, or begging.

At the end of the nineteenth century, the colonial *casarões* (big houses, mansions) in Pelourinho were largely owned and inhabited by rich families. With the end of slavery, industrial changes, and migration to Salvador, the elite moved to new neighborhoods. The mansions were left behind to decay, soon to be taken over by working poor families and people involved in underground economies.

By the 1930s, Pelourinho was a completely pauperized space. In the following decades, tourism, prostitution, and drug dealing would coexist within its borders. At the end of the 1960s, about 60 percent of women living there worked as prostitutes (Bacelar 1982). And by the 1980s, cocaine and crack became widely available to the poor, replacing marijuana and speed as the main source of addiction and commerce.[3]

Pelourinho has been declared a world heritage site by UNESCO.[4] And in the mid-1980s, a number of Afro-Brazilian musical groups (such as Olodum) and other nongovernmental organizations (NGOs), began to operate there, attracting local and international attention to their work on Afro-Brazilian identity. They were a key component in renewing Pelourinho's cultural façade and economic vitality.

In the beginning of the 1990s, the government of Antônio Carlos Magalhães (ACM, Bahia's decades-long oligarch) decided to reengineer Pelourinho as a highly policed center of tourism and a site for cultural events like carnival (see Collins 2003; Cerqueira 1994). During my fieldwork in 1996–97, more than five hundred *casarões* were being restored, and most of Pelourinho's poor and marginalized inhabitants were being relocated to peripheral slums. For many, the compensation money soon disappeared, and they had to either squat in other abandoned buildings or live on the streets. Although more circumscribed and disguised, prostitution and the exchange and use of drugs remained fundamental to Pelourinho's economy and lifestyle (Kulick 1998, pp. 19–43; Mott and Cerqueira 1996).

Dr. Tarcísio Matos de Andrade, a physician, carried out an ethnographic and clinical study with one hundred intravenous drug users who used to buy, sell, or take drugs in Pelourinho (1996). This study was part of a needle exchange program. Andrade met with the drug users in their rooms and shacks, on the streets, in the "pharmacy"—that is, the place where one gets a clean shot by a drug injection expert—and in a local health post run by the city of Salvador. All research participants agreed to take an HIV test. His study population was young, with an average age of 24.9 years and 22 percent in the 11–18 age group. With 53 percent unable to produce a signature without great difficulty, the population was also largely unschooled, and with only 4 percent earning regular income, it was largely unemployed. Fifty-six percent, in contrast, earned money through illegal activity like robbery. Fifty-six percent lived in the Pelourinho area, and just under half lived in single rooms or public spaces.

The results of the HIV testing were extremely disturbing: 58.6 percent of the men and 56.7 percent of the women were seropositive. The highest proportion of HIV incidence (68.2 percent) was among the age group of 11–18 years.

Several of Dona Conceição's patients took part in Andrade's study. One such patient, Calixto, first learned he was HIV-positive at the age of twenty in 1993, while participating in the study. "The little money I had went to drugs," he told me. Without the support of his family—"my father also lives in the streets"—he looked for assistance at Caasah and was eligible for a monthly food basket for eight months. Calixto explained that he did not want to submit to regular medical screenings and lab exams: "They always asked me, 'Where is your exam?' And that I should do this and that. . . . It was too much control. And where in the world does a doctor want to deal with a drug addict?"

Dr. Andrade argues that the AIDS condition of these individuals was largely determined by violent experiences of socialization. He reported

Pelourinho

Pelourinho

on his informants' exposure to physical violence and also their infliction of it on their immediate kin, as well as on their own bodies (1996, pp. 95–100):

> All this happens in everyday life, it is in the accounts of the neighbors and of the mothers themselves ... tearing a baby's penis while misplacing the diapers' clips ... parents burning children's tongues on lamps as a way of punishing them. Child sexual abuse is commonly reported. An 11 year-old girl had an injured leg and arm. The stepfather raped her and threw her down the stairs. Mothers report that it is not worth giving birth to daughters for at 10 they are already prostitutes. A 36 year-old woman was asked about the scars in her left arm: "It was myself. When I am angry with my man, I cannot do anything to him, so I do it to myself."

"I set myself on fire"

Consider the life story of Maria Madalena, a former prostitute in Pelourinho, whom I had first met in March 1995 during a Wednesday communal meal and reencountered two years later at Caasah. Torben Eskerod and I were setting up a place in Caasah's backyard to carry out the portrait–life stories work presented in the introduction, and Maria Madalena was the first to volunteer to have her photograph taken. She approached in a silky robe, with lots of gaudy jewelry around her neck. She sat in the chair, removed her robe, and exposed her naked body. Torben said that he just wanted to photograph her face, but she said, "This is how I want it." She had scars on her face, her abdomen, her wrists, and her upper legs, and the right side of her body had been burned.

I sat next to Maria Madalena and asked her about her life. She began by denying having AIDS: "All the exams show that I have sugar in my blood, I have weak blood. In the hospital São Jorge I got better but not of the blood, so I had to get blood transfusion. Something happened to my legs, both of them are kind of paralyzed ... I do not know what it is. The hospital's social worker brought me here. I do not have AIDS. People here told me I have AIDS, but not the doctors. I do not have anywhere else to go, so I stay here."

Travestis and mulheres da vida

With disconnected sentences, Maria then shifted to the abuse she suffered and to her own death drive in the world of prostitution:

> I used to live in the Ladeira da Montanha, it was women's stuff. . . . Life there was as it is; we drank a lot, played, danced, fucked. My man cut me in the stomach. He is dead now, the police killed him. I set fire on myself. I broke my legs, my shoulder, my arm. . . . My head opened. I looked down and threw myself from Elevator Lacerda [the elevator that connects uptown and downtown Salvador]. My man, my father and mother were dead, I did not care for living anymore. One day I drank bleach and acid, but I did not die. I also threw myself into a deep well. . . . My family is dead.

I asked her about the extent to which AIDS had been a concern for women in the Ladeira. But she vehemently denied knowing anything about the existence of AIDS. "I never heard of AIDS there. Never, not even on TV. I never got sick. Every month we had to do a checkup for venereal diseases." She recalled, however, a mysterious disease that attacked women there: "We called it *cocheira*."

Cocheira is the place where carts and horses are kept. "The person neighs and coughs, and gets very tired"—like the animal, horse or other, I thought, which is beaten up and powers the vehicle that transports people and goods for both business and pleasure. *Cocheira* also is related to *coxa*, thigh, and to not being able to walk.

Maria Madalena now walked on crutches: "I only take pills when my legs ache. . . . If I get as wasted as they are in here I will kill myself. I already did a lot of perversity in life. I will do something that they will not even figure out what it is. . . . I have a nodule in my stomach."

I asked whether she had children.

"Yes. I had my first abortion when I was fifteen. I later gave birth to a son. A nurse at the hospital kept him and baptized him. I don't recall his name. It all happened in Itabuna. God forbid I have more children. I do regret not having a child. I see so many women killing their babies. They stick stuff in here to take the baby out [to abort]. So many children die, are left on the streets, or are sold."

Maria Madalena said that she spent the nights recalling all that had happened to her: "Men liked me; they were all telling me, 'Oh, *Baiana*, come here, I have money and this dick that you like.' I made my own

Maria Madalena, 1997

money. I bought clothing, shoes, and also put new teeth on, four were of gold. I felt that I was an artist, I already went to Pernambuco, I danced naked in clubs. I rehearsed every day. I had much jewelry, had long boots on. My pimp gave me everything, clothing, he controlled the money. He was my owner. I was dumb. I did not know how to read and write. I knew nothing."

Then, in tears, Maria Madalena spoke of the house of her infancy, in the interior, where she inherited the paternal curse:

> One day, my father took my sister and my brother to the city. My mother had run away with another man. I took care of the little children. My father left me behind, I did not know a thing. When he came back he brought three pieces of cloth, he gave one to Marilda, one to Joaquim, and showed me mine. I threw the cloth back at his feet and told him, "Father, you bought the beautiful cloth for her and the ugly one for me. You took Marilda and Joaquim with you and not me." He said, "My daughter, you will have to go into the world out there. You will suffer a lot. You will have to work hard to earn money and to buy your own piece of cloth."

"They take care of me as if I were family"

Hair combed, freshly bathed, and perfumed, Lazaro de Oliveira was on his way to Caasah's infirmary but conceded that he had a few minutes to spare for the portrait. Before sitting down, he pulled his pants down, pointing to an abscess that had turned into a cavernous hole on his left buttock. Lazaro had been recovering in Caasah for six months when a venereal infection reappeared and overwhelmed him again. At the hospital, a substitute doctor who treated him had tested a new cauterization technique on Lazaro's foot warts, which did little to slow his declining health. The foot soon became infected: "It is like a stone now; I do not feel it anymore." And when a Caasah nurse tried to relieve Lazaro's pain, she accidentally botched an injection, which in turn produced the major abscess he showed us.

Despite these trials, Lazaro said that he received care for the first time in his life while at Caasah: "My medication is on time. The nurses bathe

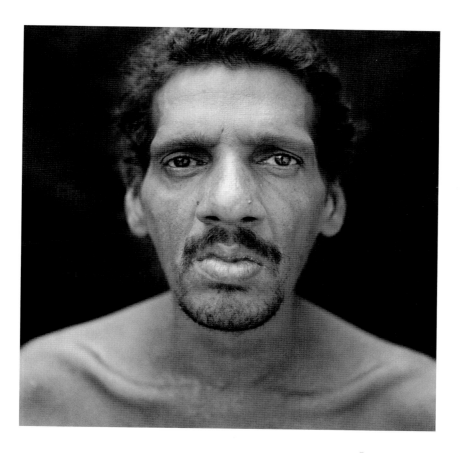

Lazaro, 1997

me, clean the wound, smear cream on my whole body, and put me to bed. They watch the house all night. They take care of me as if I were family."

Like so many there, Lazaro grew up in the streets of Pelourinho "all by myself." He meant it literally. His father left the house on Lazaro's fifteenth day of life, he said, and his mother abandoned him when he was six years old (although she returned a few years later). He spent the rest of his childhood shuttling between various households and the street, battling various drug addictions.

Things began to change when he became amorously involved with Ionice, the woman who raised him in his mother's prolongued absence. Lazaro sold fruit and helped Ionice in her liquor and cigarette booth. "I did not stop with all drugs at once. That's not how things work. I became calmer and committed myself to staying inside the house. I began to forget those bad things. I was entering another world." Lazaro and Ionice had two children.

But, in 1994, Lazaro found out that he had AIDS. Depressed, he returned to his mother's shack but became the object of selective neglect: "They separated cups and plates for me. I was not allowed to sit at the table; I had to stay in my room. People were afraid of me. I became very disoriented. There were moments I wanted to leave and never return again." In 1996, his mother dropped him off at Caasah, "like a cadaver, with a wasted body," Lazaro recalled. "I was undernourished and couldn't walk." Occasionally, Ionice visited him.

When he is "oppressed," Lazaro told us, "I go to my corner and weep. Then I calm myself." As we ended our conversation, he noted that he was content with "the little God has given me."

•••

I learned that Lazaro had died in Caasah in October 1997. The family refused to pay for the funeral, and he was buried in a nearby cemetery for indigents. As for Maria Madalena, she was asked to leave Caasah a few months after we had met—it was due to "violent behavior," I was told. She had gone back to the Ladeira da Montanha, where she was housed by Mãe Preta ("Black Mother"), a former prostitute who had begun her own shelter for the homeless in a ruined mansion that the city hall had made available.

Waiting for a meal at Mãe Preta's shelter

"Maria had nobody in life. Poor woman, she was vain, drank a lot, and had poor judgment. . . . She was fat when she came from Caasah. But it did not last. We put her in a cab and sent her to the state hospital, where she died. This must have been in early 1998," Mãe Preta told me when I visited in December 2001. At that time, the matron of Salvador's homeless was caring for fifty children and distributing more than two hundred meals a day.

Technologies of Invisibility

What I observed in the field strikingly contradicted the representation of AIDS in official data. As I met with technicians at the Bahian epidemiological surveillance service, I asked them to verify whether some of Dona Conceição's patients who reported being treated at the state's AIDS unit were registered in their database. The professionals were at first reluctant to look at the records, but once they did, they admitted that these individuals were nowhere to be found.

Together with two local epidemiologists, I then planned an analysis of the Bahian AIDS services, encompassing both epidemiological surveillance and medical care. I wanted to know how the AIDS of poor and marginalized patients took form in routine exchanges with the local public health services and what ultimately turned them into absent beings. My colleagues needed information for a report that the state's Health Division had commissioned on the course of the AIDS epidemic in Bahia. We agreed to collaborate and to produce data together (see Dourado et al. 1997a).

At the Bahian epidemiological surveillance service, we analyzed official documents, reports, and registration books, and we carried out semi-structured interviews with technicians and administrators about their AIDS work. In addition, we interviewed health professionals (physicians, nurses, social workers, and psychologists) at all public health care and laboratory services of the state (University Hospital, Hospital Caridade, State Hospital Luis Souto, and Central Laboratory).

From this work, we were able to identify clinical, bureaucratic, and political problems that make the AIDS services inadequate to meet the increasing and complex demands of AIDS in that context. The units and institutions that we studied provide AIDS data to the surveillance service as part of an information-gathering network based on mandatory reporting. We found that the Bahian surveillance service fails to keep abreast of the AIDS cases detected in hospital and laboratory units.

We moved from statistical analyses in the official databanks to our own tally of AIDS deaths, both in the state's AIDS referral unit—where the poorest are directed for specialized or emergency treatment—and in Caasah's community-run AIDS service. As I discuss later in this chapter, we counted 571 AIDS deaths at the unit between 1990 and 1996 and discovered that only 26 percent of these AIDS cases were actually registered by the epidemiological surveillance service. The majority of the unregistered AIDS cases involved people whose diagnoses coincided with their first and, in most cases, final hospitalizations.

As our work progressed, I understood that this "absence" was not just a result of a precarious surveillance service that could be remedied with a simple technical fix. I found evidence that the problem was basically rooted in three factors: the operating logic of a public health care system that circumscribed service delivery; a local form of medical sovereignty and ethics determining which patients were worth treatment; and the mode in which AIDS subjects understood their disease and related to medical services. The realities we unburied cast doubt on the value of available data, making analysis of these data through sophisticated techniques a potential waste.

In his book *The Taming of Chance*, Ian Hacking identifies scientific and technical dynamics that intermediate processes by which "people are made up" (1990, p. 3; see also Hacking 1999). Categories and counting, he argues, define new classes of people, normalize their ways of being in the world, and also have "consequences for the ways in which we conceive of others and think of our own possibilities and potentialities" (1990, p. 6). Hacking views categories and statistics as "making up people," but I am concerned here with how technical and political interventions make people invisible and affect the experience, distribution, and social representation of dying.

In my ethnography, I discovered that bureaucratic procedures, sheer medical neglect, moral contempt, unresolved disputes over diagnostic criteria, and unreflexive epidemiological knowledge mediate the process by which poor and marginalized AIDS patients are made invisible. During the course of this study, I began to call these various practices "technologies of invisibility." These technologies routinely intersect with discontinuous medical care and drug dispensation.

A System of Nonintervention

Several health professionals with whom I talked synthesized the situation of the Bahian AIDS program in the question: "Who is today responsible for AIDS in Bahia?" As of 1996, the position of AIDS coordinator was still vacant, despite ample funds from the national AIDS program for institutional development. This reflected the fact that "AIDS is of no political interest in Bahia," Valeria Batista, an immunologist working at the AIDS unit of the Luis Souto Hospital, told me.

Dr. Batista said that the Bahian AIDS program had shown some signs of efficiency in the early 1990s "because of the good will and political influence" of Dr. Eliana de Paula. "But as is always the case with public policies here, there was no continuity. AIDS is an uphill battle. We are all exhausted." With no direct support from an AIDS political body, doctors had to engage constantly in internal administrative disputes regarding the prioritization of AIDS in their respective hospitals. As they struggled to maintain a minimum infrastructure of assistance in a public health care system of already notoriously poor quality, they were overwhelmed by more and more complex demands (epidemiological, biological, and psychological) from a growing number of AIDS patients.

Immunologist Luis Alberto Costa, also working at the Luis Souto, told me that he doubted the capacity of the state's epidemiological surveillance service to record and analyze the data being generated at the various services treating AIDS patients. All the health professionals I interviewed said that they are effectively registering all AIDS cases, their mandated responsibility. However, because of the limited ability ("strategic," a few ventured to say) of the surveillance service to enter AIDS cases into the databank and the overwhelming number of incomplete forms, Dr. Costa believed that "a significantly high proportion of AIDS cases are being missed."

Dr. Costa was right. A series of problems were occurring at the epidemiological surveillance service at various levels of data collection and analysis. During our first visit there, we actually noticed that dozens of unregistered AIDS case files were just lying around in the office. We asked a top-ranking official to demand that the technicians immediately put them into the computer system. By doing this, the technicians increased the number of AIDS cases registered by a full 55 percent.

Besides an inadequate computer database and a poorly skilled group of technicians, we noticed a lack of dialogue between technicians working on AIDS at the city level and their counterparts at the state level and, furthermore, that political partisanship thwarted joint projects. The director of the service complained of his difficulty in keeping a stable technical team at work in a context where public officers were frequently devalued and demoted. Two of the technicians we interviewed considered themselves "martyrs of the bureaucratic machinery." As one of them put it, "No one with a sane mind would accept this job. . . . But with all the unemployment out there . . ." Given all these operational and financial difficulties, "there is no way we can develop a competent analysis of the data gathered," she concluded.

The fact is that the central administration had never audited the service, and there were no measures in place for the service to evaluate its performance. Deficient flows of information and feedback between the health services and the various levels of government impeded the production of an accurate technical profile of the AIDS epidemic in Bahia and compromised the design of prevention strategies as well as the effectiveness of the limited assistance policies that were taking place.

In our investigation we also found that the standard practices of epidemiological surveillance set forth by the national AIDS program were in conflict with the instituted practices of the local surveillance service, which routinely delayed notification. The very filing of the epidemiological notification questionnaire exemplifies the logic of how the Bahian surveillance service functioned. Given the steady changes in biomedical knowledge about HIV/AIDS and the criteria for treatment protocols, the health professionals' understanding of clinical AIDS increasingly conflicted with the criteria adopted by the Ministry of Health. Such disagreements had, indeed, led to more under-reporting: many doctors reported cases that could not be framed according to the AIDS criteria of the Ministry of Health. As a result, these cases were not registered by the personnel of the surveillance service.

The registration practices of public health care professionals were incongruent with those of technicians at the surveillance service: AIDS cases seen in the public services, even when they progressed to death, were often not acknowledged by the state. In such situations, the personnel of the surveillance service took an ad hoc pragmatic stance. Because they were unable to keep track of cases that resisted the Ministry of Health's categorizations, they opted simply to exclude all the questionable cases from the databank.

In short, the surveillance service had not developed a classification system that incorporated the diagnostic criteria at work in the daily activities of AIDS public health care services. Neither was the system responsive to the technical difficulties of notification. All this made the a posteriori detection of questionable AIDS cases impossible. Entrenched administrative practices generated a disjuncture between the AIDS epidemic as it was experienced by patients and health professionals at the public health care level, on the one hand, and the epidemic as it was represented by the local informational system, on the other. It was this disjuncture that informed and directed policies of, basically, nonintervention.

Infectious Diseases Research

The University Hospital has been treating AIDS patients since the first cases emerged in the mid-1980s. Patients in need of hospitalization are typically admitted to the Research and Assistance Unit for Infectious Diseases, which is responsible for all cases of infectious and parasitical disease, including AIDS. It has also served as a reference service for other endemic diseases such as calazar, or leishmaniasis, and enterobacteriosis. The clinical team of the unit is well known for its research and numerous publications in these areas. It is also known for internal discord between two opposing camps over resources and prestige.

At first, the infectious disease unit hospitalized only two AIDS patients at a time, "but now 75 percent of the fifteen beds are occupied by

AIDS cases," Dr. Carlos Moura, the director of the unit, told me in early 1997. "Another key development is the increasing number of women being hospitalized; now we sometimes have more women than men in the unit." Over the years, however, the number of beds available in the unit has remained constant; furthermore, the hospital has no emergency treatment. As a result, hospitalization was commonly limited to those AIDS patients in critical condition but not at immediate risk of death. Patients who needed emergency treatment had to look to the AIDS unit of State Hospital Luis Souto or the emergency service of Hospital Caridade.

Epidemiologist Maria Inês da Costa Dourado coordinated a study that profiles the 154 AIDS patients hospitalized at the unit between 1989 and 1991 (Dourado et al. 1997b). The average male age was 32 years, and the average female age was 27 years. Of the patients, 84.4 percent were men (recall that as of 1996 this scenario had dramatically altered and, according to Dr. Moura, there were already more women than men being hospitalized). The majority of the patients were Afro-Brazilians (63.2 percent). In general, these patients worked in low-paying jobs or were unemployed. According to the protocols, sexual contact was the most common form of HIV transmission: 65 percent as reported by men and 70 percent as reported by women. Fifteen percent of the men and 10 percent of the women reported intravenous drug use. In terms of sexual behavior, 56.3 percent of the men reported homosexuality, 24 percent reported bisexuality, and 17.7 percent reported heterosexuality. Two women said they worked as prostitutes.

In 1985, the hospital had opened its one-day-a-week outpatient AIDS service. "The demand grew so quickly that we had to open a second day, and now we already need a third or fourth day," said Dr. Moura. With an average of fifty patients seen per day and approximately four hundred seen per month, the waiting room was perpetually overcrowded and the medical staff overstretched. The unit had been able to accommodate some of the high demand for hospitalization with a "day hospital" created with funds from the national AIDS program, added Dr. Moura. "Three reclining chairs are available for patients who need some form of continuous treatment, without having to be hospitalized. They come in and leave in the same day."

Dr. Moura explained the hurdles the infectious disease teams had to overcome among the medical community in order to treat AIDS: "At first we could not even identify the blood of AIDS patients—the lab technicians would throw it away. With the intervention of the Health Ministry, matters improved in this area: every blood sample was treated as potentially infected. We also made it clear that AIDS patients needed specialists; we insisted on their presence. We appealed to the code of medical ethics and made them see that there was a human being in front of them needing their help. The hospital also gave those workers the minimum biosafety working conditions."

The residency program in infectious diseases, said Dr. Moura, was booming: "This specialization was strengthened with AIDS, the technology deployed, and the possibility of a direct contact with patients." Low-income patients were increasingly drawn to the unit's research apparatus, as it gave them access to diagnostic and medical resources and special medications that were not yet readily available in other public health services. Interestingly, in spite of its research know-how, the service had not yet developed a registry of AIDS patients to track causes of hospitalization and death.

The demand for hospitalization was increasing at an alarming rate, continued Dr. Moura. "Last week we had no vacancies at all. We had to send five patients in an acute state to the Luis Souto Hospital. They were not accepted there and were returned to us. It is inhuman. They begged us to at least give their families a death certificate." A death certificate, explained Dr. Moura, might help the family to procure help with funeral expenses and also to claim a pension. He refused to speak further about the destinies of those five cases, sent back and forth without care, but added that health professionals ought not to be blamed for this chaotic situation: "Private hospitals throw AIDS patients into public hospitals—after all, it is a patient who has no money. And the public health care system is not built to provide access for all of these patients, period." Aside from a stricter triage of patients, the unit had no choice, concluded Dr. Moura, but to send patients in critical condition home, "to die at home, or at Caasah."

While working at Caasah, I frequently heard complaints about carelessness in public hospitals. One nurse attributed this carelessness to inadequate space and funding for AIDS, as well as the administrative priorities of the hospital. He implied that poor and marginalized AIDS patients were only useful for the specialized teams inasmuch as they helped professionals increase clinical know-how and helped train doctors for private clinics with rich AIDS patients. In fact, many immunologists working at the University Hospital and other public health AIDS services also maintained private practices for rich patients seeking anonymity.

Medical Sovereignty, Local Bioethics

Hospital Caridade was founded in 1853, from its beginnings specializing in the containment and treatment of the so-called *pestes* (plagues) that perpetually threaten communities around the region: meningitis, tetanus, leptospirosis, hepatitis, rabies, and diphtheria. The hospital's administration limited hospitalization to a narrow group of diseases in order to avoid overpopulation in its three units. The rationale was to keep rooms available for epidemic emergencies. Public hospitals often refused to treat patients suffering from the above-mentioned diseases, instead sending them to the Caridade. For example, almost all new meningitis cases in Salvador are treated there.

Only in the 1990s did the hospital begin hospitalizing a few AIDS patients. In most cases, these patients were dying and unable to find a room in any other hospital. Activists and health professionals had criticized the hospital for not extending its assistance to AIDS patients. After all, the Caridade had special units, a medical residency program, and highly specialized clinical personnel in infectious diseases, and it was located near Caasah. In December 1996, the state's Health Division gave in to much public and political pressure, decreeing that the Caridade would be officially accredited as an AIDS service (funding would be provided by the Bahian government and the Health Ministry).

In fact, in November 1993, the hospital's director recommended that Caridade shift away from its function as a quarantine institution. From then on, the hospital was to "treat all patients with infectious and parasitical pathologies as defined by the international classification of diseases of the World Health Organization [WHO]." This administrative move immediately led the hospital to treat victims of a cholera epidemic that was affecting the poorest.

Adroaldo Silva, a local physician and epidemiologist who had once worked there, felt ironic about this development, suggesting that "the hospital shifted from its role of public hygienist to symbolic container of socially generated epidemics." However, as Table 3.1 shows, this administrative move also led to a significant increase (around 100 percent) in the number of AIDS patients treated between 1993 and 1994. By 1995, AIDS hospitalizations again decreased 40 percent. Dr. Silva told me that this was not due to a decrease in demand but to the removal of the director from office and a return to the old policy. This, in turn, was linked to a continuous lack of political will and medical and financial incentives to include AIDS in the clinical routine of the hospital. In fact, "poor and marginalized AIDS patients are not considered worth treating," the physician told me, "and the government is not forcing hospitals to take more patients in, as that would require expanding the care infrastructure."

When they were guaranteed hospitalization, AIDS patients were typically treated at the Ambrósio Souza unit, which also housed the ambulatory service. The space was limited, making it impossible to keep patients under observation for a longer period of time. According to one critical social worker, the service was organized "in order to not provide adequate primary care." Here the ambulatory service also functioned as triage. Doctors decided which few patients to hospitalize and treat and which to discharge, most likely without further referral.

During a focus group meeting in January 1997 with several doctors and nurses working at the Caridade, I asked why so few AIDS patients were being treated there. The main reason, they said, is that, given the hospital's mission, "rooms must be kept available for emergencies . . . we cannot over-hospitalize." Additionally, they pointed out that a vacant bed did not mean that they could accommodate an AIDS patient: "An AIDS patient cannot stay in the same room with a patient who has meningitis, TB, diphtheria, and so on." They also referred to the lack of a multidisciplinary team like that of the University Hospital: "AIDS patients are too complex to be treated here." As a result of the hospital's deficient laboratory service, specialized exams had to be done in other hospitals: "In the case of AIDS, there is a need for many exams, immediately."

Furthermore, both doctors and nurses alleged an understaffing problem: "Currently 30 percent of our nurses are doing some form of administrative work." Only two immunologists working there welcomed the possibility of treating AIDS patients on a more routine basis, as they said

TABLE 3.1
Meningitis, Leptospirosis, and AIDS Hospitalizations, Hospital Caridade,
1992–95

	1992	1993	1994	1995
Meningitis	1,152	1,137	1,194	1,327
Leptospirosis	262	213	342	316
AIDS	6	35	70	45

that this would help them to expand their research and teaching capacities. New partnerships with the state and the Health Ministry could also potentially enhance the hospital's technical apparatus.

Caridade's staff described themselves as the most committed in Bahia's public health care system. However, they were not immune to the lack of initiative so commonly found among public health professionals. They complained of low salaries (a doctor earned about $500 per month), work overload during night shifts, emotional distress from treating acutely ill patients without adequate diagnostic and therapeutic tools, and an inefficient bureaucracy at the state level. As one doctor put it, "We are skeptical about any top-down directive, as it basically means more work for us and we are tired of having to self-sacrifice." Some health professionals told me that they feared lawsuits by patients and families: "If something goes wrong, it is easier for them to blame us than the inadequate hospital infrastructure and poor administrative decisions."

The doctors I interviewed did not all agree on criteria for admission and refusal of AIDS patients. Their attitudes ranged from always refusing hospitalization to admitting AIDS patients selectively. Through these interviews, observation of medical interactions, and research on the patients' protocols, I was able to outline some of the implicit criteria and practices determining AIDS hospitalization.

The fact that the Caridade was not officially open to AIDS patients until the beginning of 1997 had led to the development of an unofficial practice of admission of AIDS patients who had a specific clinical profile. The Caridade generally accepted patients who had not found a vacant bed at the University Hospital or at the Luis Souto AIDS infirmary. Quite often, these patients were sent back to the alternative services one more time; in the event they were not hospitalized on this second try, they finally would be admitted at the Caridade. One wonders how many patients gave up or died in transit.

Patients who had clearly defined pathologies—particularly those with respiratory infections (including suspicion of pneumocystosis, cryptococcal meningitis, and cerebral toxoplasmosis)—had a greater chance of being hospitalized.

The Caridade also admitted patients who were at immediate risk of dying but who were not yet in terminal condition. The most common reason given for this hospitalization was that the patients could not be transported to another hospital (without risking death on the way).

Patients who presented undefined symptoms (mostly internal pathologies that the hospital could not identify because of a lack of diagnostic machines) were generally not hospitalized. The most typical was a patient suffering from a very long spell of high fever, whose ailment had not been previously identified by the other services he or she visited. Patients with cancer were denied hospitalization. Those who were terminal as well as those who were comatose or had dementia were also denied hospitalization.

These implicit practices were not formalized, but they guided ethical reasoning and medical interventions nonetheless. Dr. Esdras Cabus Moreira and I found a similar phenomenon among doctors assisting patients in critical condition at the intensive care unit of a general hospital in the region (Moreira and Biehl 2004). Here, too, the "acceptance of death" was part and parcel of a local form of medical power and bioethics. We documented medical conduct such as the adjustment of respirators to less rigorous control levels; decreases in therapeutic drug dosage; and the staging rather than actual attempt to resuscitate patients that doctors had already deemed "irretrievable."

These medical decisions to let die were made in spite of ambiguity over prognosis and were explained to nurses and family members as being the result of technical measures. Such practices did not correspond to the hospital's official bioethical directives, but a constant medical rotation system and a lack of regular deliberative meetings facilitated their persistence. The fact is that particular institutional logics and economic calculations allied with an all-pervasive medical form of sovereignty kept these practices unspoken and thus "normal."

At the Caridade, the practices related to AIDS hospitalization actually followed the hospital's general philosophy, that is, to avoid treating patients with "atypical" pathologies. To justify such a discriminatory practice, hospital officials argued that it kept beds available for potential epidemics of "typical" ailments. Similar to what happened at the University Hospital, Caridade's doctors also did not admit patients in terminal condition, deeming them unsuitable for any beneficial intervention. Were they hospitalized, the reasoning went, they would only occupy space and resources needed for other potential patients.

These official practices of denying hospitalization ultimately participated in the maintenance of a hidden AIDS epidemic. Many of the poor patients arrived at health facilities while "wasted" and in terminal condition, only then to be officially diagnosed with AIDS. By refusing hospitalization to terminally ill patients, the hospital kept these cases from being registered in the surveillance service. Even if the patients were already registered as AIDS cases, their subsequent deaths would also be invisible in the registries.

Triage

Salvador's main public hospital lies in the hilly slums surrounding the city. The State Hospital Luis Souto has a specialized AIDS unit that provides care for patients coming from all over the province. This state-run unit is the main source of AIDS information for the epidemiological surveillance service. Until 1987, AIDS patients were treated in the emergency room, barred from any of the specialized units of the hospital. Increased pressure from patients and family members led the administration to create quarantined rooms in 1987, making sixteen beds available for AIDS patients. With the growing demand for assistance, an outpatient service was created in 1992. Hospitalization has been progressively restricted to patients in a critical state. As of 1997, both laboratory assistance and the workforce of technical personnel remained unchanged. Moreover, the hospital administration had approved no additional technical support to address the new clinical, psychological, and social complexities faced by AIDS patients.

As of March 1997, the unit was run by a team of three immunologists, two residents, a chief nurse, nineteen nurse technicians, a nutritionist, a physical therapist, a psychologist, a social worker, and several interns. I met individually with some of these health professionals, and they had no qualms about saying that their work was "uncoordinated," meaning that medical interventions were targeted and driven by individual initiative rather than a result of a jointly deliberated and supervised treatment plan.

The professionals I interviewed were pessimistic about possible changes. "AIDS is not a priority for the administration. . . . Our salaries are low," one of the immunologists told me. "We have to get many jobs in several hospitals to make our living." Another immunologist mentioned that other services routinely "dump unwanted patients here," also admitting how difficult it was to establish productive collaborations with other specialists in the hospital. "Discrimination among doctors seems to be the rule," the doctor told me. "We get the homeless patients and those who the surgeons and the psychiatric units reject."

At Caasah, the community-run AIDS service, patients who had already been hospitalized at the AIDS unit told me things like this: "When I am ready to die, please, don't take me to the unit." "They don't give us sheets." "I saw people starving." "Someone donated hot showers for the unit, but they were installed somewhere else." "What we want as patients is minimum human consideration. Medical ethics must change."

Valquirene, a Caasah resident, described her passage through the AIDS unit as "the worst thing that has ever happened to me" (see introduction). She discovered she had AIDS in May 1996. "I ended up weighing only seventy pounds. My family brought me to the unit, and I tested positive. I always implored the nurses to call my mother and my sisters, but they began to dope me, they gave me tranquilizers." Valquirene recovered a bit and was sent back to her mother's house: "My mom expelled me from home. She rented a shack for me. But I went everywhere in town, skinny as I was, to the stores, to the church. My family was ashamed of me and they started to spread the news that I was crazy. They gave me sleeping pills, and sent me back to the Luis Souto."

Valquirene "hates" medical professionals, whom she also blames for her mental disorders. She sees her depression originating not in disease but in the combination of family exclusion, psychiatric diagnosis, and overmedication. At the Luis Souto, says Valquirene, she was devalued both as a person and as a patient:

I do not know what actually leads a person to become a doctor. Maybe it's social position, money, some perversity, but it is not care. I had a doctor who I thought was good, and when I was ready to leave the unit she and my mother decided to put me in the psychiatric hospital. They said that I was crazy because one day I stood up from my bed and tried to reach the bathroom but fell, could not stand up, and shit and threw up and urinated on the floor—I got completely dirty. Only the next morning the nurses came to put me in bed, and left me there dirty. Later that day they just cleaned the body and kept my hair sticky with dirt. I spent six days without showering because I could not stand up on my own, I was too weak. Then a Pentecostal nurse finally took me to the shower; it was a sacrifice. I entered the hospital without soap, and they had no soap at the infirmary. The nurse had to get some detergent from the kitchen. From then on I got scared to lay in bed and asked the nurse to help me to put the mattress on the ground. I took strong drugs that made me constantly dizzy and I was afraid to fall again, so I slept on the ground. The doctor and my mom decided that I was crazy.

Valquirene was transferred to the Dorneles Psychiatric Hospital, where she was treated with antipsychotic medication. Finally, a psychologist saw that she had been incorrectly diagnosed and helped her to move into Caasah.

A nurse I talked to confirmed the precariousness of the AIDS services there: "The outpatient service is immoral. It is assistance based on pure charity. It must be photographed." Regarding the unit: "There is a lot of space, but there are not even chairs available. It has never been so bad. In the beginning we had all kinds of basic medicine; now sometimes we have even to ask Caasah to donate medicines to us. They have free samples there. Quite often the patients themselves have to buy their medicines."

According to one of the immunologists I talked to, this lack of improvement or even minimum maintenance of the hospital's material and technical infrastructure was linked to a "decrease of political and administrative support by the state's AIDS program." It was also seen as a "retaliation" against the doctors by the hospital's administration: "We had some serious confrontations with the administration. We told them that the lack of resources and support was immoral . . . our critique backfired."

One of the immunologists working at the unit cynically explicated the hospital's operational logic: "AIDS is a disease that does not deserve direct investment. It is fatal disease; the person will die anyway . . . furthermore, he is a faggot, a drug user, a marginal, so why should I treat this individual? He must die; there is no reason for me to invest in him."

This doctor and her colleagues were tired of being "scapegoats" for this tragic situation: "The state's AIDS program is a disaster. We as individual professionals are supposed to keep this functioning. But how? When medical interns come to visit, we take them to the rooms, introduce them to the patients, describe the situation, and say which therapeutic course should be followed . . . and that is it. What will be done with that patient? Nothing. Because there is nothing that we can do. We don't investigate the case, we don't treat. Do you understand?" The question that followed was shared by many: "Where does the money go?"

In 1995, the Health Ministry paid the state of Bahia the highest sum per hospitalized AIDS patient in the country: $1,150.58. In 1996, the amount decreased to $989.05, but the Bahian patients were still the most expensive in the country.[5]

The two doctors I interviewed said that they had decided to stop "saving the system." As one of them put it, "We are tired of playing the role of martyrs, having to handle all the social responsibilities tied to the patients' diseases. We are not minimally infrastructured to treat them. And whatever we do, we save the system and not the patient." Some nurses were in fact critical of the doctors, adding, "It is not just a matter of changing the infrastructure. The professionals, particularly the ones who do not work directly with patients, have to be equipped to value human life." One nurse saw no other way than "to change the mentality of the hospital."

Given the poverty and marginality of most of the patients, the unit's social worker was overloaded with complicated social and economic issues. And on top of that, she said, "I have to ensure that the patients receive minimum medical care. It's just too much."

The social worker described a common profile among the patients: many have passed through several local health services with symptoms suggesting AIDS, but they went undiagnosed. "They usually say that in the past year they have felt very weak, were no longer fit to work. Then they decided to look for a local health service. Many had diarrhea for months, a general anemia, weakness, respiratory problems, even tuberculosis." There were also the patients who came from the interior: "Since getting the results takes long there, most of the times they arrive here extremely sick. Many times families use this occasion to simply abandon the patient."

A resident described the low morale of the patients: "There is so much prejudice and rejection out there that the physical and psychological effect is a tremendous burden. The patients are reduced to a point where they don't have any more strength to fight for their rights and their lives." Those who actually leave the service better, said the resident, "by and large do not maintain a regular follow-up schedule at the ambulatory."

Quite often, added the social worker, even the spouses and children of those who have died of AIDS in the unit do not return for the results of their own HIV tests. She pointed to more signs of a hidden AIDS: "Many of our patients have been drug users, and they say that they know of other persons with similar AIDS-related symptoms, with ongoing exposure to risk of infection by HIV, who have not undergone laboratory or clinical screenings."

The poorest and most marginalized AIDS patients get lost while trying to access care. "A few days ago, the Oncology Service called me," said the social worker, "and asked me to locate one of their patients who also happened to have AIDS. I answered that unfortunately this patient could not be located since he lives in the streets. By coincidence, his brother was hospitalized here. I told them that I would leave a little note under his brother's pillow, even though the street patient is illiterate. . . . Oh God, we have such a surplus of people in these conditions."

This contingent of HIV-infected and diseased persons at the margins of the public health care system may have gone unnoticed in the official AIDS registries, even when its existence has been acknowledged by health professionals. Many live in a state of *apparent invisibility*; as such, they are not officially part of the AIDS epidemic.

The Social Life of Death Certificates

To get a better grasp of the process through which AIDS victims disappear from the public registries, my colleagues and I gathered the death certificates of patients at the AIDS unit of the State Hospital Luis Souto and at Caasah. The data at the AIDS unit, obtained from the social service's registers, revealed 571 reported deaths at the unit between January 1990 and October 1996. Caasah's archives, in turn, showed 104 deaths between May 1993 and October 1996. After verifying that all deceased patients were serologically confirmed as HIV-positive, we collected data on demographic characteristics, numbers and dates of hospital admissions, lengths of in-hospital stays, risk behavior, and dates of death. We then checked how many of these AIDS cases were registered by the Bahian epidemiological surveillance service (recall that, as a result of our investigation there, technicians had updated the databank).[6]

Of the 571 now-dead patients counted at the AIDS unit, only 150 (26 percent) were registered. All were still reported as "alive." Figure 3.2 follows registration patterns over time. Annual registration rates varied from 6 percent to 35 percent, stabilizing in 1992 at around 30 percent (immediately before the first reported decrease in AIDS incidence; Dourado et al. 1997a). One can argue that this relatively stable 30 percent represents the limited AIDS population that state services are prepared to serve.

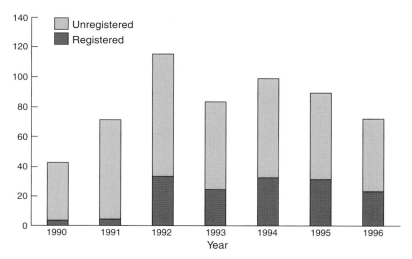

FIGURE 3.2 AIDS case registration by the Bahian Epidemiological Surveillance Service. Deaths in the AIDS Unit of Hospital Luis Souto, January 1990–October 1996

We were intrigued: What made some of these AIDS cases officially visible and the majority not?

Table 3.2 takes a closer look at the patients who died at the state AIDS unit. Just over half died during their first hospitalization, implying that the majority of these people gained access to hospital services largely to die. About one-quarter of the patients reported injecting drug use, and 80 percent were men, half of whom disclosed their sexual orientation. Of this group, the majority reported being bisexual or homosexual. The remainder, who refrained from categorizing themselves, may have feared discrimination for partaking in sexual practices frowned upon by society. Therefore, they might also be assumed to associate with sexual identities considered "deviant."

Figures 3.3, 3.4, and 3.5 consider whether any of these characteristics are associated with case notification to the Bahian epidemiological surveillance service. In each figure, the bar farthest to the left represents the reference category. Each of the remaining bars tells how likely patients in other categories are to be epidemiologically visible, relative to the reference

TABLE 3.2
Selected Characteristics of AIDS Patients

Registration Status	
Registered	26.3%
Unregistered	73.7%
Number of Hospitalizations	
One	52.0%
Two or more	48.0%
Intravenous Drug Use	
No reported drug use	73.0%
Drug user	27.0%
Gender	
Female	20.7%
Male	79.3%
Sexuality (Males Only)	
Heterosexual	21.2%
Bi-/homosexual	32.0%
Unreported	46.8%
N	571

Deaths in the AIDS unit of State Hospital Luis Souto, January 1990–October 1996.

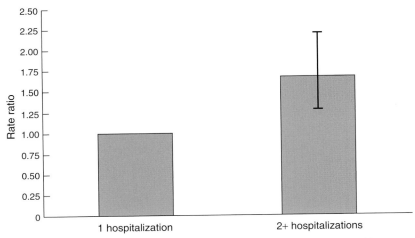

FIGURE 3.3 Registration by number of hospitalizations (Note: Vertical line represents 95% confidence interval. $N = 571, \chi^2(1) = 13.56, p < 0.001$)

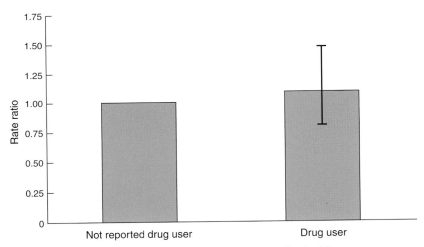

FIGURE 3.4 Registration by reported drug use (Note: Vertical line represents 95% confidence interval. $N = 571, \chi^2(1) = 0.30, p = 0.584$)

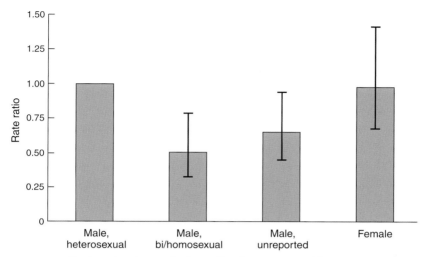

FIGURE 3.5 Registration by gender/sexuality (Note: Vertical line represents 95% confidence interval. $N = 571, \chi^2(3) = 14.45, p < 0.003$)

category. If a category has a rate ratio of 1.5, for instance, then patients in that category are 150 percent as likely to be registered as patients in the reference category; a rate ratio of 0.5 implies that they are 50 percent as likely. Because these statistical calculations are prone to misrepresentation due to random variation, the figures also include 95 percent confidence intervals, demarcated by fine vertical lines. For each category, we can say with 95 percent confidence that the true rate ratio lies within the interval.[7] If a confidence interval overlaps the value of 1 (on the vertical axis), the rate ratio for that category is not considered significantly different from 1, implying that the rate ratio is not representative of a general pattern.

Although reported injecting drug use does not appear to affect case registration (in Figure 3.4, the confidence interval clearly includes 1), the number of hospitalizations, gender, and sexual orientation all have strong, statistically significant associations with the likelihood that a case is visible in the state's surveillance system. We should note that the unimpressive result concerning drug use may reflect reporting bias, as intravenous drugs are illegal and stigmatized. However, Figures 3.3 and 3.5 bring to light several noteworthy patterns.

According to the estimate shown in Figure 3.3, compared to patients who died during their first hospitalization, patients who died during a later hospitalization were two-thirds more likely to be registered. One likely implication of this finding is that only by being fully engaged with the health care system—visiting the hospital multiple times, for example—can AIDS sufferers guarantee their visibility in the official data. Similarly, insofar as those who die on their first hospitalization may be poor, marginalized, or otherwise disadvantaged, we see a certain profile of the unrepresented emerging.

Figure 3.5 supports this idea that marginalized populations are underrepresented in the state surveillance data. Men with unreported or socially stigmatized sexual orientations had significantly lower chances of being registered than their heterosexual counterparts. Those who self-identified as bisexual or homosexual were fully 49 percent less likely to be registered than those that reported heterosexuality. Women were equally as likely to be registered as heterosexual men, but they had greater chances for visibility than men overall—the overall rate ratio relative to men, not shown in the figure, was 1.44 (confidence interval = 1.07 to 1.94), implying a 44 percent greater likelihood of registration.

•••

We also gathered information on 104 AIDS deaths at Caasah between May 1993 and November 1996 (Table 3.3). Of those deaths, 31 percent died at the Luis Souto AIDS unit, 10 percent died at the Hospital Caridade, and 1 percent died at the University Hospital. The remaining 61 patients (59 percent) died at Caasah. In addition, the deaths included two men for every woman, resulting in a male/female sex ratio substantially lower than the official ratio for registered cases, which stood at 4:1 in the mid-1990s, for both Bahia and Brazil more generally.

Limited information on 18 patients (17 percent) prevented a complete comparative verification with the database of the epidemiological surveillance service. Of the 86 cases at Caasah for which we could precisely identify names, 63 (73 percent; 61 percent overall) were not registered at the epidemiological surveillance service. Only 23 AIDS cases (27 percent; 22 percent overall) were registered. Out of 23 registered cases, only 4 were reported as dead, 18 as still "living," and one as "unknown."

•••

In setting forth a distinct representation of the AIDS epidemic, this comparative analysis challenges local and national data that indicate that

TABLE 3.3
Selected Characteristics of AIDS Patients in Caasah

Registration Status	
Registered	22%
Unregistered	61%
Unidentifiable	17%
Sex	
Male	70%
Female	30%
Place of Death	
Caasah	59%
Luis Souto AIDS unit	31%
Hospital Caridade	10%
University Hospital	1%
N	104

Deaths among Caasah's HIV-positive residents, May 1993–November 1996.

incidence has stabilized or even decreased, at least among the urban poor. At stake is an unregistered AIDS epidemic whose subjects live in social abandonment and in "post-clinical" conditions (i.e., despite making their way through and out of some form of AIDS treatment, they remain unaccounted for). This medical invisibility is not restricted to the AIDS epidemic and its local and regional management. Local epidemiologists report that during the 2000 Salvadoran dengue epidemic only one in one hundred cases were registered, that more than 40 percent of deaths in the state of Bahia have "no known cause," and that maternal mortality in the northeast of Brazil is typically adjusted more than 200 percent to account for under-registration.[8]

The AIDS unit's data included no social indicators such as level of education or income. But according to the unit's social worker, these patients are the poorest, surviving primarily through marginal drug and sex economies: "They live in the gutter. Sometimes strangers send them here in a taxi; others are brought in by the police. They come in dying; they have bad skin lesions. The ones who recover just return to the streets, where they die. They seldom come back for a follow-up. It is unrealistic to demand that a person who lives on the street adhere to treatment. They never heal. There must be thousands of people in the same situation."

The under-registration of AIDS cases by the surveillance system suggests tacit norms of intervention that are aimed at a specific target group: a self-registered seropositive population.[9] Specialized health care is provided to those who dare identify themselves as AIDS cases in an early stage of infection at a public institution, who autonomously search for continuous treatment, fighting for their place in the overcrowded services—those whom I call *patient-citizens*. While health care interventions related to the national AIDS policy help some of them, local health professionals and communities allow others to die unaided.

As our research revealed, there are also implicit medical practices that help circumscribe AIDS patients considered "apt" for hospitalization. This specific demand for and supply of care is the core reality that appears in the optimistic epidemiological reports showing a decrease of AIDS incidence in Bahia. The individuals who cannot be framed within this limited infrastructure of care and this self-selective conception of public health remain outside official registers, unable to receive life-extending intervention. The majority of these unregistered cases are only identified as seropositive when already "wasted." In these routine exchanges, many of the abandoned with AIDS simply die.

In other words, the poorest and most marginalized with AIDS are socially included through a public dying, as if their deaths had been self-generated.[10] These abandoned only become partially visible in the public health care system at the end of life and are then traced as "drug addicts," "robbers," "prostitutes," or "noncompliant," practices and labels that allow them to be blamed for their dying. They remain at the margins or absent from nongovernmental interventions. In the end, there are no records tracing their personal trajectories, and the complex social and economic interactions that exacerbate infections and immune depressions remain unaccounted for. Most likely, a large group of potential users of AIDS public services do not even seek assistance, medical or pharmaceutical. The short-term care of these dying marginalized patients is relegated to a mostly sporadic street charity like that of Dona Conceição.

My colleagues and I wrote a report to the Bahian Health Division informing them of the existence of this hidden AIDS epidemic. I learned later that this report was simply shelved. It was within this kind of unreformed and publicly discredited regional politics that the ARV rollout came into effect; it is in these local force fields that the sustainability of the "AIDS model" remains in question, that a local triage-like state gains form, and that social death continues its course.

AIDS Therapies and Homelessness

I returned to Salvador in early August 2000 to work with Dona Conceição and her street patients. Dona Conceição told me that by the end of 1997 she had been able to register her group as an NGO and a so-called entity of public utility, making her eligible for public funds. But without the necessary know-how or technical advice, she had been unable to write project proposals; as a result, she never received any governmental funding. With the aid of local priests, neighbors, and her extended family, she was now supporting a school of seventy children in the district of Pernambués, some of whom had been diagnosed as HIV-positive. She also continued to provide minimum assistance to homeless adult AIDS patients.

The shift in the focus of Dona Conceição's work had provoked uproar from several of her previous dependents, who were now being weaned from her support. Some told me that Dona Conceição used them to become an NGO and then discharged them, just as NGOs generally do with the people they supposedly represent. Others said that she chose a specific group of "favorite" patients, whom she still helped, and others proposed that she was embezzling funds and making personal gains from her donation campaigns—which, from my perspective, was not true at all.

Many were now also appealing for support from transvestite Naum Alves, who was operating a Dona Conceição-like service. Until 1995, Alves had lived and worked in Caasah. The story goes that he was expelled for openly challenging the authoritarianism of Caasah's president. He then squatted in a casarão in the Pelourinho and began lodging some of his transvestite friends who were already presenting AIDS symptoms. Alves had also been linked to a local AIDS NGO, where he worked as a "community prevention educator." With his many contacts, Alves began connecting his friends to medical services and welfare procedures. For any form of assistance, he charged some percentage of the goods received (in terms of money, food, clothing, or medication).

I worked with the roughly thirty people still under the care of Dona Conceição, who had been forced out of the now-renovated Pelourinho by the police. They were living on a cement platform adjacent to the city's main soccer stadium. People slept on the cold floor with no covers other than newspapers. It had rained the night before my first visit there, and the "residents" were all soaked. Many looked undernourished, had skin lesions, and complained of flulike symptoms. But then, as twenty-five-year-old Carisvaldo Batista put it, "we push life forward anyway." Dona Conceição, her son-in-law, and a local priest brought them breakfast once a week. She introduced me to the group, and by the end of that visit, they agreed to let the "Gringo" (as they called me) come back, which I did for the following two weeks.

People there gathered in two opposing camps. "We down here are from the good," explained Tania Bastos, a thirty-four-year-old woman also nicknamed "Sheriff." "Those faggots up there are always getting into fights. They use broken bottles to cut each other. I just want peace." Tania, who was four months' pregnant, was referring to a group of transvestites who clustered around a hut made of plastic bags. "They are all *bichada* [rotten by AIDS], but still do prostitution. They are no good. They eat their food, and we eat ours." Tania and her younger partner, Orlando Lima, had organized a storage room in a corner, where they kept some of the food Dona Conceição and her friends provided. "The one who parcels things eats more," said Tania, who cooked on a dangerously improvised alcohol-fueled stove for herself, Orlando, and a few others. Eventually, the group of six transvestites also joined the circle to pray before breakfast was distributed.

Tania had a bad cough but was adamant that she was free of TB. She wanted me to get her cough syrup and vitamins: "The winter has been pretty bad. I get too much wind at night." Dona Conceição had helped her to get prenatal care. "I first got pregnant when I was twelve years old, but the child died," she said and shifted the subject to AIDS: "I took AZT for some time, but it did not do me any good, so I stopped." Tania got AIDS, she said, by sharing needles. She learned of her HIV status by participating in a study carried out in Pelourinho (most likely the one run by Dr. Tarcísio Andrade, mentioned earlier in this part): "My whole gang is dead." Like most of her fellow street patients, Tania denied using drugs, but, as I learned from Dona Conceição, "they all still smoke crack." She

Tania's habitat

Homeless AIDS patients

also told me that Tania's partner "was previously married to Priscila, a transvestite who died of AIDS, poor thing." Tania and Orlando had been together for five years.

Orlando, who had by then joined the conversation, told me that he had already been tested for HIV three times, but, afraid, he had never gone back to get the results. With herpes all over his body, Orlando also coughed a lot. He seemed drugged and, among other things, said that he was a "proud father-to-be": "I am the legitimate father of this child. I already had a child with another woman, but that's past. I want this child to have a future, health, hope. I came to the streets knowing what was waiting for me. I studied until sixth grade. I do little jobs, here and there, some of my relatives help me when they see me. They know that I am not an evil person. It is the devil that pulls me in, and I cannot resist." Tania still mentioned doing regular AIDS-related medical follow-up but could not name her doctor or her clinic.

•••

Several of the street patients I talked to said that they had begun picking up free ARVs at the hospitals, but they had stopped using them. As Roberto Costa put it, "medication alone will not solve anything." The thirty-year-old added that "we must take things lightly, as if AIDS did not exist. Here, public health is for the miserable ones to die." He had buried his wife the week before. She died of AIDS: "When they called me at the Luis Souto, she was already in the freezer. . . . Caasah paid for her coffin."

His friend, Luis Gomes, said that he didn't believe in the efficacy of ARVs: "My medicine is food, beans in my belly." He joked that "whisky is more helpful than these drugs." He also spoke against the side effects of ARVs: "The faces of people who take these drugs change, the skin peels. . . . I have a friend whose body went back to normal after he stopped taking them. Now and then I get a fever, a diarrhea, a headache, but that's all."

Nobody there sold ARVs, I was told. Roberto, Luis, and others to whom I talked did not even know how expensive these drugs were. They had no qualms about saying that they simply threw drugs away. Here, medication was not considered for its commodity value. The free dispensation of ARVs had succeeded in halting the development of a black market, but a culture of adherence was far from here. In this context of scarcity and marginality, only the strengthening of circuits of care (of which Dona Conceição, Caasah, and their "former" families were part), I thought, could actually help to restore the medical meaning and potency of these therapies—at least for some.

Three young men told me that they still "did the agenda" (*fazem programa*), meaning prostitution. In the words of twenty-two-year-old Abel: "The Baron stops the car and picks us up. I can get ten *reais* by giving him a blow job. The other day I fucked a chick for five *reais*." Many showed me their positive HIV test and said that they use it "to beg," like Ciro, who took me to the side and said, "I lost my other documents. I am not ashamed. When I ask for help I show the test and many help." He also showed me the test of Joselita, his wife, and asked her to join us.

Joselita had just been discharged from the Luis Souto AIDS unit—no new beds had been added and the service remained precarious at best—where she had been treated for TB. "I go once a month to the University Hospital to see Dr. Vania. I was taking AZT and another drug—I forgot the name." She said that medicines were always available, "they work, they are for free . . . I am trying to recover," she said, with a low voice, desperate to get a room to inhabit. Joselita's first husband had died of AIDS, and her HIV-positive daughter, she said, would soon be placed in an orphanage. Ciro was direct: "I need money to buy a cooler to sell popsicles. Can you, please, give me a few bucks? But make sure that nobody sees you giving me the money."

•••

Twenty-four-year-old Rita de Souza—"I was a drug addict"—spoke of the violent ways AIDS recasts the nature of social ties:

> I have a family, a father, a mother, siblings, and three children, but I have a problem: HIV. They kept me locked in, and the neighbors stopped talking to my family. My mother said that if I stayed with them, this would make my children suffer even more. So I decided to come to the streets and to leave my children in a better situation. Dona Conceição has known me for more than five years. She knew my husband too. He injected. I began doing it with him. . . . he was murdered. He was drunk and did not shut his mouth. He told people he had AIDS and got into a fight. I don't know if they killed him because of the little money he had, or because they thought he would infect them.

Dona Conceição assured me that Rita was the only one there who regularly followed her AIDS treatment. "I am healthy now, but I cannot get a job," Rita told me. "Now they request HIV testing before they give you any kind of job." Resigned to her joblessness, Rita now also uses her HIV test result to beg. Dona Conceição helped her to write a letter explaining her situation, which she shows to people, door to door, alongside the HIV test. "People have no obligation to help, but they help with food, clothing, a few bucks. Society helps us more than the government does."

Rita spoke of Caasah as if it were a site of AIDS governance: "I am registered there and get a monthly food basket. . . . Half of the cans we get are of rotten food. They keep the best food for themselves and for those who live there. My husband was a founder of Caasah, and when he died, the president did not even let our friends who live there come to the funeral. . . . They always said that they would help me get my disability status, but until today I am empty-handed."

Rita said that she had already been hospitalized four times at the Luis Souto Hospital, mostly because of recurrent TB. Her favorite doctor is Dr. Nanci, she said, "a mom for me. She explains things as they are. 'If you don't stop taking drugs and alcohol, you will die, period,' she tells her patients." For reasons I could not understand, Rita had been shifted to the ambulatory unit at the University Hospital. She said she was well treated there and that she was taking a combination of AZT, DDI, and 3TC: "I am doing this for my children," she said, with tears in her eyes. "I will tell you the truth: one of my children also has HIV."

It is basically impossible to combine life in the streets and AIDS treatment, said Rita. Taking ARVs was an incentive for her to leave the streets. She told me that she now only joined the "street tribe" (as some referred to the group) on the days Dona Conceição visited. "Caasah did not accept me back. I know too much, I talk too much. They want people who don't know their rights and who don't know how to express themselves. People like me would do a revolution there." Rita also refused to submit to the "inhuman conditions at Mãe Preta's shelter. She and her people pick up food from the supermarket dump; they wash it with vinegar and then cook it. I am not a pig. God forbid, to live from garbage."

Most of the time, Rita slept at a "house of passage" in Cajazeiras: "It is two hours by bus from here." The homeless I interviewed hate this state-run facility. They say that they are mistreated and forced to work without pay there. Rita, however, was able to befriend one of the directors, she said. "They know that I have AIDS. There, I get dinner and also milk in the morning so that I can take my medication. I help with house chores." Rita was begging Dona Conceição to help her rent a room: "What I need most is a place just for myself, a place where I can reconstitute myself." Rita's other main concern was shared by all around her: "I need to obtain my documents as soon as possible."

Most people there had no official documents whatsoever. A lot of their conversations with me were about an immediate need that they thought I could provide (like sandals, milk, aspirin, a towel, for example—and, to the best of my capacity, I tried to help them materially) and about their endless saga, from bureaucratic office to bureaucratic office, to have photos taken, to make a duplicate of their birth certificates, to get a social security number and a new ID card, to apply for disability benefits—as if they were having an interlocution with the world of the law through me. Dangerously unequal conditions give contours to these subjects' valuation of AIDS, health, and well-being. And the ways they are institutionally addressed as being quintessentially antisocial further expose their individual vulnerability and lack of political value.

•••

What I witnessed during my next visit speaks volumes to the tragic unfolding of the AIDS epidemic among the marginalized urban poor. Beyond the rubric of "impoverishment," "feminization," and "interiorization," this story points to the real ways in which AIDS spreads among them and in their interactions with other young and impoverished subjects.

Two girls, ages thirteen and fifteen, were among the "street tribe" that morning. They had been lured from the interior to the capital. Promised a job, the girls were forced into prostitution, and after two weeks they managed to run away. Two of the men I had talked to earlier found them wandering and brought them to the soccer stadium, where I was told they had either consented to or were forced into sex with several men the previous night. "I want to go home," one girl told me, shivering.

As I was talking to them, the "infancy guard" came and said that they would take the girls back to their hometown. As the police van was leaving, the street cleaners who had been there all along turned to the remaining street people, and in loud and laughing voices said: "Your fresh flesh is gone."

"Science makes people equal"

That week I met with Dr. Airton Radames, the new director of the infectious disease unit of the University Hospital. He and his group had ousted Dr. Moura. "His group never did research, they were very bureaucratic and conservative." By conservative, he meant solely care-oriented. "I don't know what is more important: to give people medication or to understand the situation in which disease takes form. . . . My philosophy is that treatment cannot be an isolated affair; health is multisectoral." But then his view of science, I thought, left little room for variation: "I defend science, and science makes people equal through knowledge."

Dr. Radames had an impressive research record indeed. Having studied and taught in a prestigious North American university, he had been able to foster several research collaboration efforts. In the process, he created his own research foundation "so that the university would not eat up the resources. Now academic and private institutions contact the foundation, which then subcontracts the research with the university." He was then running multiple clinical studies (most of them on AIDS, Chagas disease, leishmaniasis, and TB), was the editor of a Brazilian medical journal, a consultant to the WHO, and the coordinator of a sophisticated AIDS outpatient clinic under construction.[11]

The state, said Dr. Radames, was losing too much money in assisting AIDS patients at the hospitals and "a lot of medication is being thrown away, either by doctors who don't make the proper request or by uninformed patients." The new service would centralize medication prescription and distribution. Moreover, patients here would receive a more interdisciplinary form of AIDS care (immunology, nutrition, psychology, social work) than they got at local hospitals. "The government was investing too much, without results."

Technical infrastructure and medical culture had to change: "Doctors keep patients hospitalized at their whim. And as patients wait for one exam they get another infection and so on. My goal is that the new service will lead to a 50 percent reduction in hospitalization rates. At my unit, I already reduced the average hospital stay from 146 days to eight days. If we introduce quality in the service we reduce costs." Dr. Radames did not explicitly account for the treatment efficacy of ARVs as being

232 | CHAPTER THREE

determinant of its managerial success. At any rate, as I learned from Dr. Nanci at the Luis Souto AIDS unit, quite often doctors keep patients hospitalized so that "they can eat well, gain their bodies back, and so lessen the chance of discrimination out there," or, simply because "they have nowhere else to go."

I was also told by several physicians who greatly mistrust Dr. Radames and his projects that in this new outpatient clinic people were also screened for all kinds of trials run by the new partnerships between this man of science and pharmaceutical companies. The fact is that Dr. Radames and his research group were involved in some forty clinical trials, "most of them to test new combinations of drugs." Dr. Radames conceded that there was not much room to innovate in these studies: "The protocols are developed in Europe and the United States, we get them approved in our ethics committee, and randomize the patients." The service gains in terms of infrastructure building, he reasoned, "and the patient has access to superior drugs, to brand drugs that are available in the first world market."

Dr. Radames told me that the AIDS unit he coordinated now had twenty beds and that every month eight hundred patients passed through the ambulatory service. "We treat about 60 percent of all patients taking ARVs in Bahia." There was an average of thirty new AIDS cases every month, he told me. The unit's pharmacy had been improved with funds from the national AIDS program, and medication dispensation was working at "85 percent," estimated Dr. Radames. "All registered patients who want to get the medication are getting it; we have some problems with the patients who have to change their drug regimen and, sometimes, they have to wait a month or two for the new drugs." He also estimated that with the availability of ARVs, there had been a 15 percent increase in the service's population: "It was mostly middle- and upper-class individuals who were being treated in private practice. But now that ARVs were freely available, they migrated to public services."

This was a self-registered AIDS population, I noted. I then asked Dr. Radames whether he had seen a concomitant increase in demand for AIDS treatment among the poor and indigent. "Their demand for service is strained by many factors," he said, "People don't stop working because of a cough, a pain, a fever. They are afraid of losing their jobs. They treat their infections with whatever treatment they have at hand. In the city they only look for the service when they are bedridden, and in the interior they cannot even manage to get to a health post." Dr. Radames agreed that this was quite troublesome: "It is undeniable that AIDS has now shifted toward the indigent." Adding to their lack of economic and political value, these indigents with AIDS were also perceived as noncompliant, I thought. That is, they "wasted" medical resources and were not serious candidates (as were the self-registered middle-class AIDS patients) to the numerous trials that now seemed integral to quality AIDS care.

•••

Later that morning, Dr. Radames introduced me to his colleague, Dr. Airton Ribeiro, an epidemiologist and infectious disease expert who was writing a report on the state of AIDS in Bahia in the 1990s to be sent to the Health Ministry as part of a nationwide evaluation of the AIDS policy. Dr. Ribeiro had also studied in the United States and told me that he had worked for two years at the WHO's TB control program. Jokingly, he mentioned liking the work of anthropologists but that the interventions they proposed "were just too expensive." He was specifically referring to the "DOTS-Plus" approach to TB control that was being championed by Paul Farmer (1999, 2003) and Jim Yong Kim. "It is not just because they suggest all kinds of social assistance linked to the dispensation of medication, but because of the technical difficulties and costs involved in specifying each patient's pattern of drug resistance. With these specific protocols you cannot buy high dosages of medication and the patient has to wait to get the drugs. In my view, in the end, a more standard collective approach has similar rates of success."

I was curious about Dr. Ribeiro's diagnosis of AIDS in Bahia, and asked him whether he thought that it was under control. "Right now I am collecting information. I will let you know in a few months, after the report has been written." As I pushed him for preliminary results, he told me that "what I see in the charts that experts are sending me is that there has not been an explosion of AIDS cases as one might have expected. In the past two years, incidence rates have remained stable. So, it seems that the various public interventions, ranging from condom distribution to prevention among sex workers to outdoor advertising, have had some sort of impact on the course of the epidemic."

I told Dr. Ribeiro about my ongoing work on the hidden AIDS epidemic and that it appeared that only some 30 percent of AIDS cases made it in the registry. Dr. Ribeiro was cautious and said that the state surveillance service had improved its performance and that given the free availability of ARVs, doctors were now basically registering all cases: "I would say that now we have an underreporting of no more than 20 percent. *The marginal AIDS patient that you are working with is no longer invisible, he is intractable,*" he said matter-of-factly.

Indeed, a large number of my informants reported going through HIV and STD testing and medical treatments and participating in seroprevalence studies of intravenous drug users. Yet, I told him, we have to find a way to account for their condition not just as individual cases of "noncompliance" but as being symbiotic with present institutional practices. Dr. Ribeiro then said that he was not denying the existence of this contingent of marginalized AIDS patients, "perhaps unaccounted for," but that he questioned "its importance in terms of HIV dissemination."

The homeless with AIDS, I thought, were to be accounted for inasmuch as they were risk vectors. I told Dr. Ribeiro the stories of the two girls, to which he replied, "They must be infected now." He seemed to have found room in his epidemiological reasoning for this nomadic AIDS population: "Yes, they are a specific vector of infection. . . . But I am not convinced that they are the most important ones. Other social groups might be more decisive to the spread of HIV. Anyway, to diagnose is not the most important thing. It is important to treat and to educate the person that from now on he should no longer be a vector."

To illustrate his point, Dr. Ribeiro mentioned the stories of two men he had recently seen in the clinic. The first man had hidden his serostatus from his wife for over six years, and now the couple and their young daughter were being treated for AIDS. The second man had had more than a hundred partners since learning he was HIV-positive: "This means that information didn't matter. Last week, we heard that he died. Thank God he died. We don't know if it was a drug overdose, suicide, or from AIDS. But his dying did good." The moral of the story was that social scientists had an important, if accessory, role to play in the control of AIDS: "You guys can help us to understand why communication breaks down, to help us think new ways of making people take on what we say."

I told him that an anthropology of AIDS should operationalize a less rational model of personhood and that I was concerned with identifying the ways social forces, including service performance, influenced disease distribution and outcome. I quoted one of my favorite Paul Farmer maxims: "A failure to understand social processes leads to analytic failures, with significant implications for policy and practice." As the astute and unapologetically cynical intellectual he was, Dr. Ribeiro replied, "I see Brazilian society as already very complex, but not with all its problems solved. And I don't know the extent to which we want to solve these problems and then create all the social problems that Americans have. They have by and large solved technical problems, but the meaning of life is gone."

Enough of this. I needed his help. In a manner similar to my previous study, I now wanted to ascertain the presence of the homeless AIDS subjects I had been working with in the streets (I had done a mini-census) in the surveillance and medication distribution system. Dr. Ribeiro agreed that it was important to find an epidemiological way of accounting for this biosocial reality and put me in contact with the state's new AIDS coordinator. In the end, she denied access to the relevant files, alleging that this would violate patients' right to anonymity, despite the signed consent forms I brought from Dona Conceição's patients.

Brasília

At the end of August 2000, I went to Brasília. In a conversation with Dr. Paulo Teixeira, then coordinator of the national AIDS program, I mentioned what I recently found in my fieldwork. "It is a portrait of Brazil," he said. "I am not happy with the work being done with AIDS and poor populations. We have to identify a working strategy." The program was planning interdisciplinary workshops to reflect on AIDS and poverty and to develop new analytical tools, he said. The poorest of the poor with AIDS, I thought, are brought into policy as a future challenge, and their destiny is at the mercy of new discourses that will advance little else but research.

Dr. Teixeira again made clear the state's position of basically deferring care to community and pastoral organizations: "We have to develop regional strategies in association with social movements." And he added: "To work with these people is not the same as working with the elite in São Paulo, but effectiveness is also possible. If I get 20 to 30 percent of effectiveness with these people this is already a very important step." As I heard him, I was reminded of the 30 percent rate of AIDS registration in the Bahian AIDS unit, of the unconsidered and renewed ways in which the local state circumscribes populations for service.

"We have to go after this," repeated Dr. Teixeira. "The Brazilian experience shows that it is possible to intervene. In this context that you are describing, everything is more difficult, but it is also possible to do something. The excluded, the marginal, also has the capacity to care for his life, to protect his partner and family. With them, you can also effectively use and incorporate modern technologies."

I was told to write a report on my work on the hidden Bahian AIDS epidemic to the national program, which I did. I also sent a report to Dr. Ribeiro in Salvador. But I never heard back.

Chapter Four

Experimental Subjects

AIDS-like Symptoms

"If I were HIV-positive," Sky said, "my husband would leave me, perhaps even kill me." The twenty-five-year-old woman, a shop attendant by profession, looked scared and was afraid to get tested. "But I cannot endure not knowing my true condition," she told Ana Outeiro, a psychologist and counselor, at the Bahian Center for HIV Testing and Counseling (CTA) in October 1996. The Bahian CTA was created in October 1994 as a subdivision of the state's epidemiological surveillance service. Here, clients have free access to confidential HIV testing, accompanied by pre- and post-test counseling.[1]

With changes in the national AIDS policy and funds from the World Bank loan, HIV testing centers have dramatically proliferated throughout the country. CTA services are committed to the early detection and treatment of HIV-related distress and disease and to the promotion of safe sexual and drug-related behavior. As of 1997, there were already more than one hundred CTAs nationwide, established as partnerships that included the national program, regional and municipal governments, and universities. Local health services, nongovernmental organizations (NGOs), and schools are encouraged to refer asymptomatic persons for HIV testing.

Upon arrival in a CTA, clients are generally asked to fill out an epidemiological questionnaire, hear a lecture on the scientific and clinical aspects of HIV and AIDS, undergo pretest individual counseling and blood collection for HIV testing, and return for the results and final counseling. Clients are also allowed twelve free condoms per month. Usually, CTA units work as surveillance systems that keep track of local tendencies of the AIDS epidemic and are supposed to guide the planning of health care services (see UNAIDS 2000). From October 1996 to March 1997, I carried out ethnographic research at the Bahian CTA. I was aided by Ana Outeiro and by Denise Coutinho, a private practice psychoanalyst.

Sky did not mention any high-risk behavior during her pretest counseling session. But she was allowed to be tested anyway, and the results came back negative. A few months later, Outeiro ran into Sky in the streets of downtown Salvador. Sky told her that after getting the results of the test, "I finally had the courage to leave my husband."

Bear, another person we met at CTA, reported swollen lymph nodes and frequent diarrhea. "It can only be AIDS." The thirty-five-year-old accountant told Outeiro that he had been in a safe-sex relationship with a woman for two years. "Trust put me at risk. One day I was drunk, and we had sex without a condom." Only recently, after they broke up, Bear had learned "the whole truth" about his partner's past. He confided, "She was bad. She slept with other men while she was with me. I suspect that one was an intravenous drug user." Bear was "sure" that his ex-girlfriend had infected him, but his test turned out negative. In the following months he returned several times to CTA to get free condoms.

Sky and Bear did not procure treatment for their malaise in psychological or medical services. Rather, they went to CTA, where they demanded and found a new truth about themselves—a truth extracted from their blood—that at least for some time seemed to help them to settle their dramas. Through this negative biotechnical truth, Sky freed herself from a conjugal tie that she perceived to be "killing" her. By having his symptoms defined as non-AIDS-related, Bear disconnected himself from a "bad" relationship.

During our work at CTA, we identified a high demand for HIV testing by low-risk clients, largely working- and middle-class, experiencing identity crises, depression, phobic tendencies, and complaining of AIDS-like symptoms. Like Sky and Bear, a large number of clients were not so much interested in receiving prevention information as they were in being tested and in knowing their truth, so to speak. Most of the clients were seronegative and returned for second and third tests. They were testing an "imaginary AIDS," we thought.

To better understand how such a phenomenon was taking form at this novel public health site, we developed a pilot study that was approved by CTA's administration. Outeiro asked clients who were assigned to her, for either pretest or post-test counseling, if they wanted to participate in the study. Of those asked, thirty-seven clients agreed to participate. Outeiro registered these clients' accounts and her own observations following the counseling sessions. She then joined Coutinho and me in reviewing the clients' individual risk assessment questionnaires, as well as her and other counselors' remarks on their files. We also verified the laboratory results of the clients' HIV tests.[2] Afterward, we discussed the collected accounts—and we were startled by their repetitive quality. As we tried to disentangle the counselors' impressions from the clients' narratives, we also began to recognize the inductive property of the acts of counseling and testing in recasting symptoms, fantasies, and self-understandings.[3]

Medical anthropologist Elizabeth Miller has identified a similar phenomenon in Japan: "AIDS neurosis is an illness phenomenon in which the person suffering is convinced that he or she is HIV positive, despite negative test results, and a range of nonspecific physical symptoms and phobic and neurotic tendencies are manifested" (1998, p. 402). She begins her essay "The Uses of Culture in the Making of AIDS Neurosis in Japan" by quoting an AIDS hotline counselor saying, "Japanese are at much greater risk for developing AIDS neurosis than they are of getting AIDS" (p. 402). Thus, Miller focuses her study on the social construction of this diagnostic category and on how cultural stereotypes and symbolic cosmologies are imbricated in the making of AIDS neurosis, a "uniquely Japanese phenomenon." Her important analysis nonetheless fails to address the cultural work of science and technology, that is, how epidemiological risk assessment and HIV testing and prevention actually inform this widespread experience of AIDS neurosis.

In the previous chapter I elaborated on how poor and marginalized people with AIDS became "absent things" in epidemiology and medical care. Here I look at how the middle and working classes are cast into risk categories and how these profiles are personally experienced. In particular, I am interested in two questions: first, how clinical epidemiologic expertise and HIV testing technology are made available to specific "market-able" populations; and second, how the concepts of rational-technical health management and biologically based identity are localized and articulated in self-knowledge.

At CTA, I argue, psychological processes are being scientifically and technically manipulated. Truth production, repetition, and fantasy become the materials and media through which new individual and group identifications are made up. As I elaborate on the access, reification, and affective absorption of biotechnical truth, I highlight the alteration of moral landscapes and the engendering of a *technoneurosis* in this testing center. These processes have coincided with the consolidation of a public image of AIDS as addressed and contained. I conclude this chapter with some critical reflections on how clinical trials carried out by local doctors and international drug companies have recently integrated HIV testing at CTA.

HIV Antibody Test

The Bahian CTA was located in downtown Salvador, in a building adjacent to the Workers' Health Center and the Center for the Study and Therapy of Drug Abuse (Centro de Estudos e Terapia do Abuso de Drogas, or CETAD). The University Hospital was within walking distance. CTA had six counselors (three social workers, two psychologists, and one sociologist) and two staff assistants—all of whom were women. The service operated jointly with the Center for Prevention and Control of Sexually Transmitted Diseases; although the two services had different working routines and separate personnel, they shared an administrator in Dr. Mirta Nogueira. Dr. Nogueira, a biochemist, told me that she ran training meetings for the health professionals of both units, "to update everyone on the latest AIDS scientific developments." The services also shared a laboratory and its staff of two biochemists and four laboratory assistants. The blood collected for the HIV test was analyzed at the Bahian Central Laboratory (Laboratório Central, or LACEN).

At the front desk, each of CTA's clients chose a pseudonym and received a coded number for identification (pseudonyms are maintained in this chapter). The next step was to complete an epidemiological questionnaire, which assessed the client's knowledge of modes of HIV transmission and AIDS-related diseases, and asked him or her to report patterns of risk behavior.[4] According to Rita Moura, social worker and coordinator of the counseling team, "It is a way of starting to figure out who they are, so that we can identify which category of risk exposure they belong to." After completing the questionnaire, the client had to participate in a group educational activity. Here, counselors discussed topics related to modes of HIV transmission, natural history of HIV infection, means of prevention, as well as diagnostic technologies.

Personal contact with the counselor followed. In a private session, counselors ascertained the clients' motives for being tested: "What are the reasons? It is a moment of reflection and recollection; the person will decide if he really wants to take the test or not, and why," said Moura. At this moment, client and counselor located the time of the potentially morbid act that motivated the person to come to the center and translated that act into a risk factor. A combination of fear and of staging ensued, as counselor Marlene Avila saw it: "The person generally does not want to self-identify. It is very difficult to put desire and social values into risk exposure categories." When asked why they exposed themselves to risk, the answers were, according to her, "always very empty, such as 'at that moment I did not think,' 'I had no condom,' 'it did not work,' 'I forgot.' I tell them that I want concrete answers and that at that moment they chose to get contaminated, because prevention was not important to them."

Counselor and client then formulated an individually oriented epidemiological plan. At stake was building a different grammar about a specific risk f/actor (factor and actor begin to coincide), educating instincts, and formatting safety-conscious behavior. Counselors also assessed the psychosocial implications of disclosing a potential HIV infection, and devised strategies to ensure that clients would return to get their test results. "It is not just a matter of taking the test and then it is done; we prepare the person to come back," said Moura. The client was alerted to the fact that he or she was only entitled to three free tests. Blood samples were then drawn at the STD laboratory next door. The client was asked to call back in a few weeks to schedule a post-test interview.

At CTA, one could argue that individuals do not belong to a specific group or to the masses, but to a technical procedure—the individual and group subject are technically the same. Testing helps the individual to find (at least momentarily) a new ideal of the ego and this experience makes him or her part of epidemiological profiles (anonymous and epistemic populations).[5]

Two dynamics have to be anthropologically charted here: (1) the ways in which material technologies such as epidemiological knowledge and HIV antibody testing are combined with confessional procedures and are then recontextualized as new social (public health) technologies; and (2) the role of this technoscientific experience in formatting ideal agency and in recasting these clients' libidinal and moral economies. Moura put it straightforwardly: "I ask them: 'Was your desire for pleasure, to be satisfied, stronger than your right to continue to live?'"

Certainty: Closing the Past

Mulata was a twenty-year-old high school student; she also worked part-time as an office assistant. Mulata told Outeiro that she needed to do the test because her boyfriend had already showed her his negative HIV result, and "I should also show it to him." She was very nervous and said that she couldn't sleep. Until now the couple had been using condoms. So what was she actually hiding from? What did she want to show her loved one?

Mulata admitted that she had had only one prior sexual encounter in which she lost her virginity. "I told that man about my fear of getting contaminated, but he assured me that he was not infected, that he had already done exams, and so on. He also told me that he would only have sex without a condom because I was a family girl and because I was disease-free." The young woman started to weep and said that she trusted that man because he was about the same age "as my father."

Mulata received a negative test result. Relieved, she said, "Now I can close the past." Arguably, Mulata's testing experience was successful inasmuch as it engendered the following: (1) the client found an anonymous and cathartic site in which she could deposit her secret and leave behind suggestions of incest; (2) she acquired a new truth (biology) that rendered past risk behavior void and that was meant to work as an identification for present ties; and (3) she crystallized a new affect—the testing experience initially increased, then decreased her anxiety.

Uncertainty: The Window Period

Love was a twenty-three-year-old clerk who identified himself as heterosexual. He came to CTA for his second test in 1996. "This past year I only had one sexual partner." Why then did he return? "I think she is cheating on me." While getting his negative result, he "confessed," said Outeiro, that he had had unprotected sex with his partner three months prior to be being tested. "We forgot the condom," he told the counselor. Thus, because he had an "open window," his second test was annulled. Outeiro told Love he "would have to repeat the test," which entailed going into a risk-free regime (using condoms or abstaining from sex).

The window period refers to the HIV-1 seroconversion lag (Alcabes et al. 1993), encompassing the time from acquisition (infection) to sero-conversion, that is, until any HIV-1–specific antibody is detectable in the serum. The period that extends from the point of seroconversion to the onset of AIDS is generally known as the "incubation period." The true date of HIV-1 infection is rarely known, so the lag period from infection to seroconversion is usually approximated.[6] On average, this window period is much shorter than the incubation period and appears to be two weeks to three months in length (Cooper et al. 1985; Neisson-Vernant et al. 1986), rarely lasting more than seven months (Horsburgh et al. 1989; Longini et al. 1989). However, lag periods of up to forty-two months have been reported (Ranki et al. 1987), and although they are very rare, they cast doubt upon the scientific validity of HIV-1 test results. Testing centers routinely adopt a standard three- to six-month window period (from the subjects' last risk exposure or situation) as a quasi-scientific reference for assessing the subject's "true" seronegativity and for timing the original morbid act to be reassessed as a risk factor.[7]

HIV infection is commonly diagnosed by the detection of antibodies (anti-HIV) by Enzyme-Linked ImmunoSorbent Assay (ELISA), or aggluti-nation. ELISA was patented in 1985 and first used in screening the blood supply, permitting positive or reactive units of blood "to be discarded or put aside to be used for research purposes" (words of one researcher of the U.S. Centers for Disease Control and Prevention [CDC] in Rugg et al. 1991). Reactive results are confirmed by Western blot (immunoblot) or further specific tests such as competitive ELISA, which, when evaluated quantitatively, allow for the differentiation of HIV types and subtypes (Gurtler 1996).

A parallel can be drawn between public health measures and social control in the era of HIV/AIDS, on the one hand, and the corresponding practices during the early modern epidemics, on the other. In his book *Discipline and Punish,* Michel Foucault (1979) elaborates on the new forms of collective and individual control that emerged from the management of the plague in seventeenth-century Europe. During the quarantine, for example, each individual was located, examined, and assigned a place among the living, the sick, and the dead in a specific way. There was a strict partitioning of the plagued town; each quarter was governed by a supervisor, each street surveyed by an inspector. No one was allowed to leave the town, stray animals were slaughtered, families had to make use of their own provisions. New sanitary policies required that every living individual appear at the window of his or her house daily. Called by the authorities, the individual showed himself or herself and reported on what was happening inside the house. In noting how the sick and the dead were counted, Foucault wrote: "Everyone locked up in his cage, everyone at his window, answering to his name and showing himself when asked—it is the great review of the living and the dead" (p. 196).

These strategies and actions gave form to a disease politics: "The relation of each individual to his disease and to his death passes through the representatives of power, the registration they make of it, the decisions they take on it" (p. 197). In this conception, modern government has become increasingly concerned with men and women in their relations, their links, their imbrications with epidemics and accidents, wealth, resources, means of subsistence, and other things such as habits and ways of thinking and acting. A new and automatic functioning of power was coming into existence, inducing a state of conscious and permanent visibility and bodily control.[8]

CTA is an ordinary site where new scientific rationalities and institutions of government intervene in the course of biological *and* psychological processes. Here the question of when to appear at the window and be counted by the government becomes both an external and an internal negotiation involving the production of truth—in this case, a biotechnical truth. Individuals have the window opened on themselves, so to speak, when they freely appear at CTA for risk assessment and the education of instincts.

Consider Lion, a thirty-nine-year-old small businessman who said he had already been tested for HIV in the United States. However, he reasoned, "Since I was still in a window period, I am uncertain about the validity of that result." Lion said he needed another "AIDS test." He had been in psychoanalysis for five years, with no improvement. "I am always anxious." He had had three sexual partners in the past six months but insisted that he had not been in a risk situation—"only a broken condom twice." Lion was tested.

Outeiro told us that as soon as Lion got his second negative result, he pulled the cellular phone out of its case, and told the good news to his mother, to his former wife, and to someone else. After his public airing, he recalled that one of his partners "had sex with others without condoms." He reverted to his initial configuration: tested but again with an open window.

Lion was also asked to return for another testing after his current window closed. Before leaving, however, Lion declared that he was going to produce another situation of risk, real or imaginary: "At an unconscious level, there are forces that lead a person to risk, and this is linked to a lack of self-esteem. Besides, considering the physical factor, contact without a condom is much better; the heat of the vagina is irreplaceable."

The stories of Love and Lion show that the window period is successfully tooled by both counselors and clients. The quasi-scientific temporality of the window period is used by counselors to annul the results of previous tests, to place clients in a kind of safe-sex quarantine, and to induce them to return for another testing. Thus, CTA's apparatus and experience provide clients the possibility of organizing subjective ideas into scenarios that are scientifically and technically legitimated and to which instincts become prosthetically affixed (if only in that space and time). As the window period is integrated in the clients' fantasies, it also frames, in complex ways, the repetition of low-risk situations.

It would be misleading here to refer to the unconscious as "a world of prelinguistic and pretheoretical phenomena" vis-à-vis sociopolitical and material effects, as Evelyn Fox Keller does in her attempts to understand the subject of science qua scientist (1992, pp. 3, 9). In this context, rather than trying to get rid of the subjective dynamics of endless truth production and fantasy, science is coextensive with them. If, on the one hand, as Lion suggested, the unconscious seems to impose itself on science (as it is being performed) in the form of an action that disrupts objectification, then, on the other hand, CTA's conflation of confessional and bioscientific know-how successfully integrates unconscious dynamics into manageable categories that are part of experimental forms of subjectivity. In the process, a different local AIDS reality is generated. As CTA's director states, without acknowledging herself as a vector of it, "All this that is happening. . . . I think that in a few years we will have a very different history, something very characteristic of these times."

A Population of Doubts

At the Bahian Central Laboratory (LACEN), the director, Dr. Iolanda Menezes, told me of an impressive increase in demand for HIV testing since CTA made free testing available to the general population in 1994. During the first years of the epidemic, persons who were suspected of having AIDS were tested at the state's blood bank. Since 1988, LACEN had performed HIV testing for patients already diagnosed with AIDS in the public health care system and for the general asymptomatic population. The University Hospital, with its own laboratory, did HIV testing for those patients who needed diagnostics to begin specialized treatment. According to Dr. Menezes, "Initially, testing was by and large restricted to high-risk groups and people who already had AIDS."

A colleague and I reviewed LACEN's records (then non-computerized), counting the number of individuals tested for HIV infection during 1990–96, as well as the number of seropositive exams in this period. Interestingly, alongside an increase in the number of people being tested for HIV, the prevalence of seropositivity among tested individuals had diminished (Table 4.1). After 1994, these data included HIV testing done at both hospital AIDS units and CTA. Arguably, the increased availability of free HIV testing at CTA led lower-risk individuals to demand more tests. LACEN's data do not represent the general population's seroprevalence or individuals' risk of HIV infection, but they instead reflect a specific and novel AIDS population, the population of what I call an *imaginary AIDS*.

CTA's clients did not need a referral to be serviced. "The demand is spontaneous," the director, Dr. Nogueira, told me. This assertion of spontaneous demand, however, ignored CTA's initial aggressive strategies of advertisement and recruitment. According to counselor Rita Moura, at first most clients said that they had heard of CTA in the media or were sent there by public health services, "but now the majority comes here recommended by friends, family members, by someone who already passed

TABLE 4.1
HIV Tests at LACEN, Bahia's Central Laboratory, 1990–96

Year	HIV Tests	Seropositive (%)
1990	962	19.5
1991	1,014	17.1
1992	1,585	17.0
1993	2,768	11.2
1994	3,073	14.3
1995	4,882	13.4
1996	4,902	10.3

through our service. Now publicity is mouth to mouth. The demand has increased dramatically. We don't have enough space."

Dr. Nogueira referred to CTA's testing constituency as persons without symptoms of "actual AIDS." The staff made this demarcation explicit. As Moura said, "In the beginning, public health services sent us people who already had AIDS-related symptoms. Then we contacted these services and told them that this was not our objective." The logic was that "people who have AIDS symptoms look for the state's medical units. There they are registered, get tested, and undergo follow-up treatment. Our clientele experienced risk, they are full of doubts, and they want to know if they were contaminated or not; many are just curious and want some information. They want to change their values." Moura conceded that, "quite often, there are people who are already experiencing things like skin rashes, diarrhea, wasting without being infected." Office assistant Zenaide Silva also said she saw many people in despair: "For example, if a girl begins having sex, she immediately loses weight. . . . The other day a woman was eight months' pregnant and the husband started to feel many things. He was getting crazy thinking that the child was infected."

Both CTA's director and the counselors with whom I spoke agreed that, generally speaking, this "population full of doubts" had some economic and social stability. Dog, one member of this population, was a twenty-five-year-old medical student. He said that he came to CTA because a professor suggested it to him and also "out of curiosity." He had had only one sexual partner in the past six months: "We have been together for three years." They didn't use condoms because of a "mutual trust." The testing would be a way to ratify that trust: "I really need to know her serology." According to Moura, "The demand for testing by people living in marginality is much lower. CETAD sends us some cases of intravenous drug users, but there are not too many. We predominantly serve people with high school and college degrees." In this case, the low-risk behavior of this specific population facilitates a change in values.

Tedania was fifty years old and worked as a biologist. She came to get her negative result: "It took me two months to have the courage to come back," she told the counselor. Tedania was content with the result but immediately asked to be retested: "I might have an open window." She talked compulsively about her exposure to risk (no condom use), about the "nonsexual life" of her son who is in Europe, and about the instrumentality of fear: "I think that condom use is a cultural question. It is like fear of the soul that one learned from childhood. The pressure of sex is enormous. Things do not change from one day to the other. It is a process. Perhaps spirituality is a way to develop risk-free practices. The younger ones have more fear of AIDS. My son is eighteen years old and has not yet had sex. He is still virgin because of fear."

For Tedania, a divorcee, "life without sex is miserable." But she was trying to get adjusted to the new AIDS reality, "a disease that shatters emotions, it changed the world." In this process, old Oedipal structures seemed to reappear in new scenarios: "Now I like to go out with younger men, because they use condoms. The older ones have not gotten used to condoms, they were not educated in that way."

As I talked to people on the periphery of Salvador, I noticed a generalized lack of information about CTA and also a lack of desire to undergo HIV testing (Biehl 1996). Many times I heard the expression, "I don't want to be discriminated against." For some people, HIV testing was seen as a threat inasmuch as it could reveal secrets (sexual and drug-related) to families and neighbors; they knew how to keep those secret practices functioning within a certain routine and a sense of normality. Many people were afraid that news or gossip about being seropositive would scare away clients in the informal economy or jeopardize their possibilities of being employed at all. At stake was also an unwillingness to submit to new behavioral controls inherent to "successful" AIDS prevention. Moreover, slum dwellers made it clear that they knew other persons (relatives, acquaintances, or neighbors) who had either died of AIDS or were living with AIDS without undergoing laboratory screenings or public clinical assistance. They suggested that this proximity of real AIDS scared them—"it's horrible"—and that they therefore preferred, as one informant said, "to go on with living without knowing it."

What Is Socially Visible Is an Imagined AIDS

Between October 1994 and July 1995, approximately 2,000 persons used Salvador's CTA services. Using admission questionnaires filled out by 744 of these clients (randomly chosen), Sérgio Cunha and colleagues (1996) produced a profile of 40 percent of CTA's population during the first year of operation. Fifty-five percent of the clients were male, 45 percent were female. Regarding sexual practices, 76 percent of these clients were reported to be heterosexual, 7 percent to be bisexual, and 9 percent to be homosexual. Ninety-two percent of the clients came from Salvador and the remaining 9 percent came from the interior.

Most of them had some level of formal education: 35 percent had finished elementary school, 38 percent had finished high school, and 15.5 percent had college degrees; only 2 percent of the clients were illiterate. In contrast, Salvador has a 25 percent rate of illiteracy among its adult population, and only 16 percent of the Bahian population have studied beyond elementary school.

Additionally, of the CTA clients, 72 percent were regularly employed, compared to only 51 percent of the Bahian population (as of 1990) (Secretaria da Saúde do Estado da Bahia 1996, pp. 15–18). The rate of seropositivity in the population profiled by Cunha and colleagues was about 5 percent, approximately one-third of the rate of seropositivity reported by LACEN in 1994–95.

Results from our pilot study, carried out with 37 clients, appear in Tables 4.2 and 4.3. In their admission questionnaires, 23 clients mentioned that they learned of CTA's services by word of mouth, that is, from relatives or friends; 6 clients learned through the media; 4 learned through public health services; and 4 learned through AIDS NGOs. Thirty-two clients answered that the main reason that brought them to CTA was "the desire to be tested"; 4 clients answered that they "needed AIDS information."

The demographic profile of our study group was very similar to the profile found by Cunha and colleagues: 21 were male and 16 were female. This was also a young cohort, with 22 clients 16–25 years old; 8 clients 26–35 years old; and 7 clients 36–53 years old. Thirty-three of these individuals were single.

Data on occupation and income revealed that 27 clients were employed and that 10 clients were high school students. No client ever referred to himself or herself as unemployed. Some reported professional status, with jobs such as office assistant, hairdresser, vendor, biologist, cook, radio talk-show host, security officer, accountant, health professional, maid, teacher, carpenter, businessman, and government clerk. The sample also had relatively high levels of formal schooling. No client was illiterate; 8 clients had finished elementary school; 21 clients had finished or were in the process of finishing high school; and 8 clients had finished or were in the process of finishing college.

Regarding sexual orientation, 32 clients reported heterosexuality, with only 5 claiming homosexuality or bisexuality. Thirty clients said that they were involved in stable relationships. Data collected on risk practices showed that these clients' practices were in fact "low-risk," as Outeiro and other counselors put it. The most commonly alleged risk factors were inconsistent use of condoms (26 clients) and, primarily, doubt about the serostatus of current or former partners (30 clients). Only 2 clients confirmed illegal substance abuse.

Regarding risk exposure, 9 clients said that their last exposure to risk had been more than a year ago; 28 said they were exposed to risk during that year; 14 said they had been exposed to risk in the past 6 months (potentially placing them in the window period); and 5 had been tested for HIV before. The counselor asked 14 of the clients to return for a second or third testing. All 37 sample members were confirmed HIV-negative.

•••

While investigating the Bahian AIDS services, I often heard public health officials referring to CTA as an "exemplary service." In a 1996 state-sponsored AIDS conference, an officer praised CTA for having an efficient management, for being freely available to the general population, and for providing "precious" epidemiological information to the state's

TABLE 4.2
Demographic Profile, CTA Pilot Study

	N	%		N	%
Gender			**Employment Status**		
Male	21	57%	Employed	27	73%
Female	16	43%	Unemployed	0	0%
			Student	10	27%
Age					
16–25 years	22	59%	**Education**		
26–35 years	8	22%	Elementary School	8	22%
36–53 years	7	19%	High School	21	57%
			College	8	22%
Marital status					
Single	33	89%	**Sexual Orientation**		
Married	4	11%	Heterosexual	32	86%
			Bi-/homosexual	5	14%

TABLE 4.3
Questionnaire Responses, CTA Pilot Study

	N	%		N	%
How Learned of CTA			**Risk Exposure, Timing**		
Word of mouth	23	62%	More than a year ago	9	24%
Media	6	16%	Previous year	28	76%
Public health services	4	11%	Previous six months	14	38%
AIDS NGOs	4	11%			
			Risk Exposure, Type		
Primary Reason for Visit			Inconsistent condom use	26	70%
Testing	32	86%	Uncertain partner's serostatus	30	81%
Information	4	11%	Illegal substance use	2	5%
Previous Testing Experience			**Test Result**		
Previously had HIV test	5	14%	Seropositive	0	0%
No previous HIV test	32	86%	Asked to return for retesting	14	38%

Health Division. In fact, the evaluation and distribution of the CTA's data is one of the director's main goals: "We are training people and bringing epidemiologists in to help us to analyze this epidemiological data, so that we can maximize notification to the surveillance service and state's AIDS coordination. I tell my people that you do not live without a history." In the meantime, CTA had not succeeded in bringing seropositive clients into regular clinical monitoring. For example, from October 1994 to June 1996, only twenty-two of CTA's seropositive clients had actually been treated at an AIDS service.

As I mentioned in chapter 3, the Bahian measurement and management of AIDS was largely limited to and aimed at a very select group of "self-reported" AIDS subjects. Specialized health care was provided to those who dared to identify themselves as AIDS cases in an early stage of infection at a public institution and who fought for treatment in the overcrowded services. CTA's practices were an extension of this public health machinery; they also generated a reduced representation of the complex factuality of AIDS in Bahia. One could argue that the AIDS sufferers excluded from AIDS statistics held something in common with the subjects of this imaginary AIDS, who ironically had easy access to services and were statistically included: at some level, both realities were technically and administratively engendered.

CTA stood for HIV/AIDS prevention policies that were directed at individuals in the formal economy, who were not necessarily at high risk of infection and who "spontaneously" sought testing. Poor and marginalized people suspected to be HIV-infected remained mostly absent at CTA. The presence of AIDS in the media and in massive prevention campaigns, the introduction of the condom to relationships, and the confluence of all this with the real AIDS deaths that people face in their extended networks had created a new anxiety. CTA was the place where this "AIDS anxiety" was acted out and repeatedly tested as negative. In this way, CTA ended up serving a specific segment of the population, whose demand to be served was produced by the service itself. CTA's reality was thus part of a fictitiously contained and technically controlled local AIDS. In this case, the socially visible AIDS was an imagined AIDS.

Occasionally, CTA became a site for the desperate purging of the daily misery and violence that places many at high risk of HIV infection. Star, for example, was a twenty-one-year-old woman who first came to CTA in January 1996. It took her exactly a year to come back to receive the negative result of her test. "Very skinny, worn out," read her protocol.

Star lived in the interior. She said that she had elementary schooling and that she worked as a maid. She wanted to be retested. In the past six months, Star had two sexual partners, she said, and at least one might have put her at risk of HIV infection. She then confided that "three months ago I had a sexual relationship with a truck driver." She opened a window by not using a condom "because I wanted to have a child." If she could, she said, she would have "a child every year."

Yet Star enacted another vision of her future. Distressed, she told the counselor that that morning "after peeing, I washed myself and introduced a water hose in the vagina to clean the inside, I hurt myself. I bled, and saw two pieces of flesh on the ground. It happened today." Most probably, Star used the HIV testing service to check if she was pregnant; that same day, she probably wanted to ensure a negative pregnancy test result by attempting an abortion through a technique common among poor prostitutes. As the session ended, the issue of prevention was reinforced, and Star was sent to get another test.

Risk and Prevention Models

What happens at CTA Bahia is part of larger international initiative to prevent HIV infection (see Epstein 1996; Patton 1996, 2002; Treichler 1999). Following the commercial licensing of ELISA in 1985 and its use for blood screening, the CDC recommended that individuals in high-risk groups be given "the opportunity to know" their HIV serostatus as a means of enhancing risk-reduction efforts (Valdisseri et al. 1993). A U.S. nationwide alternate test site (ATS) program was initiated by the states, with guidance from the CDC and with regional and federal support basically aimed at preventing infected individuals from infecting others.

As the program grew, U.S. state legislatures passed laws to require mandatory reporting of persons found to be HIV-positive, and a civil liberties controversy developed over the purpose and value of the testing. In response, the CDC and other organizations began to emphasize the importance of a counseling service that included HIV risk-reduction education preceding the HIV antibody test. ATSs were renamed HIV counseling and testing sites (CTSs), and the testing service became increasingly a prevention strategy under the banner of "information dissemination on risk reduction and behavioral change" for both uninfected and infected clients. This intervention was based on a rational decision-making model, in which knowledge of the potential negative consequences of one's behavior was seen as sufficient to influence that person's behavior. By the end of 1989, with the introduction of anti-HIV drugs such as AZT, the focus of the service changed (at least officially) once again: now to early detection and referral of infected individuals to therapeutic intervention and medical monitoring.

By September 1989, HIV counseling and testing was provided in more than 5,000 cities in the United States. By the end of 1990, the CDC was spending nearly $100 million per year to provide this service, and more than 2.6 million tests had been performed at publicly funded testing sites. In 1991 alone, more than 2 million HIV antibody tests were performed at publicly funded sites, of which approximately 2.8 percent were positive (Farnham et al. 1996). In 1993, CDC guidelines were published in the *Journal of the American Medical Association*, emphasizing that HIV pretest counseling should include a personalized client risk assessment. If at first counselors were called on to explain to those being tested the meaning of HIV test results, including the possibility of false positive or false negative results, now they were asked to develop a "client-centered" managerial plan of instinct control and risk-behavior change based on a potential HIV-positive result.

In 1991, Donna Higgins and colleagues published a review of 50 studies worldwide (mostly in the United States and in Europe) on the effects of HIV counseling and testing on risk behaviors. The authors reviewed 17 studies on homosexual and/or bisexual men, 12 on intravenous drug users, 11 on pregnant women, and 10 on other heterosexuals at high risk of HIV infection. Based on all these studies, the authors concluded that overall HIV testing and counseling "does not have a direct impact on reducing risk behavior" (1991, p. 2427). Considering sexual behavior and psychological outcomes, Jeannete Ickovics and colleagues (1994) also reported on the very limited consequences of HIV testing and counseling for seronegative women in the United States. Furthermore, Dawson and colleagues (1991) found little evidence that having an HIV test played a substantial role in reducing risky homosexual behavior in England. Interestingly, these authors reported but did not elaborate on the fact that, in many cases, the rates of risk behavior actually increased after clients received their negative results.

In all of the above-mentioned studies, the impact of the HIV testing and counseling technology was measured in terms of its efficacy in reducing high-risk behavior. The studies failed to address the appropriation of HIV testing and counseling by the clients and the other impacts of testing, such as the repetition of risk exposure. Additionally, demand for testing and retesting was on the rise among "low-risk individuals," while new categories of "high-risk individuals" increasingly excluded themselves from these services or used the test to various other ends. In 1993, CDC researchers publicly acknowledged that individuals at potentially high risk for HIV infection—such as "adolescents, blacks, and clients served in family-planning and STD clinics"—had "lower return rates for HIV posttest counseling" (CDC 1993).

In their qualitative study in Australia, Deborah Lupton and colleagues explored why "low-risk" individuals decided to undergo HIV testing. The authors argued that the test had become a form of "social currency" and "ritual" in that society: "Some people may not feel personally at high risk of infection from HIV, but have a test for reasons other than advocated in official policy statements. The people in the study drew upon reasons which included pressure from parents or lovers, the desire to give up condom use, the need to display mutuality, as a symbolic closure or commencement of a sexual relationship, and values concerning responsibility" (1995, p. 179).

Lupton and colleagues concluded that the costs of testing for HIV antibodies would continue to escalate "as long as the test is being used in these ways" (p. 179). Clients have to be taught not to misuse resources.[9] This cost-benefit analysis operates with the assumption of an ideal testing service and of a deviant, if creative, user of it. In the end, the study loses sight of the myriad ways in which testing service and client influence each other's performance.

In spite of all these side effects and affects, HIV counseling and testing have remained one of the keystones of national AIDS prevention in the United States, Europe, and Australia (Phillips and Coates 1995). They have been widely exported as prevention models (through international health and development aid packages) to poor countries that began their public health efforts against HIV/AIDS late.

Libidinal Order

The first Brazilian CTA unit was created in 1989 in the southern city of Porto Alegre. According to Carlos Passarelli, former national coordinator of CTAs, "our program was inspired by the testing centers operating in San Francisco, and was basically aimed at guaranteeing the secrecy and anonymity of the individuals who wanted to know their HIV status" (personal communication, January 2000). In 1992, the Brazilian AIDS program was redesigned and the founding of CTA units became a central feature of the new World Bank–funded prevention strategies. Since 1993, on average, more than 20 new services opened each year. As of March 1998, there were 102 CTA units functioning in Brazil, with 34 new ones in the works.

The installation of these services followed the pattern of official AIDS incidence in Brazil, spreading outward from its epicenter in the southeastern context of São Paulo and Rio de Janeiro (Saraceni et al. 1996). It also followed the diverse regional modernization of public health services. For example, as of 1998, the state of São Paulo had the largest concentration of CTA units in the country (25, with 9 being installed)—in 1995, the state accounted for more than half of the total AIDS cases in the country (CN 1997c, p. 145). As of 1998, the southern state of Paraná, which has one of the best public health systems in the country, had very low AIDS prevalence, but had 11 CTA units functioning. Meanwhile, there were no HIV testing and counseling services provided in the northern states of Amazonas and Amapá or in the Bolivian frontiers of Acre and Rondônia, where AIDS surveillance remained precarious and specific AIDS public health services were almost nonexistent.

At the time of this study, CTA units were established in the following way. On the demand of a provincial or municipal government, the national AIDS program undertook the financial and technical procedures involved in the creation of CTA services. Around $30,000 was immediately made available for building renovation and the acquisition of laboratory equipment, anti-HIV ELISA kits, educational material, and computers. Local governments were responsible for the daily maintenance of each unit, and the national program ensured a regular flow of new technical procedures and epidemiological data, meanwhile periodically evaluating the services.

The Bahian team, for example, was trained by technicians who flew in from Brasília. "We had to learn quickly how to minimally handle the complex demands of the service," counselor Rita Moura told me. In practice, professionals with different backgrounds were administratively balanced in the routine of the service (all partake in activities such as counseling and teaching). Counselors confided that they competed among themselves "not to be the one who tells a person a positive result." Why? "We have to open the envelope and read the 'bands.' The lab technicians should do that. And what if we did a wrong reading of the band?" The result is an equalization of competence among the social workers, psychologists, and laboratory professionals.

The creation of a CTA unit generally implies a technical upgrading of local laboratory services. In fact, according to Dr. Iolanda Menezes, the large-scale HIV testing at the Bahian CTA brought many benefits for LACEN: "AIDS helps us acquire a more modernized infrastructure, both in terms of laboratory operation and in terms of logistic support, such as a van." She specifically mentioned that "our service is in the process of being computerized and our professionals are being trained. In the future, we also hope to use this infrastructure to carry out scientific studies." The serological analysis done at LACEN followed the recommendations of the Ministry of Health, including two ELISA tests (recombinant ELISA, lysate ELISA) as part of a triage process, and then two immunofluorescence or Western blot tests.

All CTA counselors reported that LACEN was very slow in returning the testing results. As Moura put it, "This gives a lack of credibility to the service and generates revolt among the clients, not to mention the added anxiety, fear, and extended depression while they wait for the results." According to Dr. Menezes, the delay in returning the results to CTA is directly linked with the high demand for LACEN's services statewide: "We do both clinical analysis and public health exams. The demand is immense. Sometimes we issue 1,200 exams per week. We are the only centralized public laboratory in Bahia." At the time of this research, it took about thirty to forty days before test results were ready to be delivered in the post-test interview. When CTA first began offering the testing, the results were announced within ten to fourteen days. As counselor Lucia Silva told me, "This delay induces the development of mental illness in persons who are already stressed."

According to anthropologist Sean Patrick Larvie, the model that informs Brazil's new prevention programs "suggests that health—defined as the absence of illness—is a universal and invariant value as well as a powerful motivator of individual and collective action" (1997, p. 102).

Larvie followed the training of community leaders in Rio de Janeiro and their work as HIV prevention educators, identifying an emphasis that he calls "psychological prophylaxis." This emphasis is part of a change in public health that locates "the nature of the problem as well as the possibilities for its solution, within the minds and instincts of individuals" (pp. 99, 100). The key idea is that risky behavior can be rationally manipulated through correct and neutral information, based on scientific principles. These prevention initiatives are representative of the circumscribed ways in which the reforming state both fulfills and rewrites its public health contract.[10] Here, selective populations are reorganized and addressed, and the medical domain is increasingly reduced to a question of individual improvement and to community care so rarely in place (Biehl 2005). CTA's experiment is part of a push in the public health world toward technological sophistication and self-service, moving away from bureaucracy and traditional clinical work in the name of cost-effectiveness.

Gemini, another of CTA's clients, was a nineteen-year-old high school student. For her, CTA's experience amounted to instinctual literacy. Gemini's questionnaire read: "I have had sex with only one partner and he has been using condoms." However, during the pretest counseling she disclosed that she had unprotected sex. She blamed her libido. "I need to learn to control myself. We kiss, and then things get hotter and hotter. When I realize I have already given myself away. When one likes the other person, one thinks that he does not have the disease." Her education at CTA was caught between the disappearance of face-to-face relations and the vital necessity to decode appearances. "I must learn to see how appearances deceive." As if the testing experience could guarantee the end of a phantasmic rule. She seemed to have rationalized CTA's lessons well: "It's an illusion to think that one lives in a good moment now, the consequences can come later. If a couple does not want to use a condom, both must do the testing. Life is at stake."

Here we notice not simply the shift from clinical care to the client's objective accumulation of facts, but a technoscientific tinkering with subjectivity (Fischer 2003). Discourses and interventions of clinical epidemiology and biotesting have already turned the concept of individual risk into a parallel practice (Haraway 1991; Petersen and Lupton 1996; Castiel 1999; Barata 1996). We are increasingly living in epistemic environments, argues Naomar de Almeida-Filho, "peopled with fictitious beings, recognizable by their individual probabilities to acquire disease and die, non-subjective subjects, epidemiological profiles" (1992, p. 150).[11] CTA's rationality goes further: these individuals are educated to become—or at least to think of themselves as—administrators of their minds and instincts, and they have in the actual HIV antibody test the means to verify this "bioadministration" of themselves.

Meanwhile, the advertisement strategies of the Bahian CTA have successfully restricted the inflow of people suspected to be presenting symptoms related to real AIDS. In our sample, only one person had been referred to the testing site by a local public health unit. This individual, Sun, was thirty-four years old, was homosexual, and worked as a cook. He said that he would not have come to CTA if the health professional had not "scared me." Sun then took his shirt off, and showed a large area of his body covered by herpes: "I get this urge to fuck and then I forget this business of the disease. Pleasure makes me forget the risk to get infected. Now I am thinking more of risk, because I am watching prevention campaigns on television."

Science and Subjectivity

What happens in institutional and human terms at CTA cannot be fully understood through psychology, knowledge/power dynamics, or interpretive anthropology alone. Here, new forms of knowledge, technology, the unconscious, and social experience combine to engender an auto-bioadministration. An encounter between Michel Foucault and Jacques Lacan helps shed light on how material interactions actually mold subject and moral responsibility—there is no autonomous self.

In 1969 in Paris, before beginning his lecture "What Is an Author?" Foucault told his audience that the neurotic's position better qualified the open-endedness of his presentation (n/d, p. 30). His comment resonated with the subtitle of the talk, "the return to a . . ."—where "a" referred directly to Lacan's work on the *objet petit a*. For Lacan, the *objet petit a* is the object cause of desire, that which constantly remanages its inscription in the body as it answers in the place of the impossible truth of the neurotic. It corresponds to Freud's idea of a "representative of the representation of the drive." Eager to hear Foucault's thoughts on his work, Lacan attended the talk.

Foucault began and ended his lecture asking, "What does it matter who is speaking?" (1998, p. 206). He argued throughout that the author's name legitimizes certain discourses within our cultures and societies, and that this process has the effect of constraining meaning-making practices. Foucault also marked an important difference between the founding act of a natural science that "can always be reintroduced within the machinery of those transformations which derive from it" (p. 218) and the sciences of discourse—for instance, those of Marx and Freud—which feature "an inevitable necessity . . . for a return to the origin. This return, which is part of the discursive field itself, never stops modifying it" (p. 219).

Overall, Foucault was interested in the modes of circulation and appropriation of discourses as they are modified within each culture. Analyses of this type, he said, called into question the absolute character and founding role of the subject, pointing to the subject's system of dependencies. That is to say, the subject is found in the materiality of epistemic thresholds and transformations (1972). "In short, it is a matter of depriving the subject (or its substitute) of its role as originator, and of analyzing the subject as a variable and complex function of discourse" (1998, p. 221).

During the follow-up discussion, Lacan sided with Foucault in criticizing those who suggested that Foucault's work advocated the "negation of man in general." "With or without structuralism," said Lacan, "it is not a question of the negation of the subject. It is a question of the dependence of the subject, and that is something completely different" (in Eribon 1996, p. 150). On several occasions, Foucault actually noted the importance of Lacan's work on this matter: "The subject has a genesis, the subject is not originary. Well, who said that? Freud, certainly; but it was necessary for Lacan to show it clearly" (in Eribon 1996, p. 147).

Lacan mentioned this encounter immediately thereafter in his own seminar (n/d). Again, he said that Foucault had appropriately valued the originality of a function internal to discourses, a function that evokes "an effect of splitting and of laceration that is proper to everyone" (n/d, p. 90). Lacan has argued that, contrary to science's ideals and practices, something in the human subject "knowingly lies . . . without the contribution of consciousness" (1991, p. 194). Functioning at the level of the unconscious is *the return to a*: "there is a knowledge which says 'somewhere there is a truth which does not know itself'" (n/d, p. 90).

Paradoxically, says Lacan, "we no longer have anything with which to join knowledge and truth together but the subject of science" (1989, p. 17). Thus what uniquely characterizes capitalist forms of subjectivity is that truth has become our labor: we love truth and deploy knowledge to both further enjoyment and proliferate symptoms: "There is no discourse that is not of *jouissance*, at least when one expects from it the work of truth" (1994, p. 74).[12] The reoccurrence of the experience that "perhaps this is not the true meaning" points to these workings of truth in the subject.

We moderns do not elude a relationship to changing forms of truth. "The fact is that science, if one looks at it closely, has no memory," argues Lacan. "Once constituted, it forgets the circuitous path by which it came into being" (1989, p. 18). In psychoanalysis, however, says Lacan, knowledge is deprived of its absolute position in the subject: "What do we call a subject? Quite precisely, what in the development of objectification, is outside of the object. . . . The ego acquires the status of a mirage, as the residue, it is only one element in the objectal relations of the subject" (1991, p. 194). Or, in Lacan's rewriting of Freud's motto, "Wo es war, soll Ich werden"—"Where it was, there must I, as subject, come to be" (1989, p. 12).

Consider what Eyeglasses said and did. He was a forty-two-year-old single, bisexual man, known among CTA's counselors as "the one who wore thick glasses." A college graduate, he worked as a city government clerk. Eyeglasses mentioned that he had two male sexual partners during the past six months. He considered encounters with both to be of "some risk," although he did not engage in anal sex, and arrived at post-test counseling extremely anxious. He complained of stomachaches and diarrhea: "It must be related to the disease. I keep searching for symptoms in me. I am scared, I exposed myself a lot." He confided that, "last week, an acquaintance of mine died of AIDS." The moral of the story was: "When we see people dying with AIDS, we start to take better care of ourselves."

Eyeglasses told the counselor about his problems with condoms, control over his drives, and trust of strangers: "I do not know why I do not use condoms all the time. It's a mistake. Man's head is a very serious business. I have to control myself. At the moment I must understand that

I have to use condoms. I think that condoms hinder my type of sexual relations. I do not think that penetration is important. The preliminary things are much better, and if I keep using condoms I will have to have at least three to five condoms in one relation." Governmental AIDS prevention campaigns were generally ineffective, he said: "They must use the body, and incite fantasy."

Eyeglasses agreed, however, that the counseling and testing experience had an important emotional effect on him: "Since I came here, I have been much more coherent with myself. I began to have love for my life. Counseling was very important because I started to channel all the guilt." But the experience of blood drawing for the HIV test was the decisive moment. Eyeglasses described the experience as if it were the conclusion of the preliminary pleasure-giving practices he looked for in his sexual adventures. "The tension is overwhelming. In the end, we have to give our blood, and then things are no longer at the level of hypothesis."

In other words, with the blood drawing, the rule of the phantasm seemed over, at least for a while, and biotechnical truth production yielded satisfaction. "The test helps me understand that I might have it and might die. It is difficult, but the test reveals that. So, I prevent." The counselor asked Eyeglasses to return for a second testing because he had an open window.

At CTA, the scientific process of objectification has successfully moved from the scientist, the inaugural maker of a new system of knowledge, to the individual who acts out the system such that everything subjective is reduced to "error." A "progress" in forming subjectivity has been spurred: as the client's field of risk-free consciousness is enlarged, a biologically based self-understanding displaces his or her speech and keeps his or her desire for truth prosthetically in place, ideally foreclosing the subject of psychoanalysis as a mirage of the past.

It is worth pointing out that this reforming of human possibilities and relations via technology is contemporaneous with what Sherry Turkle identifies as the movement from a psychoanalytical to a computer culture—"a new culture of simulation" (1997, p. 22; see also 1991), and of a growing biologization of human experience (Haraway 1991; Luhrman 2000; Rabinow 1996; Young 1995; Dumit 2004; Fischer 2003; Rajan 2006). In the article "Our Traumatic Neurosis and Its Brain" (2000), Allan Young argues that the new taxonomies of mental disorders—as well as the scientific and clinical apparatuses in which they are imbricated—are actually invested in deleting the term *neurosis* altogether from research and practice, "dropping it into the waste-bin of psychiatric history" (p. 2). However, as Young insightfully shows, this process is linked to a purge of the psychosexual version of neurosis and to the transfiguring of traumatic neurosis into post-traumatic stress disorder, or PTSD. Experts are "passionately interested in discovering biological features particular to the disorder and its defining process," such as hypocortisolism (pp. 5, 14). This experimental "remnant" or "epistemic thing" (Rheinberger 1997) is to become a measure against which subjects will be able to define their "true" pathological status.

Foucault's vision of an experimental mode of subjectivity at the end of "What Is an Author?" seems to be well under way: "I think that, as our society changes, the author function will disappear, in such a manner that fiction and its polysemous texts will once again function according to another mode, but still with a system of constraint—one that will no longer be the author but will have to be determined or, perhaps, *experimented with*" (1998, p. 222; my translation, emphasis added).

The testing of an imaginary death, a new truth, and a surplus of enjoyment derived from this technical experience is what ties subjects like Eyeglasses to CTA. While CTA's "epistemological drive"[13] is operationalized, other interpretations that might concern the subject are condemned to a further obscurity or pain to be retested.

Dangerous Worlds of Intimacy

A continuous procession of tested subjects returned to CTA. In our pilot study group, the counselors asked fourteen clients to return for a second or third testing, either because they were in the window period during the previous testing or because they exposed themselves to another (low-) risk situation while waiting for the results. Rather than interpreting repeated testing as a sign of prevention failure, I saw it as a productive outcome of the tie established between this testing apparatus and the client.

Artemisa was a twenty-five-year-old man who worked as a radio talk-show host. In October 1995, he showed up at CTA complaining of "an incredible anxiety." He said that he had had unprotected sex with a woman of unknown serology and suspected that she might had been an intravenous drug user. He did the HIV testing that same month, while still in the window period. In January 1996, he came to get his negative test result. He agreed to go into a safe-sex regime and to return in a few months for another test.

Artemisa was retested in November 1996, but during post-test counseling he confessed that he had put himself at risk once: "It is the business of the moment, a desire to fuck. She was someone that all of my office mates were lusting for; she gave me clues that she liked me." He recollected the risky act: "It happened in the bathroom of our company. I thought that water from the shower would go into the condom, therefore I did not use it. At the time I did not think about contamination. It only happened once."

As he received his second negative test result, he plotted his next fantasy with the counselor: "Is there any possibility of getting infected if I have sex with a virgin, a younger woman, without using a condom? She told me that she had never had a blood transfusion." Artemisa was asked to return for his third testing.

A great number of heterosexual men used HIV testing to experience an imaginary sexual liberation (by being technically "zeroed"), resulting in a "scientifically" legitimated subsequent reenactment of risk behavior. In this way, men used CTA's procedures as a fantasy-space through which they were thrown back into the libidinal order where the ego was inscribed. Many men also used the testing as a way to have access again to sex without a condom, once "normal" and now a rare and fetishized occasion. They acted as if now, through CTA, there was the possibility of a "true" sexual relationship—a fantasy that is plotted out in the exchange between the client and the representative of the testing institution. In this process, matrixes of sexual domination and deadly practices toward vulnerable Others are newly inscribed.

Snake was a thirty-six-year-old married man. "I only have one mistress," he said. "I always use condoms with her. She lives with a guy, but she says that they don't have sex." Snake did not use a condom with his wife, "because I trust her." Snake was a self-assured narcissistic male: "The two women only fuck with me." How could he then be contaminated? Did his wife contaminate him? Or was he looking for proof that the second partner was betraying him? Snake did not respond.

The circuitous logic he presented at CTA was that as he betrays his wife he feels betrayed by the Other; and as he is imaginarily betrayed by the Other, he might as well have been betrayed by his own wife. In this reasoning, the possibility of Snake infecting his wife was never considered. He was sent for a second testing.

Women, in turn, often used their symptoms and the testing experience to deal with the socially condoned sexual servitude to which they were condemned. Highly oppressive relations were both reproduced and newly faced at CTA—recall the story of Sky at the beginning of this chapter. Several other women in our group experienced similar impasses. Their voices were faint, lost in an orthopedics of "right" answers and unspoken fears. They were filtered through the counselors' zealous clinical-epidemiological framing. In the end, this framing left us with little more than glimpses of dangerous worlds of intimacy.

Equator was twenty-four years old and worked as a maid. She spoke little and in a confused manner. Equator was afraid that her current partner might harm her if she was found seropositive: "He already hit me hard once." They didn't use condoms: "I never saw him with another woman." However, "it's true that I caught him once having sex with my sister-in-law." Equator authorized her partner to get the result in her name. In this case, the woman did the testing for the man.

Mango, a forty-nine-year-old store attendant, showed up for her post-test counseling in a profound depression. In the pretest session, Mango mentioned to Outeiro that her partner had had sex with "a woman who apparently died of AIDS." She told the counselor that "I trusted him. It is much more pleasurable without a condom. But I do not blame him, because if one gets contaminated, the only one responsible is the person herself."

The testing experience and the long wait for the result had integrated themselves into a deepening of Mango's confusion and pain now expressed in skin lesions: "I got crazy waiting. Look at my legs, see these wounds. I am so afraid." She was content with her negative result, but she then admitted that she had recently had unprotected sex with her partner again. Outeiro told her to close the window and then come back for a second testing. She agreed to use a condom for the next three months. She left with a narrative of purity: "Women are naïve. They fuck without a condom because they think that the partner is as pure as they are."

Technoneurosis

Consider the outcome of Oxygen's experiences at CTA. She reported a pain that found no rest, from test to test. Despite three previous negative test results, Oxygen had returned yet again for testing. She said that she had been extremely depressed: "I am very nervous, tired, I sweat a lot, and have insomnia. My eyes are yellow. My throat aches." She had been "happily married" for over ten years.

The woman pointed to skin lesions on her face. "I have no strength left at all. My whole body aches." Then she told Outeiro what she had already said in other pretesting sessions: "A year ago I was raped." The rapist supposedly told her he had AIDS. No police reports were ever filed. All the counselors who spoke to the woman at the (not so) anonymous testing center agreed on an interpretation: "She jumped over the fence [committed adultery] and is dying of guilt."

Now even Oxygen's husband was displaying AIDS symptoms, she said. The woman already saw two psychologists, but "I did not like it. The first said that she disagreed with me, that I had nothing; the second was too quiet, she did not say anything." The need of a test was entangled with the possibility of Oxygen's own voice and actually seemed to replace it. Outeiro told her that there was nothing else they could do for her at CTA. She had used her testing quota. "But I need another test." Later that day, Outeiro saw the woman next door, waiting to be examined at the unit for sexually transmitted diseases.

I want to mark a social and subjective swerve at CTA. The confused and painful experience of Oxygen was somewhat technically engineered. This testing apparatus played a determinant role in the emergence of a socially visible imaginary AIDS. The psychosexual neurotic "fate" of clients like Oxygen was not simply purged in clinical epidemiological frames and biotesting, but was instrumentalized and co-produced as "normal," returning to social reality as *technoneurosis*.

CTA's testing mediation was aimed at "zeroing" the client, that is, finishing off the "deadly" fixation to the object of his or her fantasy—all the while reframing instinctual urges within epidemiological risk assessment, a new truth, and conversion to deeroticized safe sex. During this testing experience, which was both performatic and biologically verified—a bildungsroman of sorts—the institution and the client produced a new discursive formula for him or her to handle with psychological dynamics and patterns of risk exposure. This experience provided the client with a "double" that had scientific voice and legitimacy (see Cohen 1998, p. 269; Freud 1955; Petryna and Biehl 1997). CTA's rational-technical intervention guaranteed the control of the sexual variants explicated by clients, validated their symptoms as "untrue," and paradoxically induced their literal return (risk subjects, fantasies, and symptoms).

This was a two-way street. As these new risk subjects were trained to overcome their morbid acts by new knowledge of themselves (now imaginarily emptied of the menace of death), they made more and more "natural" their capacity to reinvent their subjectivity (now with science and technology). In the process, a client like Oxygen came to depend on the secret significance that HIV testing was supposed to carry. This might have foreclosed access to the singularity of her dramatic experience and to the formulation of desire. In this way, the CTA experience directly impacted the course of mental health.

In sum, in this chapter, I have shown how working poor and middle-class subjects interpellated the AIDS services that were made available to them and how they used the testing experience to recycle power relations, sensibilities, and ways of being. Many of CTA's low-risk clients used the testing and counseling experience to handle intersubjective dramas (family and love relationships such as loss of virginity or adultery) in the void or vanishing of traditional social ties. Others used the service to formulate or disguise publicly new definitions of sexual orientation (e.g., to affirm or deny homosexuality), to move out of oppressive situations, or to cope with shifting gender roles. Some simply exercised free access to a "modern" and well-infrastructured public health service, so rarely available to them. Overall, I saw CTA's clients trying to deritualize domestic relationships and to actualize programmatic fantasies of a supposed autonomy.

Some of the immediate effects of the supply and demand of biotechnical truth at CTA are the consolidation of a technoscientific ethos of governance, the strengthening of fantasy as a regulator of social reality, the new inscription of patterns of social and sexual domination, and the client's addictive self-tooling. As the categorical imperative of science—to go on knowing—is embodied (as in Oxygen), the subject remains searching for evidence that might open up the future. It remains to be seen how massive and socially transformative this form of self-governance is—this is the very nature of its experimental quality.

"They own their bodies and are responsible for their actions"

When I returned to Salvador in August 2000, CTA had moved to a three-story building, newly constructed by the Bahian government in the Garcia District (a ten-minute walk from the previous location). CTA was now "a state-of-the-art testing facility," the proud director, Dr. Mirta Nogueira, told me. "The government recognizes that CTA has a very positive impact on society, and that we responsibly administer public resources. It gave us the best possible physical and operational conditions so that we can keep expanding and delivering the best possible service. The health secretary knows that it is better to invest in prevention now than to deal with AIDS treatment later."

The facility had a large auditorium for prevention lectures, numerous counseling offices, special rooms for meetings with clients found HIV-positive, a library (with largely empty shelves), and a technologically advanced laboratory. CTA was now doing ELISA tests and also confirmatory tests for syphilis, resorting only occasionally to LACEN, the Central State Laboratory. "Our staff has doubled, and we need more help. More and more people come for testing," added Dr. Nogueira. "The health services are referring all pregnant women to us. We are also getting an increasing number of heterosexuals from all social classes." She credited this high demand "to recent governmental efforts to address HIV vertical transmission and to continuous advertisement of free testing in schools and communities. AIDS keeps spreading not just here, but in the whole of Brazilian society, and the media must keep talking about it."

CTA had also developed a special service for rape victims: "We are making the 'morning after pill' available. We are also carrying out workshops with young people, educating them about their bodies, drugs, and sexual violence. AIDS has changed public health. It has made evident that disease is preventable. Our job is to help people understand that they only get ill if they want to. They own their bodies and are responsible for their actions."

CTA's waiting room was packed the day I visited. Biochemist Maria Delgado told me: "In the old CTA, we ran twenty tests per day, but now we are running an average of one hundred tests per day." Counselor Rita Moura chipped in, saying, "The majority of our clients still fall into the same profile: they have some basic education and are heterosexual. But we also see a considerable increase in people of lower socioeconomic status asking to be tested. Our problem now is finding ways to translate all our technical information into a language that they can understand."

Moura added that the dominant mode of HIV transmission remained sexual and that a large number of clients were indeed "repeaters": "An average of forty people [out of one hundred] per day come back for a second or third test." The demand for testing rose threefold in the months following carnival and the São João festival, she said. She then referred to the pilot study we had carried out there: "The same thing that you documented in your study continues to happen. People come complaining of AIDS symptoms. They seem to know the basics of prevention. And they use the test to zero themselves, as if they could then reenact risk."

Moura told the story of a man in his thirties who had come in with a suitcase: "He had been here before and asked for an HIV-positive report so that he could be hospitalized. He complained of diarrhea and lack of appetite. He was unemployed and had fought with his mother with whom he still lived. A lot of our work goes into helping these individuals to discover what it is that is making them return to us, and this goes from dealing with their sense of guilt and lack of self-esteem to issues of social vulnerability to their need for a truth." Indeed, HIV/AIDS stands in for other forms of anomie and abandonment. All the technical information and infrastructure, reasoned Moura, do not necessarily help, "unless we personalize the intervention."

There were also strictly economic reasons to get tested. A growing number of local employers were asking for HIV antibody tests as part of job applications, and CTA (like all other testing institutions throughout the country) was no longer enforcing anonymity. According to Vera Menezes, a recently hired social worker, "By enforcing anonymity, we were denying access to testing for many people who needed it to get a job. With so much unemployment, maybe one person doesn't want to show the test out of fear of discrimination, but twenty others are willing. Those with some money do the test in private labs. We were thus indirectly excluding the poor from the job market."

In spite of the increased testing demand, the overall rate of seropositivity found at CTA remained around 5 percent, Delgado told me. She added the caveat that "this is a spontaneous demand by a very heterogeneous population; we cannot scientifically infer whether HIV infection has slowed down or not."

Most of the HIV-positive cases were in the 20–35 age group, and the ratio of men to women was now 1:1. In July 2000, for example, 9 men and 9 women tested HIV-positive, out of an overall test population of 432; in the previous month, 3 men and 4 women were found HIV-positive. The critical professional added, "Personally, I believe that HIV infection is on the rise in Bahia." She based her suspicion on the fact that "we are having higher positive results for STDs. This suggests to me that condoms are still not routinely integrated in sexual practices."

A 1999 national survey by the Health Ministry showed that 48 percent of the population had used a condom in their last sexual relationship.[14] The national AIDS program cited this statistic and the free distribution of more than 12 million condoms per month as indicators of its success with safe-sex campaigns: "In the mid-1980s, a study showed that 5 percent of the population used condoms, the most effective method to prevent HIV/AIDS and STDs. Today, the scenario has significantly changed. . . . The Brazilian youth has already systematically adhered to condom use" (MS 2002, p. 16).

Indeed, condoms have become widely available in Brazil, and their use has certainly gained prevalence, particularly among younger people, since safe-sex campaigns began. But as my informants at CTA told me in 1996–97 and during this last visit, condom usage is far from a systematic reality. Considerable variation remains among certain groups and populations, but this reality is obscured in the various surveys and reports. Married women often report difficulties in negotiating the use of condoms with their partners, and people over forty do not seem to have systematically adhered to condom use. The use of condoms is mostly sporadic, one could say. In the locally inflected worlds of sexuality and male dominance, specific circumstances redirect prevention and lead people to abandon condoms.

CTA's counselors complained that in the past year, condoms had not been available on a regular basis. "It is horrible, but we have been forced to limit to whom to give condoms," stated Moura. "Before we gave twelve condoms per month to whoever came to be tested, but now we are giving them out only to those who are HIV-infected or those with STDs." Menezes alleged: "Some people say that there is no more money from the World Bank loan; others say that the national program is wrongly buying into the social marketing idea . . . the idea that people will more conscientiously use the condom if they pay a price for it, even if minimal. According to this reasoning, condoms should become a basic food basket item, like rice and beans." Unfortunately, added Moura, "We have no other option but to send people to their local health posts. But to tell you the truth, I am pretty sure that they don't have condoms to distribute either."

I told Moura and Menezes that I had also been working with homeless AIDS patients under the care of Dona Conceição and that several of them used their HIV test results to beg for aid. Like the working poor, they were also using the test to deal with the financial side of life. When asked whether they were undergoing treatment, some even told me "at CTA" or "at CETAD," revealing their actual dissociation from AIDS care. Moura was quite explicit: "It is true . . . we are getting fewer intravenous drug users to do the testing. And it is not because there are fewer people injecting drugs. . . . When we operated next to CETAD, they were directly referred to us."

CTA still espoused the idea of a spontaneous demand for testing. It was not structured to search actively for HIV infection where it was most prevalent. No specific initiatives had been designed to address intravenous drug users, for example. This stood in stark contrast to the state's policy of mandatory HIV testing for all previously untested pregnant women—those who tested positive were immediately placed on AZT. Vertical transmission was beginning to decrease, mainly as a result of this active search and quick form of testing.

Due to their antisocial status and behavior and their lack of political value, it seemed, drug users were not specific targets of intervention. Most likely, these marginalized subjects only looked for HIV testing after having experienced severe disease processes (see chapter 3). As far as I could infer from my street informants, CTA was failing its mandate to follow up cases that tested positive with prevention and early treatment measures. This was particularly striking in light of the lack of anonymity in the testing procedure, which should have facilitated the location of the HIV-positive person.

Then and now, the model AIDS service remained at odds with the real course of the epidemic, at least as far as the marginal sectors of society were concerned. Its "standstill modus operandi" was in tandem with the local AIDS clinical infrastructure, which remained precarious at best, neither prepared nor wanting to receive and treat these abject subjects. Moreover, the rich data gathered by the service were left unanalyzed, at least in any systematic fashion, and the counselors' critical insights about limitations and possibilities by and large met administrative indifference.

Clinical Trials

The administration had other subjects in mind. CTA's new headquarters and operations were part of an AIDS Research and Assistance Center that was under construction next door. Funded by the Bahian government, this outpatient clinic was designed to treat about 230 patients per day and it would be run by a multidisciplinary staff that specialized in HIV/AIDS (immunologists, gynecologists, pediatricians, and pharmacists, among others). "This will be a first-rate facility," Dr. Nogueira told me. "After testing positive here, the person will be immediately referred to the AIDS center. There, he will be registered, get a card, receive medication, and access medical and psychological care."

What the director did not say—and what I later learned from local doctors—is that these new AIDS patients would also be targeted for pharmaceutical testing. Dr. Radames, one of Bahia's most influential medical scientists, had conceived the AIDS center and was coordinating its construction. Through his many international contacts he had made his research unit at the University Hospital part of an international clinical trials network (the state's new AIDS coordinator, Dr. Lizane Arruda, was part of his team).

The AIDS center spoke volumes about the ways market interests influence the form and course local public health institutions take. Portrayed as a state-of-the-art and cost-effective outpatient care unit, it was in fact a hybrid public-private scientific enterprise filled with conflicts of interest. Here, persons newly identified as HIV-positive at CTA (their social status and their ability to adhere to treatments now known) would be screened for all kinds of clinical trials these local scientists were running for international companies. As a result of novel economic and political maneuverings and amid murky value systems, a new experimental population was coming into existence at CTA.

"All big pharmaceutical companies are inundating us with proposals for studies," explained Dr. Airton Ribeiro, the epidemiologist and infectious disease expert who supervised all trials run by Dr. Radames's team (see chapter 3). Their team was then running two phase II studies and more than thirty phase III studies. These were large trials (three hundred to six hundred patients) outsourced throughout the world: "Our studies generally involve groups of ten to twenty patients. They are designed to test the efficacy of ARVs and of drugs for opportunistic diseases. Dr. Radames works the ethical approval process, and we do the clinical handling of patients and fill out the forms the companies give us. Laboratory exams are centralized in São Paulo."

Trial subjects were recruited through an informal network of infectious disease specialists working in both public and private hospitals. These doctors (many were co-authors of the studies) referred newly identified seropositive individuals to the University Hospital. The partnership between CTA and the AIDS center would greatly facilitate subject recruitment. "Our studies are normally carried out with patients who have not been yet treated," added Dr. Ribeiro:

It is a matter of proving that a new drug is more effective than an older one. And for this we need treatment of virgin patients. Doctors tell these patients that we have a new drug being tested and that according to protocols from the national AIDS program they are not yet at the point of having to take anti-HIV drugs. They explain the potential benefits of early treatment and that they will have access to all international branded drugs. . . . "Would you like to try this?" Of course, doctors are told to always make clear that this is not a cure and to strengthen the idea of prevention, that participation in the trial might halt opportunistic diseases from emerging. If patients say yes, they are sent to the clinical investigators who make the final decision. If all is okay, we send the person the informed consent form, draw blood, and do all kinds of exams. The patient waits to be randomized. In our experience, 60 percent of patients asked say "yes" to the trial.

Such trials and subjects are also meant to test and shape local markets, as anthropologist Adriana Petryna has documented (2005). Companies are here articulating ways of making up consumers who, in the near future, will demand that the state include these latest drugs in accepted treatment protocols and purchase them as they officially enter the medical market. "Through these trials, we are helping to change medical culture," reasons Dr. Ribeiro. "With the drugs that are available and the new ones you can sustain the health of an AIDS patient. We also see that patients who are enrolled in trials take better care of themselves." As for the continuity of treatment: "It depends on what is stipulated in the consent form. Patients in phase II trials don't have this possibility, for drugs are still under evaluation. In some of the phase III trials patients are enrolled until treatment fails. . . . But in most cases, there is a time limit."

In the face of corporate interests and the shadowy mingling of political and academic institutions, ethically troubling scientific practices are replacing public health and medicine for affected populations. As far as I could tell, neither AIDS NGOs nor the academic community were openly discussing the risks and benefits of these developments, which—here as in other zones of crisis—equated drug trials with authentic care for the disadvantaged (Petryna 2006).[15]

There is no efficient pharmaco-surveillance system in place in the country, a local pharmacist told me: "Drugs are easily accessed over the counter, and there is no monitoring of therapeutic failure. Doctors are increasingly trained by pharmaceutical companies and they simply dispense medication. Patients do not receive adequate information about dosages and side effects, and they remain unaware of why they rotate from one treatment regime to another. People are dying via medication, and these are not cases of suicide. This is the reality in which trials are okayed and carried out."

The National Poisoning Information System (SINITOX) reports that since 1996 medicinal products have become the main agents in human poisoning cases (ahead of venomous animals and agricultural pesticides).

Of the 79,366 cases reported in 1998, 28.2 percent were medication related. Of the 72,786 cases reported in 2000, 30.4 percent were medication related.[16]

Even if the trend in globalized AIDS drug trials can be rationalized (as potential benefits apparently outweigh harm), this by no means justifies "risk dumping," nor should this pharmaceuticalization unfold without a continuous assessment of its effects on public health budgets and on concepts of care.

Chapter Five

Patient-Citizenship

"On the plane of immanence that leads us into a life"

While democratic political rights are being realized in Brazil, public policies are increasingly shaped by market interests and dynamics—the AIDS policy is emblematic of this trend. As I showed in chapter 1, lifesaving therapies had become widely available by the end of 1996, a result of civic mobilization combined with the interests of a reforming state, international institutions, and the pharmaceutical industry. The state, one could say, is now locally present in the form of medication dispensation. The antiretroviral (ARV) rollout, however, coexists with historically entrenched mechanisms of social exclusion that continue to shape the course of life and death for the country's most vulnerable citizens. In this chapter, I explore how impoverished AIDS patients fare in the long term as they engage with this large-scale intervention. Which forms of institutional and individual care materialize in contexts where pharmaceuticals and scarcity exist side by side?

AIDS therapies are boundary- and institution-making technologies. The distribution and use of ARVs make certain populations visible to the state. These drugs are the means through which civil organizations take on and improvise the work of medical and political institutions; that is, they play an important role in defining the field and scope of policy implementation.

Poor and abandoned AIDS patients self-select for social and medical regeneration in grassroots care units called *casas de apoio* (houses of support), which are spread throughout the country. The Health Ministry considers these houses a "great ally of the state," as they guarantee the "patients' right to health," "rescue their full citizenship," and ensure "community reintegration" (CN 1997b). On the ground, however, the situation is more complicated. Houses of support mediate the relationship between AIDS patients and the precarious public health care infrastructure. They address the paradox that ARVs are available but public institutions are barely working.

Over and above this population's history of political exclusion and human devaluation is an urgent problem: how to achieve the self-discipline and care required for AIDS-related treatment adherence in a context of multiple scarcities. To access these grassroots units of care (many of which are religiously inspired) and to get what they are legally entitled to, people must break with their old habits and communities and behave in certain ways. Volunteers and patients together seek out services and take the initiative to make AIDS therapies work. Being in a non-stigmatizing environment and part of a collective treatment routine greatly enhances ARV adherence among these patients. My work at Caasah was specifically intended to uncover these new public formations and the phenomenon of patient-citizenship. How does the consumption of pharmaceuticals alter these patients' self-conceptions and political possibilities?

Caasah began functioning in 1991 in a two-story house in downtown Salvador, the first AIDS hospice in the state of Bahia. Most of its first residents were prostitutes, both male and female, and drug addicts—people who either had been abandoned by their parents at a very young age or had left their families to live on their own in the streets, often squatting in abandoned buildings in the city's historic district (see chapter 2). In 1992, Maria Luiza, Caasah's founder, learned that a famous Catholic philanthropist was planning to convert an abandoned maternity ward owned by the Red Cross into a home for the elderly. She decided to take it over first.

On the night of August 5, 1992, two of Caasah's AIDS patients approached the building's security guard, chatted with him, and offered him *cachaça* (liquor). After he became drunk, a fleet of cars appeared, unloading AIDS patients and dozens of packages with provisions. The squatting of Caasah became a local sensation and national news.[1] The homeless who slept in the backyard were allowed to stay.

Grassroots institutions such as Caasah are at once precarious and strong, and their work can be quite effective. As they operate on the intense register of survival, they come to the foreground as life guarantors. For many, these are sites of salvation—as thirty-four-year-old Evilásio Conceição referred to Caasah: "For me, this is the house of God." And because these care units work, they garner support from various levels of government. The placement of the poorest and sickest AIDS patients has become a communal undertaking. Over time, national and regional policies have incorporated Caasah and similar initiatives, qualifying them as

Caasah

proxy public health care services—they circumscribe afflicted populations, triage services, and deliver care.

ARVs are now embedded in these pastoral sites; novel ideas of citizenship and modes of subjectivity travel and gain currency among those who use or who refuse to use them. This is not a top-down form of control—one could call it a market-based biopolitics. Pharmaceutical companies are themselves engaging in biopolitics, gaining legitimacy and presence in both state institutions and individual lives through ARVs. The government is not using AIDS therapies and houses of support as "techniques . . . to govern populations and manage individual bodies" (Nguyen 2005, p. 126). Poor AIDS populations acquire temporary form through particular and highly contested engagements with what is made pharmaceutically available. The political game here is one of self-identification, and it involves a new economics of survival. Desires are fundamental to life chances, unfolding in tandem with a state that is pharmaceutically present (via markets) but by and large institutionally absent.

Interviews and daily logs capture the events at Caasah and are the building blocks of this ethnographic account. Most of Caasah's residents did not read or write. These are their texts. Our encounters were in many cases a first-time opportunity for them to test the social recognizability of such texts. The anthropological endeavor was an opportunity for them too. As I gathered ethnographic material and wrote this book, I became aware of how the speech of Caasah's coordinators and residents was imbued with a calculated sense of what they wanted to reveal or not about themselves, as they could now, perhaps for the first time, author the outcome of their lives. Beyond what was said, sometimes predictably, I became interested in the explicit use and even inductive role such accounts had for the pursuit of health and in keeping Caasah's machinery of care in motion.

A Place of No Government

Celeste Cardoso Gomes, Caasah's current director, is a housewife and trained nurse born in 1957. She saw news of the squatting on TV and decided to join Caasah as a volunteer the next day. Her first marriage left her with three children, the first of which she had at fourteen: "It was a marriage of convenience, only for social status." At the end of the 1980s, she divorced. Conversion to Spiritism (a religion marked by belief in communication with the dead, reincarnation, and pursuit of charity) changed her life: "One day I looked in the mirror and did not like what I saw. I decided that I would become another person." She also married a younger man and decided that her work in Caasah would fulfill her earthly mission.

Having worked in an orphanage and taken care of her dying parents at home, Celeste had the needed experience. "In the next month or so, Maria Luiza called me and asked me to handle the finances, but I did not dare—it was too much of a mess. There was no control, no receipts or anything like that." This was at a time when AIDS philanthropy in Salvador was at its height. Charity flowed from religious groups, schools, and activist organizations. Army and police trucks delivered clothes and food. Rooms were filled to the ceiling with goods.

Soon this initial scenario of charity turned from haven to disorder. And a few months after the invasion, Caasah's residents felt Maria Luiza had proved to be too controlling and expelled her. As Lourival, a founding member of Caasah recollects: "You know, first Maria Luiza was the

Celeste, 2001

mother, then she became the stepmother, and, finally, the bitch. We forced her to resign. Romildo and many of us wanted to go back to the street, to have some fun, drugs, you know." Residents were extremely savvy in using the media and denouncing corruption and mistreatment (now powerful elements in arousing public sympathy). Romildo, the leader of the insurrection, called in several TV stations and newspapers. As Celeste witnessed:

> TV cameras were in Maria Luiza's face. And residents yelled at her and denounced her embezzlement—"where is my bicycle that you stole and gave to your children?"—and her previous criminal involvement in a child-poisoning case. It is difficult for me to say where the truth is. She provided everything that was needed. But at the same time there was no control. She asked me to go to the television stations and clear her name, but I had no papers to show how much money came in, how much money went out.

Celeste was elected Caasah's new president. She hired her niece Naiara, then twenty-one years old, to assist her. They have worked closely together ever since; the entire burden of running Caasah rests on their shoulders. Before working full-time at Caasah, Naiara worked in a kindergarten and coordinated a youth group at a charismatic Catholic congregation. Caasah's statutes were changed to guarantee that Maria Luiza would not be able to reclaim her position.

A few months after the "revolution," Caasah started to get bad publicity. The police caught some of the residents involved in petty crimes around the city, and newspapers reported that visitors had been caught in fights and had been exposed to HIV-infected blood. According to Celeste, "They did whatever they wanted in here. Everybody had sex with everybody, they were using drugs, selling things, there were children in the house. . . . It was chaos."

In 1992, Tiquinho, a nine-year-old HIV-infected hemophilic child, was brought to Caasah by his mother from Rio Real, in the interior of Bahia, and left there (see introduction). "Guaranteeing the future of this weak child motivated us. We could not abandon him in the midst of all these marginals," Celeste told me. In 1993, Naiara adopted the newborn of a Caasah resident, and later Celeste would adopt an HIV-positive orphan. The incorporation of those children in Celeste's and Naiara's households intensified their philanthropic commitments in a human environment that was tense and life threatening at times. There was a constant migration of patients between Caasah, the hospitals, and the streets. Medical treatments started and stopped. TB-infected patients were not isolated, and many with TB died. Some patients stole goods that had been donated to the institution, others sold their medicines. The kitchen was infested with roaches and rats. And there was no clear consensus on what was collectively good.

In those early days, Caasah was caught in political disputes between the authoritarian provincial government and the alternative city administration (a coalition of the Brazilian Socialist Party and the Workers' Party). State officials kept threatening eviction, casting Caasah as an example of left-wing anarchy. Salvador's mayor, Lídice da Mata, and the city's health secretary, Dr. Eduardo Mota, made sure Caasah would not lose its place. Indeed, city officials and local AIDS activists helped Caasah to gain legal status and become a nongovernmental organization (NGO). Funding from projects of the national AIDS program supported the institution's core maintenance and professional upgrading.

Celeste kept detailed logs of the everyday challenges of sustaining Caasah in those early years—"a place of no government." The logs speak of the conflictual character of the communal-life-in-the-making and of Caasah's incorporation into the political and activist AIDS scenes at both the local and national levels. The realistic features of these logs reveal the micro-politics and economics that secure survival and belonging. Arts of government, social ties, and affect are reconfigured through these circuits of care.

•••

June 3, 1993: Oh, my God, I am so overwhelmed with this responsibility you have given me. I am exhausted, but I will not give up, I know that I have to take the work onwards. Today at 1:15 p.m., outpatient Rildo came in with a very high fever. We are taking care of him. I asked Naum to solve the problem of the bounced check; we need the money so badly. Thank God, the ambulance finally arrived and I sent Ronaldo and Irene to get treatment. I hope they find a vacant bed in the Luis Souto Hospital.

June 4, 1993: I went from room to room to check how my children-patients were doing. I found Rildo still burning in fever. Naiara told me that she discovered that [transvestite] Raquel was smoking pot in the backyard. I called Raquel to my office and asked her to recall all the good we had done for her. She wept and promised not to do it again. An unidentified person donated vegetables. The social worker of the Court for House Arrest came to discuss Romildo's case. I told her the difficulties he is causing me.

June 5, 1993: What a sad surprise, so much suffering . . . for I am accepting into the house a mother and her 2-year-old son, just bones, and also HIV-positive. Lord, which tie did I have to these persons in the past? I try in vain to distance myself. I do not feel them as unknown.

In the afternoon high school students came and interviewed us. We also had the visit of a group from the Adventist Church. It is so difficult to endure the repeated mistakes of Naum; it is also unbearable to live with Genovildes. Oh, Lord, all this is too hard, please help me. There are times I wish I could just abandon all this.

June 6, 1993: Today is Sunday, the day is cloudy, and it seems that things in the house were calm last night. I did not eat anything. I am anxious. What can we do to take these people out of idleness? I scheduled a meeting with patients for this afternoon. I just called Dr. Cristina de Melo from the City's Health Division. Finally, there is some good news: she told me not to be alarmed anymore, for the Secretary has taken legal measures to guarantee that we can stay here. Night is falling, and I long for my family.

June 7, 1993: I am sick. I went through all rooms to see my patients. Anaildes is not well. Thank God her baby seems to be getting better. Romildo is not well either.

June 8, 1993: All patients are helping to clean the place. The city government sent us two security officers. I have faith in you, Lord, and I hope that things will normalize now. I went to the state's Epidemiological Surveillance Service, and reported that we have a patient with cholera in here. Rildo is not well. I don't have the means to give him adequate treatment.

June 9, 1993: Our financial director left. I spent the day trying to regularize our accounting.

June 10, 1993: We spent the whole day in the streets asking for food donations. The patients helped.

June 11, 1993: Romildo died last night. They took the body to the Luis Souto Hospital to issue the death certificate. I am trying to get a coffin. Shirlene told Naum that for a long time Milton had been injecting Romildo with cocaine. Where did he get the money from?

I am concerned with this abdominal pain. I will not go to the funeral. Father André will take care of everything. I went to the emergency room and the doctor requested two exams and gave me medication.

June 12, 1993: The food truck did not dare park in our yard. The drivers were afraid of the patients, who were screaming obscenities at them, what a show it was. Today a TV crew came to interview me.

June 14, 1993: Naiara told me that Joselita came back. Then Genário, her husband, showed up and started a fight. He threatened to denounce me to the city's Health Division. Later, we had to hospitalize Joselita. A city hall engineer came to inspect the building.

June 15, 1993: The security guard told me that Lindoval and Mario were drunk last night. Maria from Caritas [Catholic mission] called me and said that archbishop Dom Lucas Moreira praised me in front of a large audience, calling me a "heroine."

June 16, 1993: Last night Shirlene and Jorge fought with knives. I will have to put them in separate rooms.

I must be a president, a mother, and also a nurse. We have such a lack of professional help. Oh my God, each one of them develops the disease in such diverse forms. Raul had 105 F of fever and herpes in both feet. Irene is laying with 103 F of fever, she has a sore throat, and is spitting blood. Shirlene has severe pains in her head. And Luciano cannot move at all. Naiara took our children to the University Hospital. In spite of all these problems, I thank God for letting me pay my debt on earth. I keep trying to do my best.

June 17, 1993: Luciano, Raul, and Ronaldo are in terrible shape. Today there is a mass in Romildo's name. I must solve Mario and Andreci's problems; they really do not get along. Andreci also promised me that she would behave better and respect Luciano. I asked Naiara to buy medication for Tiquinho. Ronilson does not have a house anymore, the family kicked him out. I removed Shirlene from Jorge's room. God only knows what was happening in there. I hope I can avoid a tragedy.

June 18, 1993: The mayor's secretary called me and scheduled a meeting. Naiara discovered that Claudionor was selling cigarettes to other patients. Milton and Jorge stole goods from the storage room. Jorge hit Shirlene in the face, and Naum hit Jorge back. Some patients are spreading lies and gossip about the administration.

June 19, 1993: I expelled Milton yesterday. But as soon as I arrived this morning the patients demanded that he be allowed to return. A doctor from Luis Souto Hospital came and treated patients. We talked at length about the many problems in coordinating this place.

June 20, 1993: Today I arrived later—it is Sunday—and I found so many problems. The security guard told me that Fabinho took hypnotics with alcohol and robbed things from Mario and Rose. I was obliged to kick him out. He was given many chances but failed to change. Gilmar was also removed from the house. He was drinking *cravinho* [aphrodisiac liquor] and giving it to other patients. Lindoval and Geovan were caught in a violent fight. Geovan slammed a glass and cut himself; he was bleeding all over.

June 21, 1993: Lídia from Paratodos [For All, a philanthropic organization] called and said that our project was approved. Next month we will already get some financial aid from them. The city's engineer came again to talk about the building's renovation.

June 22, 1993: Yesterday I met with the mayor. Maria from Caritas came along. The city government finally approved a partnership with Caasah. A journalist from the *Tribuna da Bahia* interviewed us about it. This will help to alleviate some of the suffering here.

We celebrated São João with the help of volunteers from the Catholic Church. We had traditional food, cake, and soft drinks. Glaucio and Paloma [transvestites] entertained us. It was fun and all really loosened up. Tiquinho was radiant, with a cowboy hat. Today my deepest concern was with Lindoval. I never saw so much sadness in the gaze of a human being. I tried to talk to him, to help him to get all that resentment against life out of him. I confess that I fear him.

June 25, 1993: Rildo died last night. Danilo and his wife helped me to prepare his funeral. I heard that Jorge and Jovanilton (outpatients) had a secret meeting with Maria Luiza. Today I sent a revised version of the project SOS AIDS to the Health Ministry, in Brasília. I also sent the paper work to the City Hall so that we can become an institution of public utility. A reform in the building is key to the patients' well-being.

June 26, 1993: Lord, deliver me from people who say that they are my friends but who plot behind my back. They are preparing spears to wound my heart.

I will have to separate Andreci and Edson, before a killing takes place. Luciano is such a mastermind, and is always making intrigues.

TV Bahia [belonging to the conservative Governor Antônio Carlos Magalhães] came to interview me. It was horrible. They did not let me say anything about the help Mayor Lídice de Mata is giving us. The media is a true mafia.

I had another meeting with all the residents. They heard me. I read them the Gospel. There was silence. I felt the Lord's presence.

June 28, 1993: I spent the day going from hospital to hospital, checking how our patients are doing. At the Luis Souto, I saw that Raul is still very weak. Shirlene has pneumonia but is getting better. Ronaldo's health is very bad, I am very concerned about him. Joselita will be discharged tomorrow. At the University Hospital I found Joilson dying.

I am having difficulties getting condoms from BENFAM [the Brazilian Society for Family Welfare].

June 29, 1993: I have been trying to put our infirmary in order. I had to stop fights between Genovildes and Marta, and between Gilmar and Geovan. To have to live with this kind of people is a punishment from God. But I endure.

I got a call from the Gay Group of Bahia. They collected donations for us. I spent almost the whole day rewriting another project for the Ministry. I am very tired, but feeling accomplished and content—I have been able to guarantee some peace for the residents and for myself. Joselita left the hospital and asked us to take her back in. She might be spying for Genario.

June 30, 1993: I arrived and immediately went to medicate the patients. I bathed Joselita and took care of the other debilitated patients. Some patients went to the Luis Souto; hopefully the doctors will see them and give them medicines. Dona Lola brought meat, chicken, and salad. Thank God, when Genario came to visit Joselita they stayed outside the building.

July 1, 1993: I am very concerned. I am waiting for the food and medicines the city government promised us. Someone told me that there are rumors that the night-shift security guard is having sex with a male patient.

July 2, 1993: Yesterday, Luis Souto's social worker sent us another patient. Joselita refuses to take her medication. Danilo and his wife are helping with our accounting. Dona Lia visited me and gave me some *passes* [energizing touches]. I read the Gospel to the residents. I am very anxious, there is something in the air, and I do not know what it is.

Now I know, Naiara just told me that Raul died.

July 4, 1993: I found Luciano and Roberto smoking pot in the backyard. And to think that just a few days ago Luciano came back and, in tears, convinced me that he deserved another chance. Oh Lord, why are people so ungrateful? As soon as Raul was dead, the residents went to his room and took all his belongings.

Three Spiritist friends came to visit me in the afternoon.

July 5, 1993: I had to tell Anailde the sad news that her son Piuí died. Last night, patients stole the fruit that had just arrived. We only have beans left. I went to the state's Health Division. Now it is their turn to do something for us.

Sanctuary and backyard at Caasah

July 6, 1993: The city hall sent us some food. I was forced to expel Claudionor. Once again he was caught smoking pot. He threatened to kill me.

Pastoral Power

By 1994, threats of eviction stopped, and Caasah had obtained sufficient resources for basic maintenance. Caasah was being funded by two projects approved by the national AIDS program and had formalized partnerships with Salvador's Health Division. It also had strategic exchanges with hospitals and other philanthropic and AIDS organizations. Strict disciplinary mechanisms led to the expulsion of residents considered unruly *marginais*. A reduced group of thirty patients underwent intense resocialization mediated by a newly arrived psychologist (paid for by the city hall), and some eighty outpatients (*pacientes externos*) remained eligible for monthly food baskets. By the end of that year, concerns about internal violence, aggression, and drug trade and consumption were replaced by concerns for hygiene and house maintenance.

The next move involved medicalization, under the guidance of a newly hired nurse, Carlos—because he taught in a nearby private nursing school, residents respectfully called him "Professor Carlos." In 1995, he established a reasonably well-equipped infirmary post, with a triage room and a pharmacy, and hired a few assistant nurses.

Religious groups and schools visited Caasah on a regular basis—they were actively channeling donations of food and clothing, as well as providing volunteer support. Caasah's residents paid visits to public and private schools, where they exchanged firsthand AIDS knowledge for financial contributions. Technicians from the Health Ministry, accompanied by representatives of local NGOs, now periodically visited and audited the house. The city hall kept providing security personnel. The general public and the local government had assimilated Caasah as part of social life. The circumscription of some of those abandoned AIDS patients in Caasah paralleled the acceleration of other pastoral efforts, like that of Dona Conceição (see chapter 3), to address the problem of AIDS in Salvador's streets.

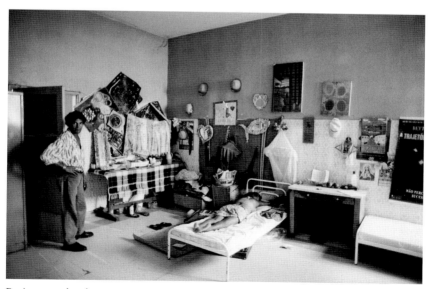

Patients and volunteers

Grassroots initiatives such as Caasah exemplify the limits as well as the possibilities of public formations at the edge of neoliberalization. As the state rationalizes its assistance interventions, we witness an increasing pastoralization of the social domain.[2] Governmental institutions do not actively search for specific problems and needs to attend to. And it is up to grassroots organizations such as Caasah to control access to welfare provisions and health care for marginalized patients. Community-run services thus compensate for financial and technological limitations and operate as social instruments of remediation.

It comes as little surprise that the support from the national AIDS program was aimed at strengthening Caasah as an institution, "so that we learn to struggle on our own, to walk on our own feet, to become self-reliant and look for other resources," as Celeste put it. Celeste mentioned that for the first three years, the government capped funding for projects like Caasah's at $70,000, and she knew how to guarantee the flow of those funds: "The next project must be less; I have to take a few items out in order to renew the project again." The funds were also used to pay volunteer workers (*trabalhadores voluntários*), responsible for maintenance (cleaning, laundry, and cooking) and the improvised infirmary, and a newly hired financial administrator. Each volunteer worker—very poor, barely literate—received $50 per month as well as food aid, a prized commodity in that world of unemployment and misery. Celeste also alternatively hired some patients to give them further incentive to change. The workers to whom I spoke were all well informed of the forms of HIV transmission and never showed much fear of getting infected. They spoke of their activity as a mixture of solidarity and employment: "I like to help. This is my job."

The project with the Health Ministry provided the institution with a minivan, which facilitated the ongoing movement of patients between Caasah and the hospitals. In fact, records of this time show that Caasah was both a challenge to and an extension of the social work services of the major hospitals dealing with AIDS patients in Salvador. The Ministry's project also allowed Caasah to offer outpatient support through monthly basic food baskets.[3] These baskets became highly prized commodities, drawing AIDS patients into a minimal and indirect welfare exchange with the state. A Ministry-paid social worker was specially hired to register the AIDS patients entitled to the baskets. Registration made

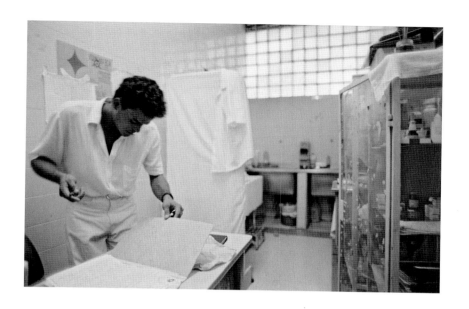

Caasah's infirmary

Leonardo, a security guard, and Edimilson

patients socially visible in the medical system and exposed them to their communities as *aidéticos*. To have access to the basket, people had to undergo routine medical checkups. Every three months, the basket allowance was renewed on the condition that the recipients presented medical reports. "Through this mechanism, we force the street patient to go to the ambulatory service," the social worker told me.

In late 1994, the social worker visited the applicants to distribute the baskets to them in their shacks or wherever they lived. Most applicants, however, had given fictitious names and addresses, as they did not want their neighbors to discover that they had AIDS. In the end, the social worker was able to put together a reduced list of around eighty basket applicants, which was sent to the national AIDS program and approved. This kind of administrative strategy circumscribes a visible AIDS group— that is, individuals self-reported and publicly acknowledged as *aidéticos*, for whom there are provisions. Meanwhile, AIDS in social abandonment remains institutionally unaccounted for. Furthermore, as I observed during my fieldwork at Caasah and with Dona Conceição in the streets of Salvador, basket access was integrated into underground economies. People learned how to accumulate baskets, how to sell them to buy drugs,

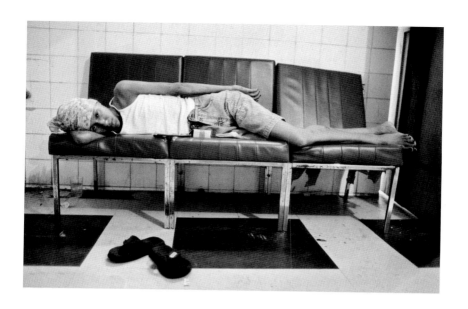

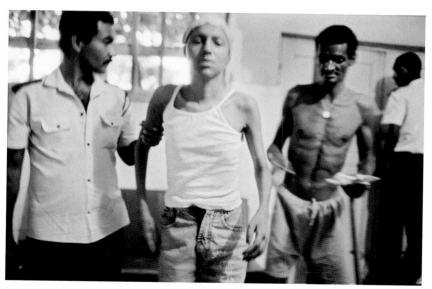

Taking a new patient to the triage room

and how to mediate other people's access to them and profit from this intermediary role.

Inside Caasah, patients and coordinators held weekly community meetings, and the psychologist offered individual psychotherapy. All patients were required to undergo supervised medical treatment at public health care services, under the strict monitoring of Professor Carlos. The inflow of medicine was maintained by donations and free samples from drug laboratories and distributors (ARVs were available by the end of 1996). An average of five to eight patients occupied the triage room, where Professor Carlos's team examined newly arrived patients and cared for patients with severe infections, who remained isolated there until their deaths.

In short, a local AIDS order was taking shape around Caasah's hybrid pastoral-governmental-medical initiative. In many ways, Caasah performed the work of public institutions and set the terms and conditions of how to be alive with AIDS or to be destined to the streets and death. By taking over the immediate care of patients and overseeing their medical treatment, Caasah became a venue of an incipient AIDS public health triage system. It mediated the relationship between AIDS patients and the extremely limited public AIDS services. In a more specific way, it selected the patients who could benefit the most from the scarce resources, as the state's AIDS unit had only sixteen beds. Caasah provided the means through which these individuals could accede to a distinct and tentative *patient-citizenship*, something completely unavailable to them in the past.

Celeste addressed this field of tensions and possibilities: "Caasah is not a hospital. The hospitals have to improve their own infrastructure, to make more beds available to AIDS patients, to offer better care and medical treatment. The government must take up its social responsibility and create shelters for the hundreds of homeless people with AIDS living under the bridges, in the parks, and on the sidewalks. Public institutions and politicians must learn that people with AIDS can live."

This practice of citizenship via patienthood (or at least a claim to it) would transform in the subsequent years—at an impressive speed—into a very focused and sophisticated practice of care for one's pharmaceutical well-being. These individuals and their AIDS community would become less confrontational with political forces, less inclined to street life, and more integrated with the life-guaranteeing mechanisms and technologies associated with the AIDS policy, local and national. This sort of with-drawal into Caasah and engagement with their own therapy could be expected to increase patients' adherence in multiple ways. First, pulling away from "street life" may lead to decreased drug usage, a predictor of non-adherence (Nemes et al. 2004). Also, preliminary results from multiple studies (Gifford et al. 2000; Hofer et al. 2004) have led authors to postulate that increased patient understanding of HIV/AIDS and its drug treatments is essential to improving adherence, whereas lack of information is a major obstacle. Many of the residents "converted" and followed alternative religious values.

Celeste was very proud of having been able to teach the *aidéticos* to take care of themselves: "We even had to teach some patients to brush their teeth. They knew nothing. When there was a fungus, they did not want to look for a doctor, they thought that an anti-inflammatory would help. . . . We showed them the importance of using medication. . . . Now they fight for their lives."

Institutional Belonging and Treatment Adherence

Studies carried out in both rich and poor settings have shown that higher adherence to ARVs is generally correlated with better treatment outcome (Hofer et al. 2004; Oyugi et al. 2004). For example, Hofer and colleagues found that an adherence level of less than 80 percent was the factor most strongly correlated with non-response to therapy. Spire and colleagues (2002) stated that adherence must be 95 percent or greater in order to get the full benefits of combined ARVs.

Interestingly, treatment adherence rates are comparable in developed and developing countries (Lange et al. 2004). Oyugi and colleagues (2004) actually found 91 to 94 percent adherence in a group of 34 patients in a poor region of Uganda, whereas adherence values around 70 percent are frequently reported in studies on resource-rich settings.

Several studies in Brazil uncovered a variety of factors associated with poor adherence to ARVs, and these findings have bearing on how we understand particular hardships and the possibilities for positive treatment outcomes in a place like Caasah. Nemes and colleagues (2004) carried out a study of 1,972 AIDS patients across 322 outpatient services in 7 Brazilian states. These authors associated the following factors with non-adherence: fewer years of formal education; injection drug use; treatment in small, poorly organized health services, especially those where patients tended to miss appointments; and previous non-adherence. In another Brazilian study, a drug regimen's interference with daily life (such as required diet changes) was found to be associated with poor adherence, as was the number of pills taken daily (Garcia, Schooley, and Badaró 2003). Patient attitudes and self-image can also impact adherence; Gifford and colleagues (2000) found that a belief in "self-efficacy," meaning confidence in one's personal capacity to abide by a certain treatment plan, is a predictor of higher adherence.

In a study at a university hospital in São Paulo, trouble integrating self-medication into a work schedule and lack of money to pay for transportation to pick up medication were other reasons cited for non-adherence (Brigido et al. 2001). In Rio de Janeiro, the issue of stigma and discrimination from family members and neighbors was found to be an especially large problem for injection drug users taking ARVs (according to self-report) (Malta et al. 2003). In this same cohort, ARV adherence increased following the implementation of a weekly focus group in which HIV-positive drug users could talk to each other; more communicative and supportive relationships with health care professionals were found to boost adherence (Malta et al. 2003).

Along these lines, Garcia and colleagues (2003) stress the importance of an "active partnership" between health care professionals and patients. Although contact with health workers could exert a positive influence on adherence for many reasons, research (Gifford et al. 2000) has shown that one benefit of such interaction may lie in the information exchange that it facilitates; for example, patients who understand that non-adherence results in viral resistance practice more adherent behavior.

Besides trust and support, the bond between care providers and patients also creates a mechanism of observation beyond the institutional walls of a hospital. This type of medical intervention, known as Direct Observed Therapy (DOT), was initially developed as a strategy to promote adherence to expensive but effective tuberculosis treatment regimens in resource-poor settings, in an effort to minimize the rise of drug-resistant strains of TB. First tested in the 1980s by Karen Styblo in Tanzania, DOT showed that the use of short-term drug regimens was possible and safe in the absence of a strong health care infrastructure (Styblo 1984). Styblo's method of maintaining patient contact throughout the entire period of treatment achieved an 85 percent cure rate, and it eventually inspired the WHO's DOT Guidelines in Effective Tuberculosis Control (Dye et al. 1998; Odgen, Lush, and Walt 2003).

DOT achieves continual patient contact through health visits by nurses and community health workers, who deliver medication and supervise its consumption until the end point of treatment. Working with TB patients in Mexico, Menegoni (1996) observed that, after the initial intensive period of TB therapy, symptoms (a persistent cough, for example) would often disappear whether or not patients were biologically cured, leading them to believe themselves cured and reducing their motivation to continue treatment (p. 394). Continuous contact and constant reinforcement of the necessity of treatment reduced the patients' inclination to terminate treatment abruptly.

Paul Farmer and his colleagues working in Haiti have shown that, in addition to the value of DOT in AIDS care, the involvement of village health workers is vital (Farmer 2003; Blower et al. 2003; Behforouz, Farmer, and Mukherjee 2004; Lyon and Farmer 2005). In one of the world's poorest regions, local health workers called *accompagnateurs* have been effectively delivering ARVs and monitoring their use, in addition to antituberculous drugs (Walton et al. 2004, p. 142).

Partners In Health pairs an HIV-positive patient with an *accompagnateur*, either selected from the current staff or hired from the community at the patient's request (Behforouz, Farmer, and Mukherjee 2004, p. 431). Most *accompagnateurs* are peasant farmers or market women, educated in the importance of patient confidentiality and emotional support, as well as the clinical presentation and management of HIV infection and TB, including proper use of medications, management of side effects, and prevention of HIV infection. Though only required to visit their patients once a day, *accompagnateurs* often return a second time (p. 431). They are also reported to share food with their patients, to babysit and to run errands, acting as neighbors as well as health workers (p. 432). In all, besides monitoring the health of their own patients, *accompagnateurs* promote safe lifestyles and refer suspected HIV and TB sufferers to the clinics (Koegnic, Léandre, and Farmer 2004, p. S22).

As of 2004, Partners In Health had tested some 8,000 patients and treated more than 1,000 using what has been named the DOT-HAART method (directly observed therapy with highly active antiretroviral therapy). Almost all patients experienced weight gain, improved functional capacity, and suppressed viral loads (86 percent on average)—what Farmer and colleagues call the "Lazarus effect"—indicating the program's immediate success (Koegnic, Léandre, and Farmer 2004, p. S22). Hospitalization rates of HIV-infected patients in the region have also declined (Singler and Farmer 2005, p. 22). According to Farmer and colleagues, *accompagnateurs* may "be the answer to the missing health infrastructure" in poor areas (Lyon and Farmer 2005, p. 470). Beyond this, Partners In Health plays a key role in holding these neighborhoods together. Its daring vision and practice of social medicine fundamentally influence the sense of responsibility and care displayed by individuals and households. Through this medical activist institution, they have a real chance to remedy affliction.

In Caasah, we see that the constitution of this unexpected political body as a care unit is integral to ARV adherence. *Institutional belonging begets treatment adherence.* The story of Caasah illustrates how a particular pastoral vision of care and demand for accountability, combined with medical technology and policy, can produce an independent circumstance that transforms social ecology and mindsets in the vicinity. Instead of succumbing to the factors that predispose them to non-adherence (such as poverty and drug addiction), Caasah residents use their "disadvantages" to create an "AIDS-friendly environment." People in Caasah have access to a social network of fellow patients and medical personnel, and at least within the compound, they need not deal with stigmatization because of their serostatus.

As thirty-seven-year-old Jorge Leal put it, "I am low-income. I have two years of Caasah. It was impossible to stay in my neighborhood. People discriminate. I chose this family because here everybody has the same problem that I have. In my mind, this is good for me. Here, there is more life for the person." He was adamant that he had never taken drugs intravenously and that he been infected by his girlfriend—"she destroyed my life." He used to work at a food market and had two daughters from a previous relationship. "I still want them to be proud of me."

This new institutional (family-like) belonging curtails some of the ordinary worries involved in having AIDS and being a "person" and a

Jorge Leal, 1997

"patient" at the same time. The will to live is one with the milieu. And this interior force of life—articulated at the intersection of abandonment, medicines, and makeshift care—shapes both concepts of personhood and degrees of political membership.

New Prohibitions

In March 1995, psychologist Ricardo Silva began holding joint meetings with Celeste, Naiara, and the residents. This group therapy was aimed at building institutional and behavioral norms. All patients were receiving incentives to participate in daily house maintenance routines. Three patient representatives were now in charge of channeling complaints and suggestions, and patients themselves wrote the minutes of the weekly collective meetings (Book of Proceedings). An entry from March 28, 1995, reads: "Dr. Silva explained that if immediate behavioral changes do not take place, if the norms we create are not effectively obeyed by all, and if individual faults are not acknowledged and punished, the institution might well have to be moved to a worse location or will even cease to exist. If this happens, this will be the worst 'loss of privileges' for all of us."

Self-transformation was a must. "Through Dr. Silva's testimony we realized that in most cases people can only evolve either through pain and suffering or through knowledge. Therefore, we have the urgent need to obtain a better knowledge of ourselves." Residents needed to renounce their pasts and build a clear consciousness of their present condition. Protecting their bodies meant protecting the institution. The body was the site of construction and destruction, and out of this tension the identity of the patient-citizen was formed. As Judith Butler notes, "The formation of this subject is at once the framing, subordination, and regulation of the body, and the mode in which that destruction is preserved (in the sense of sustained and embalmed) in normalization" (1997, p. 92). In this case, the destruction is also literally preserved through a coexistence in the triage room of dying patients with recovering patients, who had to assume the role of caretakers. Soon the residents spoke of the need to address their "psychological side" so as to justify fully why they were not taking better care of themselves and allowing the institution to help them. "The mental and behavioral profile of each patient must be thoroughly handled, so that we might learn to take personal care and to care for the group."

Besides the weekly meetings, Dr. Silva also held small-group sessions in which individual behaviors were assessed. Groups were divided according to people's previous experiences (e.g., drug addiction, homosexuality, and prostitution). "Liberty with responsibility shall be our aim," the residents wrote. "We will all profit if we develop more civilized attitudes towards each other. Unfortunately, our situation in society is very negative. We must urgently improve our image and make peace with the larger community which supports us."

Two incidents led to the rewriting of the institution's rules. The first incident concerned sex, extortion, and violence. Transvestite Nerivaldo (see introduction) had lived in Caasah since the beginning and had routinely been amorously involved with other residents. Before the first incident, he was having a relationship with Wellington, for whom he stole all kinds of goods. Then Francisco, a new homosexual patient, entered the story. He was placed in the same room as Nerivaldo and Wellington. Nerivaldo blamed the administration for disrupting his affair, and soon Francisco was caught by volunteers and visitors having sex with Wellington. In fact, Celeste and Naiara then learned that they had often performed sex for the other residents to see. "First we expelled Francisco. Then Wellington had to leave as he was physically threatening other residents. Nerivaldo followed Wellington back into the street," said Celeste. "It is sex all the time. They act like animals, it must be their culture."

A second incident involved drugs. Forty-three-year-old Marta Damião, one of the patients' representatives, learned of her HIV status in 1992 and was left at Caasah in 1993 by the Luis Souto social worker. Marta and her partner owned a food stand at the Itapuã beach and were both intravenous drug users. After he died of AIDS, "I sold everything I had. The neighborhood made it tough for me. I gave custody of my children back to my ex-husband, for I thought that I would die soon. My mother was in a wheelchair and couldn't help. My only medication was alcohol. Caasah became my home. And here I am, four years later and alive."

In late May 1995, Marta Damião was caught smuggling marijuana into the house, and the administration called the police to arrest her. A day later, she was brought back and kept under house arrest until a collective patient meeting decided her fate. During the meeting, Marta admitted to her infraction and read a passage from the New Testament, explaining that she was now "imbued by divine grace" and that she would change. "If I leave Caasah I will be reduced to rubble." Senior patients sided with her, saying that, after all, she was a founder of the new Caasah. Led by Dr. Silva and Celeste, the residents drafted rules to prevent similar acts that put the individual and the institution at risk.

All residents at Caasah signed a "contract of responsibility." The new contract read:

> Drugs, alcohol, sex, theft, physical or moral abuse are prohibited in the institution. If a resident is caught doing any of these things, he or she will be immediately removed from the premises and will not be allowed to return. No one is allowed to walk around the building improperly dressed. Each resident must clean his or her room and keep it orderly. Residents must respect the volunteers and are prohibited to enter the kitchen. All meals must be eaten inside the cafeteria; no meals are allowed in the rooms. Everybody is obliged to undergo a monthly medical check-up. During the weekdays, the residents can leave the institution with the written authorization of the president or vice-president or someone designated by them. All residents must be back to the house by 6 pm.

Round-the-clock presence of two security guards paid by the city government led to further enforcement of rules and surveillance. Guards controlled the flow of residents, intervened in conflicts, and kept a registry of all incidents. They had the power to evict people who were threatening other residents. Good behavior and adherence to medical treatments were fundamental to bargaining for ordinary concessions such as weekend outings and to activating Caasah's legal and welfare capacities (to channel claims of disability and aid to families). Rights were no longer viewed as entitlements, residents wrote in the Book of Proceedings: "Now we understand: rights are the concessions we work for and acquire."

Self-medication was also a great source of conflict. The administration used to complain that patients brought medicines from their medical visits and kept them in their rooms. Many sold their medicines to external patients to buy things like cigarettes or doped themselves with painkillers

Marta Damião, 1997

and antidepressants. Finally, all parties agreed that all medicines should go to the infirmary post and that Professor Carlos would be in charge of them. Three more nurses were hired in mid-1995. Professor Carlos established medical records for each patient and systematized the administration of medication. He contacted the doctors who routinely monitored the Caasah patients in the public hospitals, asking them to pressure patients to comply with the new norms.

Strict rules and medical routines gave residents a real sense of the possibility of improving the quality of their lives. Patients began disciplining each other. If someone picked a fight or "disrespected" a volunteer worker, the case was taken to the general meeting, which came to a collective decision over punishment. According to Edimilson, a resident in Caasah since 1992, "If the other person is doing something wrong, breaking a norm, I must point it out, because if I don't, I might fall back into that old thing as well." Caasah's therapist also challenged the residents by having them participate in admissions decisions.

April 26, 1995: Dr. Silva called a meeting for us to decide over a vacancy. Celeste reported on the inhuman conditions of homeless AIDS patients in the Pelourinho area. During a recent visit she met transvestite Raisa, who begged to come to Caasah. She lives in a room rented by Dona Conceição. Raisa is in a regrettable state, probably in a terminal condition. Given the fact that Raisa is known to create much confusion wherever she is, we decided not to give her the spot.

"In Caasah we don't just have AIDS—we have God"

Caasah had dramatically changed by 1996–97, the year I carried out my long-term fieldwork there. The infirmary space had been enlarged, and the main corridor was now crowded with nursing trainees and volunteers wearing white lab coats, carrying trays with medicine to their patients. Most of the "marginal" patients had either left or died, and a higher number of working poor and white people were now living there.

The face of AIDS in Caasah had changed. According to Naiara, the vice president, "In the beginning, there were mostly homosexuals in here;

you only found a few women and one or two heterosexuals. Now, there are only four homosexuals here. There is an even proportion of men and women. Most of the men got contaminated through drug use, and most women say they got AIDS from their partners." Celeste said that she was proud of having transformed Caasah's marginality and criminality into a life to be cared for: "The patients who wanted to attend to the norms stayed; the ones who did not want to submit had to leave. They went back to the streets; they belong there."

There was now a minimum sociality established in the house; drug addiction and interpersonal violence were more or less under control, and residents were following the austere disciplinary and medical norms. Celeste characterized her presidency as democratic and normalizing: "I was always being evaluated by them; I was audited by them, not just by society. I was also putting them in their places . . . then things began to improve, and now there is stability."

Edimilson, a founding member of Caasah, described himself as "a lost man" in his past life. At the age of twelve, he ran away from home, turned into a thief, and got into serious trouble with guns and drugs. He pointed to his atrophied finger: "This happened because another person injected me. It was supposed to have been in the vein, but it went into the skin."

Edimilson learned about his seropositivity in the early 1990s. In between sporadic hospitalizations and the underworlds of Pelourinho, he saw many of his friends dying. "But then one day a doctor told me that if I took good care of myself I could live ten more years. I put this into my head and I sought for this objective here in Caasah."[4] He was now undergoing regular medical treatment, working as a security volunteer, lecturing on AIDS at high schools and universities, and was a born-again Christian. "Caasah saved me. Today there are new things in my head. I know that if I do not take the medication, I am harming myself."

As Caasah's residents put their drives into place, so to speak, their biological condition became the locus of concentration and improvement. Many, including Edimilson, referred to the HIV virus as "my little animal," saying, "I want to let the little animal sleep in me." I frequently heard comments such as these: "The moment you fall back into what you were and stop taking your treatment, the virus occupies your place. And the virus only occupies the place because you let it." In Edimilson's own words, "We live in a constant battle." They know that they are trapped between two possible destinies: dying of AIDS like the poor and marginalized do—that is, being *animalized*—and living pharmacologically into a future, thereby letting the animal sleep and preventing it from consuming their flesh.

Reconstituting one's personhood is key to accessing housing and care and to making ARVs work in extending life span. "When I go out there into the world, I see so much suffering, and I am so glad to be living here. I do not want to lose my place here. I tell people that in Caasah we don't just have AIDS—we have God," Edimilson asserted.

Edimilson underwent monthly medical checkups at the Luis Souto Hospital and was taking AZT. "My CD4 count is low now. My doctor is preparing me to begin taking the AIDS cocktail soon. When I come from the ambulatory I take a copy of my records and the prescriptions there. Every eight hours I get the medication on time. The patients who refuse the medication are not opposing the nurses or the administration, but are harming themselves."

●●●

"When I first got here, the picture was scary, but today things are under control," Professor Carlos stated. He was particularly happy with having achieved "TB containment" (a claim rebuked by several doctors with whom I spoke later, who maintained that "new patients actually acquire diseases in Caasah," but that, after all, "it is the only available place"). Professor Carlos's team was now composed of ten paid nursing assistants (*auxiliares de enfermagem*) and twenty-five nursing interns.

A single gay man, Carlos was proud to say he had adopted the seven-year-old child of Rose, another senior resident (see introduction)—his parents were helping to raise the seronegative boy. Professor Carlos had established partnerships with three nursing schools and had opened Caasah for internships in the area of infectious diseases. One day he hoped to open his own nursing school.

For many, Professor Carlos's care literally meant survival. As Soraia, a sex-worker who had migrated from southern Brazil and had just left the triage room (see introduction), put it, "My life is in his hands. The professor is a doctor. He knows the medication I can take, and what my reactions are. If there is a pain he comes and treats me." Trust and a strong bond had been created between nurses and patients and, as far I as could

Edimilson, 1997

see, this had positively influenced treatment adherence. "Professor Carlos knows everything about me," explained twenty-two-year-old Marilda. "I pray that he might be always here."

Marilda had been discharged from the Dorneles Psychiatric Hospital when the doctors discovered that she had AIDS, and she has lived at Caasah since. Professor Carlos constantly defended her from verbal attacks by other patients who were disturbed by her playfulness: "Here I play with everybody, I talk to everybody, I entertain myself. Yesterday, I went to the church and prayed for the patients in the infirmary. Today I was a little bit dizzy, but thank God now I am feeling better and I never want to leave Caasah. I do not want to leave here to go anywhere else, only to the hospital if it is necessary, but with the professor." I later learned that Marilda's brother had most likely infected her with HIV while he was sexually abusing her and that her father had expelled her from home. "I do not want to go back there. That's all I have to say."

Between 6:00 a.m. and 8:00 a.m., the nurses went to the rooms, woke the patients up, and distributed the medication. Either before or immediately after breakfast, patients went to the infirmary to discuss symptoms or abnormalities. Those who had prescriptions were given their medications. If the nurses could not solve the problem, they prepared the patient to go to the emergency room of the nearby Estadual Hospital or to the Luis Souto AIDS unit. Meanwhile, other nurses fed and bathed the patients in the triage room. Nurses also accompanied patients to their monthly medical checkups.

Throughout the day, residents came in and out of the infirmary, having all kinds of lively conversations with the nursing personnel. I often heard that Professor Carlos had coached them on how to demand quality care and medication from public services. "We are citizens too," said Edimilson, who bragged about being in high demand to give AIDS lectures to "rich high school and college kids. . . . Who would imagine that a bandit would become a teacher?" In this case, overcoming fear of the medical establishment and actively searching for public recognition and respect seem interwoven with treatment adherence.

Caasah was more and more an extension of AIDS public services, which by and large remained unable to handle an increasing and diversified AIDS population. Indeed, patients did not come directly from the streets anymore. Only patients referred by a hospital were considered for admission.

Marilda, 1997

In exchange, hospitals gave Caasah's patients priority in hospitalization. Patients nonetheless remained objects of negligence and "medical sovereignty," in which doctors alone decided who was to be adopted and given special (meaning adequate) treatment.

"The worst abuse happens in emergency rooms," said Professor Carlos. In fact, he envisioned Caasah's infirmary becoming an emergency service of sorts, making hospital transfer unnecessary. Accomplishing this, according to Carlos, would require a doctor. "A doctor would help in emergency situations. It is difficult to get a bed in the hospitals, and we only have one van to transport patients. When we call an ambulance, they say that it is unavailable. Without a doctor, we can only administer medication that has already been prescribed."

According to the chief nurse, AIDS services "do not handle death." The regular presence of a physician would also grant Caasah immediate access to a professional who could expedite the writing of patients' death certificates. Caasah's work to extend life coexisted with a concern for a dignified death. Celeste recalled that even asymptomatic residents told her, "Please, when I am in a bad condition, do not leave me in the hospital. I want to die here."

Religion, Health, Wealth

In Caasah, patients could also generate cash. Since 1994, the federal government had been paying a disability pension to all AIDS patients—one monthly minimum wage, approximately $100. Caasah's administration was mediating the extremely bureaucratized distribution of these payments for its residents. Priority was given to those residents who showed "progress," that is, behavioral change (e.g., stopping fights and denouncing infractions of rules) and medical compliance. Patients were also encouraged, some said "forced," to use the salary to buy their own medication (then still irregularly available in public hospitals). According to Celeste, these measures were necessary "so that they would not have money to reinvest in past vices." Well-behaved patients were also allowed to help in the storage room, where they then had priority in choosing clothing for themselves and for family members living outside.

Amid all these institutional and medical changes, many residents converted, were baptized by the Holy Spirit, and began to participate in various Pentecostal and other Protestant services. As Marta Damião put it, "I had to change, to stop messing up with drugs, fighting with everybody. I drank, smoked, and did not eat. I converted because of the great need I had to stay here, so that I could live. And because I was ashamed of myself."

Clifford Geertz (1973) has elaborated on religion as that human capacity of making meaning in the face of limiting situations—religious experience is found at the point where cultural resources fail, he says, where our equipment for living threatens to break down in the face of the radically inexplicable, painfully unbearable, or unjustifiable. But what I found most fascinating in my work in Caasah was how religion moved from helping AIDS victims know "how to suffer" to providing a means through which they could construct another social identity. In addition to providing a tone or temper to lived life, religion played a crucial rule in the internalization of discipline and institutional belonging—it was a key instrument of their patient-citizenship and a platform for lives to come.

•••

"My full name is Evilásio da Conceição. I am thirty-four years old. I was born in the charity hospital, here in Salvador." Evilásio was one of the many poor, illiterate, and wasted patients who pass through the public health system but receive no adequate treatment. Whether or not they are diagnosed as HIV-positive, they disappear from the statistics and are literally left to die alone.

For three years, Evilásio hid his illness "until I became so weak that I couldn't stand up anymore." Neighbors brought him to Caasah in August 1996, and, with treatment, Evilásio quickly reversed his fate: "Today, I am strong."

A carpenter by trade, Evilásio believed that he "must have been infected by prostitutes. Right now, I have some lesions on my genitals. They say it is the worst kind of sexually transmitted disease." He had few family ties: "I have only one brother from my mother and father. There are half-sisters, but we don't get along. Sometimes I dream that my father is alive, talking to me; then I am so happy that it does not look like a dream."

Evilásio explained that he was very frightened to live in Caasah at first. "I saw many patients dying next to me, and all that smell of feces. I wanted to die too. But with the nurses' help, I started to fight against that death."

For the time being, Evilásio was regaining his health and building strength: "Some people throw medication away. I never do that. The nurses have nothing bad to say about me. We have to go to the hospital to do so many exams so that we can take the medication. I know that death exists, but as long as I struggle, it will take a long time for it to find me. Let me tell you, for me, Caasah is the house of God. Even though I suffer, God knows that I have this will to live, and that's why He has not yet taken my life."

•••

Of course, religion was part of an economy of survival—for even though life-extending medication was increasingly available, to make it work people needed the tie to Caasah to guarantee access to ongoing specialized medical care in a context of inequality and clientelism, not to mention housing and food and some form of legal accountability. But it was also more: religion provided them with an alternative value system that made them more than a social and statistical void—that is, they now had a proxy community, hope, and a real chance at life.

For example, given Marta Damião's overall transformation, Caasah provided her with an improvised studio where she did her handicrafts (mirrors decorated with shells). These objects were sold in public fairs, high schools, and churches as well as to Caasah's visitors. She accumulated her disability pension and two monthly basic food baskets for her children. With capitalization, she was able to reestablish ties to her family. Thus, there was a double movement at work in Caasah's salvific process: a subjective trajectory that redeemed the patients from their past lifestyles and put them in touch with their medical conditions and a movement of throwing patients back to the ordinary, now not through disease but through capital generated from their conditions.

The residents who were relatively well off, taking medication and vitamins and doing ambulatory checkups, repeatedly mentioned having "abandoned drug abuse." Their discourse was very religious: "I used to

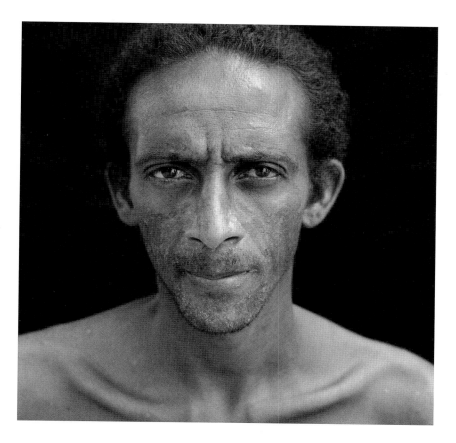

Evilásio, 1997

be in the wrong path, but now I follow the Lord." They were constantly requesting more vitamins and a better diet and told their conversion story to visitors.

Jorge Ramos, a former drug dealer and addict who had lived in Caasah since 1994, refused to get hooked on sleeping pills, he said, "even though I have problems falling sleep. I do not like these pills, because one gets dependent on them. This is a drug too. The medication I take to fight the disease is enough. I fall sleep around 2:00 or 3:00 a.m., and at 6:00 a.m. I am already up. . . . Here we need much fruit and vegetables, healthy food for us to live longer."

However, as I read the patients' medical protocols, I saw that the majority of the former intravenous drug users like Rose, Marta Damião, and Nerivaldo were now addicted to tranquilizers. These patients, as Professor Carlos told me, did not sleep without psychotropics: "They complain of insomnia and anguish. The medication turns them off, and they only wake in the morning." Without tranquilizers, these patients usually paced Caasah's central corridor, back and forth, or went to the infirmary post, where they talked to the night shift nurses until dawn. But with the medication, Marta Damião told me, "I sleep through the nights . . . I have nightmares . . . fights in here and blood flowing. But the other night I finally dreamed that I was having an orgasm. I woke up and no one was next to me though. I felt absolute pleasure in the dream."

What I found most disturbing in Caasah's process of health accumulation was that the residents' regeneration became evident in constant, if not obsessive, interpersonal surveillance and the open contempt they showed toward other patients who echoed their own deviant pasts or who posed an immediate medical threat (patients recently out of the triage room and potentially "infectious").

"Did you ever see an AIDS patient in here hoping for the Other's good?" Evangivaldo asked me. He had been quarantined because of his scabies (see introduction). Residents constantly denounced each other's faults and demanded the rigorous application of the law: "Is there a law? Where is it? Why is it not being applied?" The Other's misbehavior was also a measure of their own progress, a measure of their own change and self-control. "I am not like him." "He did it to himself, and now wants another chance." I wondered about the significance that this kind of

Rose and Fátima

morality (constituted through watching others' failures) had for treatment adherence.

The fact is that the triage room made terminal AIDS a constitutive element of everyday life in Caasah, a constant reminder of what to fight against. As Rose, a former prostitute and intravenous drug user (whose children were adopted by Naiara and Professor Carlos) put it, "I have no fear of death, but of the suffering that comes with dying. There are so many corpses that we have to coexist with in here, that I even got used to the idea of death. I dream about my friends who lived and died in Caasah. I see them alive in my dreams. There are times I want to go into the triage and visit people, but then it is too much to see what I will pass through . . . then I stay on this side, praying for God to take the suffering away."

Residents followed each other's disease progression very closely, measuring their own well-being relative to the plight of others. "That's what happens when you do not take care of yourself." "I have my own life to be concerned about." "I solve my problems myself." "I have faith in God that from here to there, scientists will discover the cure and that I will not die like this." Edimilson said that by seeing so much suffering and dying, he had gained further resolve to fight "this AIDS that is like an animal gnawing the flesh."

Ambiguous Political Subjects

In houses of support, marginalized citizens have an unprecedented opportunity to claim a new identity around their diseased and politicized biology, with the support of international and national, public and private funds. Here, immediate access to AIDS therapies and the administration of health—the micro-politics of survival—has priority over "metasocial guarantees of social order" or over political representation (Alvarez, Dagnino, and Escobar 1998; Doimo 1995; Abélès 2006a).

For the moment, let us think of Caasah as a "biocommunity," in which a selective group of poor patients fights the denial of rights and carves out the means to access them empirically. Their political subjectivity is articulated through pastoral means, disciplinary practices of self-care, and monitored pharmaceutical treatment. At work are new arts of self-governance and survival as a cost-effective patient (see Petryna 2002; Crane, Quirk, and Straten 2002).

Caasah's life-extending work challenges Hannah Arendt's rigid view of an opposition between the realm of the political and the realm of private life. For Arendt, the modern political process has progressively eliminated the possibility of human fulfillment in the public sphere, excluding masses and reducing them to the condition of *animal laborans*, whose only activity is biological preservation (1958, pp. 320–25).[5] This preservation is an individual concern; this metabolism is superfluous to the state and to society at large. "They begin to belong to the human race in much the same way as animals belong to a specific animal species" (1973, p. 302). That is, for Arendt, the primacy of natural life in modern society has foreclosed the possibility of authentic political action. Unfortunately, according to the late Iris Young, "Arendt criticized efforts to improve social life and to promote social and economic rights in the same category as the merely biological. She thought that concern for social and economic equality and inclusion was not appropriate in the public sphere" (personal communication, March 2004).

Philosopher Jacques Rancière is also concerned with Arendt's "archpolitical position" being developed into a "depoliticizing approach": "Paradoxically this position did provide a frame of description and a line of argumentation that later would prove quite effective for depoliticizing matters of power and repression and setting them in a sphere of exceptionality that is no longer political, in an anthropological sphere of sacrality situated beyond the realm of political dissensus" (2004, p. 299). This overturn, for Rancière, is clearly illustrated by Giorgio Agamben's theorization of biopolitics in *Homo Sacer* (1998). Following Hannah Arendt and Michel Foucault (1980), Agamben states that the original element of sovereign power in Western democracies is "not simple natural life, but life exposed to death" (p. 24). This bare life appears for Agamben as a kind of historical-ontological destiny, "something presupposed as nonrelational" (p. 109). Rancière gets straight to the point: "The radical suspension of politics in the exception of bare life is the ultimate consequence of Arendt's archpolitical position, of her attempt to preserve the political, from the contamination of private, social, apolitical life.

This attempt depopulates the political stage by sweeping aside its always-ambiguous actors" (2004, p. 302).

Caasah's residents set the question of what politics is on a different footing. Some of them engage the state via the AIDS policy and are able, in the process, to become a local body of citizens. The universal availability of lifesaving drugs does not reshape the bureaucratic apparatuses of the state, nor does it substantially alter the medical establishment. Novel strategies for biosocial inclusion and exclusion are, rather, consolidated under perennial structural violence. Many stay outside, on their own. Caasah's distinctive feature is its selective transformation of a diseased biology, marginal and excluded, into a technical means of inclusion.[6] Scavenging for resources and navigating through complex treatment regimes, a few constitute themselves as patient-citizens. Against an expanding discourse of human rights and possibilism, we are here confronted with the limits of the on-the-ground infrastructures whereby accountability and the right to envision a future are realized, biologically speaking, on a selective basis only, and also with the consolidation of a new political economy of pharmaceuticals.

It is within this interrelated context of local, national, and international forces that I became interested in how the project to extend life informed institutions and political agency, particularly at the margins. As I have argued throughout this book, the distribution and consumption of pharmaceuticals are significant means through which state, community, and citizen empirically forge their presence today. Nongovernmental, medical, and pastoral networks link the worlds of marginality and the state through AIDS response.

In order to make these new medical technologies work, people have to participate actively in local circuits of care. In contexts of unemployment and scarcity, for many disenfranchised patients, the mere extension of life is literally a form of work (see Petryna 2002). The deployment of AIDS therapies thus instantiates new capacities, refigures value systems, and alters people's sense of their bodies and of the future. The pressing needs of newly circumscribed patient populations for physical and economic survival dynamically inform private and public involvements; and these "affective entanglements," in turn, redefine the local terms of politics and ethics.

We can only understand the conflicting social effects of neoliberalism and decentralization by looking at the materiality of policies and related individual and communal struggles for survival. By ethnographically charting the ways policy-spaces and people operate at the margins we can also illuminate political rationality in the making. Politics here is not a sphere, but a lack, a technology, and a process all at once. In the case of the AIDS policy, medication makes people equivalent—difference lies in laboratory testing and viral loads. It is up to grassroots work to address the social determinants of patients' conditions or to make those markers invisible. Here medical commodities work in tandem with other ways of claiming citizenship, and desperate and creative interactions occasion novel public sites in which rights are group-privatized, so to speak.

Life in Caasah draws on what is available. Individuals do not operate alone. Their decision making and actions are entangled with those of other patients and the multiple arts of government at hand for this group. Contemporary conditions are pressing, and disjunctions abound. Channels of communication with people and institutions are constantly renegotiated, and newly found personal identifications cannot be taken for granted—this is what Veena Das and Deborah Poole call "the temporal experience of the state" (2004, p. 16).

Resuming Sexual Life

In mid-1997, before I left after a year of fieldwork, Caasah was in transition once more. The community ethos was definitely vanishing, there was a clear antagonism between senior residents and new patients, and the institution was redefining itself as "a rehabilitation center, a house of passage," in Celeste's words. A growing number of terminally ill patients populated the triage room or wandered around the house. Some of the recent arrivals were taking ARVs, eager to return home with the promise of minimum support (such as a monthly basic food basket, medication, and rent aid).

In a collective meeting, Celeste said that all healthy patients who had the means to take care of themselves would have to leave the institution.

"According to the statutes," she said, "this has always been the institution's mission":

> These statutes were never acted on before because the residents had no way to survive outside. But now things have changed, and there are real possibilities. Most of you receive a pension from the Ministry, you are physically well now. If your viral count is down, you can take the AIDS cocktail. You must leave so that other people can be helped. It's time for you to be free, on your own. We want you to see yourselves as normal persons, saying "I am alive" . . . with a paper calculating your monthly expenses and possibilities, and buying your own additional medication.

Celeste referred to Fátima as an example of what the institution was all about. A former prostitute, she was the first of Caasah's senior patients who voluntarily moved out, back to an apartment in Pelourinho with her two teenage daughters.

In November 1996, Fátima had represented Caasah at the seventh National Conference of People Living with HIV, which took place in Rio de Janeiro. After returning, she told all residents that the conference had "helped her to see that everywhere in the country people with AIDS are living a normal life outside institutions. And now with the AIDS cocktail we can continue our lives."

●●●

Dulce seemed to have lived much longer than her twenty-seven years. She had worked as a maid since she was a teenager. She says that she contracted HIV from her husband, who used to beat her.

Dulce learned that she had AIDS a year ago, when her third child was born prematurely and died: "I wanted so much to give birth to a girl. But it is better like this; she would have been another child with problems." Her husband left her with the two older children, who are also seropositive. "They are now living with their godmothers." Dulce's sisters left her at Caasah, her body wasting and diarrhetic.

She marveled at her quick recovery. Treated with antidepressants and ARVs, she had also been in psychotherapy: "Before, I was so depressed, isolated from humanity. But now I am not afraid. Sometimes I try to forget that I am *aidética*. But I can no longer be the same person I was."

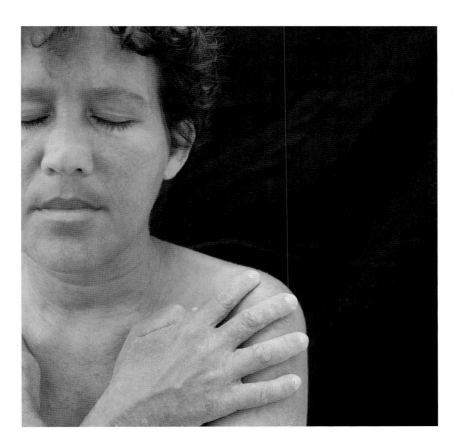

Dulce, 1997

A few days after the photo shoot, I saw Dulce again. She was enraged at the newspaper *A Tarde* for a report it had published on the accelerated spread of AIDS among women in Bahia. After interviewing both Dulce and Sara, a former prostitute and drug user, the reporter switched their identities, presenting Dulce as a call girl. She angrily told me that she had already contacted a local AIDS NGO and that she knew her rights: "They need to retract. The newspaper cannot simply say what it wants to say and slander me."

Dulce also revealed her plans for a future outside Caasah: "God permitting, I will leave soon. I want to run a little business—I am good at selling things—and raise my kids."

•••

In the meantime, anxiety, revolt, and depression were on the rise among residents like Edimilson, Rose, and Marta Damião, who had helped to build Caasah. As Edimilson put it, "This is my house. I want to see it grow so that we can have what we need. I have diabetes and problems walking. My disability money goes for medication. *I need Caasah to live.* Here people understand my rights, and I thank God for that. That does not exist in the streets. I don't want what is past."

Caasah's transition from a house of support to a house of passage was mediated by three new psychologists, who came as part of Caasah's growing partnership with the city health services. The psychologists' task was to help to develop new institutional strategies: "Caasah has to define its public role better, it cannot be a hospital any longer," Dr. Amilton Bastos told me. "Support needs to be regularized, and we can help the administration with projects and governmental negotiations. We also have to bring the patient's family into the recovery process. Caasah is still too paternalistic, and we have to decentralize actions." As recommended by the national AIDS program, professional workshops were being planned. Dr. Aurora Alves emphasized that by learning new skills patients could "show that they are productive, alive, and part of society."

Most important, psychologists were there to help each patient to handle new personal involvements, as they were now able to extend their lives and were having amorous relationships and reconnecting with family members. One of the first actions taken by the psychologists together with Celeste and Naiara was the design and implementation of a new and controversial punishment system.

Instead of collective decision making over individual cases and fates, now patients who trespassed rules (both personal and institutional, as in the case of Rose, who beat up her lover Jorge Ramos and who often had another resident inject her with a pain-relief drug) had to do the difficult work of toilet cleaning and helping in the triage room. Patients were supposed to see that as "training" but spoke of it as "labor camp," and some left, saying that they would rather work outside. In fact, the new punishment system testified to a distance already in place between administrators and patients, the mediating role of psychologists who selected problems to be administratively addressed, and the constitution of a social environment inducing self-exclusion from Caasah.

Senior residents spoke of themselves as the institution's mirror. "I am from Caasah" like one is from a family, city, state, country, or religion. "I work for the institution, I give lectures. I help in the reception. If it were not because of me last night, something bad could have happened." They were terrified at the prospect of losing this vital identification and began to use the socially minded psychologists to make their case to stay. I noticed that patients on ARVs in particular were denouncing each other to the administration (via the psychologists) as a way of having more chances to be "adopted" and to get extra time at the institution.

During one of the last general meetings I participated in, there was a tense confrontation between Marta Damião and Celeste. "Caasah helped me change," stated Marta:

> I changed so that I could stay in the institution. And now Dona Celeste wants to change the laws, and make this a house of passage. I want to stay. My future depends on a cure for AIDS and on Dona Celeste's decision. I am in God's hands. I pray a lot that she retreats from this unexpected decision. I agree that she sets this as norm for the new patients, but not for the older ones who help to bring order and progress, to consolidate the institution. This is injustice. We had to build this up from scratch. I am completely depressed, and I don't know what to do or what to think. I tell you, if death comes right now it is indeed welcome.

Much of the discussion about patients leaving Caasah was focused around their love relations and family life. The asymptomatic patients were reentering a field of activity that had been interdicted during their period of regeneration: some were already developing affective ties and wanted to have an active sexual life. Much of the administration's argument about the need for the healthy patients to move out was based on this "demand for sex." According to Celeste, the administration did not want to prohibit the patients from loving: "Together with the psychologists we are working on this question of them having the liberty they think they need. . . . We are preparing the patients to face life outside, because in here sex will not work."

But as I heard in conversations with the residents, for them, sex was not related to "liberty" as much as it was part of the will to live what they were now calling a "normal life." Sexual vitality was an indicator both of physical health and of the desire to move out of the loneliness that AIDS had made their social destiny. They wanted to have real-life problems. What Celeste called "high libidinal demand" was also tied to complex gender economies that kept informing the relations of residents in the process of medical regeneration.

Consider Jorge and Rose. Rose's partner died in Caasah while she was pregnant, and Naiara adopted her newborn daughter in 1993. The boy she had with another man (who, according to Dona Conceição, also had AIDS and lived in the streets) was now under the care of Professor Carlos. In the past year, Rose and Jorge, also a former intravenous drug dealer and user from Pelourinho, had fallen in love. Rose was a strong and outspoken person with whom nobody wanted "to mess." Yet she worked for Jorge, washed and ironed his clothing, and asked for donations to get cigarettes for him. His role seemed to be that of a dominant and jealous pimplike figure. Once a month, Jorge took Rose out for sex. As Rose put it, "I like to have a man." To get a weekend out in a rented hotel room once a month (patients called them honeymoon weekends), they had to get authorization from Professor Carlos, who reviewed the medical records and prepared their medication for the trip.

Jorge had successfully fought TB; he was taking DDI and AZT. "This house is my life. I obey the laws here. It is impossible to live with one's

Jorge Ramos, 1997

blood family because, again and again, the faults committed are recalled and accusations abound. I do not have conditions to live on the streets. I can only live within this family here. In any other place, I will destroy myself; I will destroy the new life I have."

Jorge had strategized a new subject position to make his case against what residents were calling "eviction." He claimed that Caasah needed him, a strong patient, to serve as role model for the newly arrived and debilitated ones. He had learned to accumulate health by himself, was prevention driven, and was "married" to someone with HIV:

> I no longer have the appetite for women who do not have the same problem I have. I go regularly to the doctor, I have my room and I don't mess up. I take all my medication on time. I am all right, and every day I am trying to perfect what I have so that I might live longer. Caasah needs the stronger patients like me to tell others: "Take this pill to become like me."

Beyond Direct Observed Therapy

In tandem with the country's most innovative social policy, Caasah had successfully evolved from a disassembly of marginalized AIDS patients into a community of care and proxy health care service. "State-run services need us more than we need them," Professor Carlos told me during our last interview in March 1997. Caasah provided shelter and assistance for some of the neediest, coming directly from the streets or from the deteriorating hospitals. It was a place for them to decode and recode their abandonment. A selective group had socially rehabilitated, accessed antiretroviral therapies, and regained health. Through food and medication dispensation, Caasah also played a pivotal role in the survival struggles of a number of homeless AIDS patients.

This grassroots and DOT-type health care setting reveals that treatment adherence is shaped through institutional connectivity and intersubjective work. Instead of being subsumed by all the factors that predispose these patients to non-adherence (lack of formal education or injection drug use, for example), Caasah's coordinators and residents (the founders in

João, Luis, and Torben, 1997

particular) used their lived experience and what was philanthropically and medically available to them to create the AIDS-friendly environment that is necessary to promote adherence. "AIDS-friendliness" had two components: first, here people did not have to worry about the stigma and discrimination that accompany the public disclosure of seropositivity. Second, Caasah provided a daily routine and an infrastructure that made it easier to integrate drug regimens into life on a purely schedule-based, convenience level. The ability to consume pharmaceuticals "rightly" is key to social inclusion and civic participation and people give multiple meanings to this medical experience of citizenship.

At Caasah, human forces—such as having a will, an understanding of real impossibilities, and an imagination of the future—had to be combined with other forces, such as medical knowledge, religious membership, the chemicals of therapies, adoption by a physician, disability pension from the state, and so on. In the midst of this general reorganization of an everyday life, an intense process of individuation and a spirit of competition with fellow residents motivated adherence as well. Out of these collective arrangements around lifesaving drugs (that Gilles Deleuze would call "instruments of our actuality"[7]), we see emerging patient-citizens and self-doctors, with an unprecedented expectant interiority.

Until the end of my fieldwork in 1997, many residents, such as Marta Damião, Edimilson, Rose, and Jorge, argued that Caasah was both the "family and government" they needed to make the medication work, to remain asymptomatic, and to keep AIDS under control. Not having this form of house/clinic/welfare state/NGO was unjust. They knew that "we had to build this up from scratch." Scratch was their social abandonment and death sentence. And they were now proud of their manners, of their health, of having confronted social and medical discrimination, of their life being unfinished.

I left apprehensive and with new questions. Removed from the materiality of Caasah, what would be the reach and staying power of their will to live? How would their medical investments fare as they were being thrown back into violent domestic and social coordinates? How would they turn everyday foreclosures into a space of courage? Which events would they precipitate?

Chapter Six

Will to Live

Lifelong AIDS

I returned to Caasah in December 2001. Photographer Torben Eskerod came along. We wanted to know what had happened to the AIDS patients we had worked with in 1997, and, if possible, to make new portraits of those still alive. By charting the trajectories of long-term AIDS patients— those who traversed the time before and after access to antiretroviral drugs (ARVs)—we can identify the particular pathways by which AIDS becomes (or not) a chronic disease amid poverty. How did variation in treatment quality, quotidian factors, and lifestyle affect Caasah's survivors?

Caasah was a changed institution. It had moved from the occupied building in the Dendezeiros District to a new building constructed by the state government a few miles away. Located in a residential area near the famous Igreja do Bonfim (the Church of the Good End), Caasah's new facility was gated all around. After passing through strict security proce- dures, we were greeted by a receptionist and children running back and forth between the front desk and the playground, asking us to chase after them. Caasah had transitioned from a house of support to a short-term care facility for ill patients and a shelter for HIV-positive orphans.

All former residents had left the institution, with the exception of Tiquinho (the hemophiliac child who had been raised there), who was allowed to stay. Some had died, the welcoming and "burned-out" Celeste (her own words) told us. "But many from the older group are alive. They know how to survive. As far as I know, they do their treatment. Now and then we see them here. We help them with food baskets. And when there is a medical emergency they look for our help."

Celeste showed us how well-organized and well-equipped the vari- ous offices—social work, infirmary, medical assistance, psychotherapy— now were. We sat in her air-conditioned room. The triage room had been closed, and Caasah was now providing temporary care for some thirty adult AIDS patients and housing twelve AIDS orphans sent there by the state's child custody service. "I got a marginalized place. You saw it. Today there is order. We are a reference of AIDS care and a home and school for the children," she said. "We have the capacity to take more children in."

All patients now had to sign a term agreeing to their confinement: "We want to make sure that people have no access to drugs and alcohol. They can get visits, but can only leave with someone from the institution. They are here to recover, and we want to show them another side of life, not just physical but also spiritual. After the patient leaves, he can do whatever he wants with his life, but at least he had a beginning here."

We looked at the portraits that Torben had made in 1997. "You really captured the person," Celeste told Torben, with a sigh. And denoting a certain sadness and detachment, she admitted that "in the day-to-day work we really didn't see this . . . we were always arguing so much, and thought that we knew them." She then told us how, one by one, the former residents had either left or "had been weaned from Caasah" and how they had fared in avoiding death and establishing a different way of existing:

> Nerivaldo went back to the streets and died soon after you left. He and Maria Madalena were last seen at Mãe Preta's shelter [for the homeless]. Jorge Leal died in the old Caasah, all of a sudden. Lazaro and Marilda too. Medication did no good for them. Nobody from their families showed up for their funerals, that's the reality.
>
> Rose and Jorge Ramos were having an affair, you recall, right? We were able to get a house for them from the Bahian government; it is in the Cajazeiras District. She got pregnant, and Jorge died before the baby was born. We try to help her as much as we can. The baby girl is already one year old.
>
> We suffered a lot with Marta Damião, we had to force her out. We got a house for her from the governor's wife. The older residents were always very critical of us . . . but they had an easy life here. She is now living there with her daughter, and will be a grandmother soon.
>
> For Edimilson, we also pay rent. I don't know where he lives, though, and whether he is still in the AIDS lecture circuit. Our social worker might know. Like any other normal person they have to take

care of themselves and work to pay the bills. Most of them, I think, are finding a joy in everydayness. Nadia was dating Edimilson. She got better and went back to live with her family. But she stopped taking medication at the right times and came back to us with the old symptomatology. It was too late. We took her to Hospital Estadual, and she died there.

Valquirene, I saw her in the street the other day. She looked beautiful, had a cell phone, and laughed about those psychotic episodes. She found a guy who accepted her seropositivity and they are living together. I don't know whether he is HIV-positive or not. When people like her look me into the eyes and say "thank you" I recharge for the daily battle here.

Evangivaldo married Fátima. They met here. He is hard-working. They also have a baby girl. We give them a food basket every month.

Evilásio was living outside. But he recently came back. He is not responding well to therapies. But he is happy, he is in love. He wants to finish building his shack.

Soraia . . . People who are under treatment have not seen her. Someone told me that she went back to doing prostitution, but I have not seen her.

Dulce moved in with her sister. Her mother takes care of her children. I see her often selling costume jewelry and perfume in the streets. She also comes here to get her food basket. She is doing great.

Luis is our office assistant. He helps with our accounting and bank service. He is like a son to me and an inspiration to all new patients.

Tiquinho is still with us, he converted and was baptized. We would not like him to spend his whole life in here. He needs our clinical support, but he might be better off with his family.

AIDS remains lethal. Yet, I was delighted to hear that many were alive and to learn the pathways they had taken. They were literally living beyond history. The majority had entered Caasah with a sense of imminent death and a reluctance to consider the future. The vitality they engineered via antiretroviral therapies and the care of Caasah had indeed taken root in the everyday. Through and beyond AIDS, social abandonment, and Caasah's culture of assistance, they were now back to their families or

Jorge Ramos

Edileusa

Rose

Valquirene

Tiquinho

Marilda

Dulce

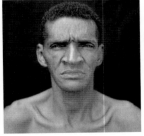

Edimilson

Nadia

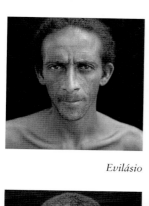

Evilásio

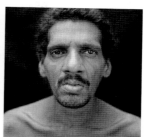

Luis

Lazaro

Evangivaldo

Soraia

Nerivaldo

Marta Damião

Maria Madalena

Jorge Leal

neighborhoods. Several had become parents, all were in the job market and were looking for whatever aid, public or private, they could find.

I was curious about the physical and mental work in which they engaged as they faced new relations, enduring poverty, and a different disease. Celeste seemed to know little about their specific struggles and the subjective work they undertook in order to frame a new lifetime and give it body. She said she would help us to contact some of our former collaborators. "Evilásio and Tiquinho know you are here and are waiting to see you." They themselves, I felt, could provide us with key insight into the human mechanisms needed to shield oneself and to make AIDS into a chronic condition vis-à-vis the violence of everyday life and the flux of institutions (Biehl 2006b; Abadia-Barrero and Castro 2006; Castro 2005; Rosenberg 2006).

Caasah's passivity toward former residents was troubling. The Direct Observed Therapy (DOT)-type environment of Caasah that had been crucial for treatment adherence had not been followed up. Celeste's recollection suggested a lack of systematic effort to search for them and to keep track of their treatment actively. Now "physicians of themselves," it was up to them to stop by, to pick up a food basket, spare medication, some *reais* to help with rent. Caasah was only one of the many stops in their struggle to survive. In the end, I thought, we might be aware of the quantities of drugs dispensed and of patients' viral loads, but much of the ancillary work needed to make life with ARVs possible remains uncharted.

Human Values

Celeste asked Caasah's new chief nurse, Elida, to give us a quick tour through the residential wing, while she finished some urgent paperwork. The children and the adults lived in opposite wings, and both shared a central plaza. The majority of patients were in their rooms, either sleeping or sitting at their desks, reading, knitting, playing solitaire. The rooms' walls were bare, and few of the patients looked at me. Two-thirds of the patients were women, Elida told me. "Most are poor. Their stories are similar. Many got infected by their husbands, lost their jobs. Most of the

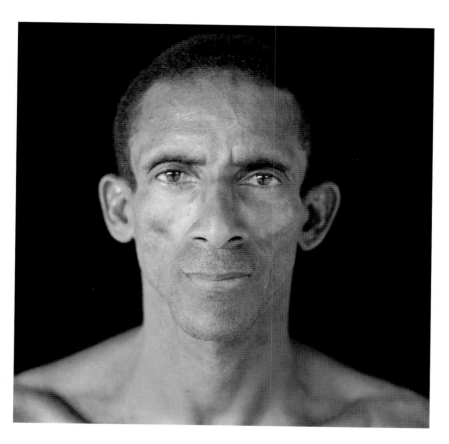

Evilásio, 2001

men here are bisexual or at some point injected drugs. All of them tried to hide their AIDS as much as they could, until they fell ill and went to the hospital, and from there to here. Some have relatives who actually come and visit—this happens mostly with women. Some make a strong friendship with the religious people who visit and hold worship services here."

The children's high-pitched voices were all that resembled everyday life, as they broke the hospital-like order enveloping the place and the people. In the old Caasah, patients were constantly moving through the main corridor and backyard, engaging each other and the volunteers, claiming rights, pointing out each other's faults, and articulating their biographies to the visitors. This constant articulation and reciting of an AIDS biography was an enterprise of health vis-à-vis the suffocating and persistent sense of death nearby. But now there was little motion. It was a time of waiting, I sensed, and no patient volunteered to talk to me. What lay beneath their silence? A strange mixture of shame, desperation, and hope, I thought, while their identities, at least for the time being, were equated to treatment compliance.

A frail Evilásio was waiting for us outside Celeste's office. His appearance had completely changed. And as we sat down, the forty-year-old man said, "Medication has not been working. My doctor has changed the treatment several times. She is getting the dosages right though. I am thin, my face is drying, but I am getting better, I feel it."

In Evilásio's account the past was always worse than now: "When we last meet in 1997, I was in pretty bad shape." He looked at his portrait: "My expression looked worse then, right?" As if one had to look constantly for signs of recovery in the body. And in order to be cared for in Caasah one had to display a healthy image of oneself: "This is the house of God. Things have improved a lot here. To be confined is better for the patient. It gives us a greater responsibility, for our own good. I thought that I was well, and I relapsed with the medication, I began drinking and smoking."

Evilásio did not want to go into details. He, rather, voiced a personal theology of health: "It is all past. I know that I have trespassed God's laws. When one mistreats oneself, this is the art of Satan. Even though I suffer, God knows that I have this will to live. That's why He has not yet taken my life. He has removed all vices from my mind. Thank God, life is no longer thinking solely about death." Doubt was ever present though: "My viral load decreasing, I can say that I am well, right?"

Like Celeste, Evilásio knew little about the travails of his former Caasah neighbors: "Some died, some are alive. Sometimes, I see a few of them here, getting their food baskets. I think that they are no longer doing bad things to themselves, but I don't know."

Evilásio smiled, as he pulled a photo out of his pocket: "This is Viviane, my girlfriend. We met here. She is much younger than I. She said she does not like to go out with the modern crowd. We understand each other. She also takes the cocktail." Viviane was all the family he had. She was living with her mother and worked as a maid. "She is lucky to have a job," he said. Evilásio received a disability pension. "It is better than nothing. But I would like to resume work as a carpenter. My dream is to finish my shack. I am saving a little money to buy the windows."

Medical Disparities

As we resumed our conversation, Celeste made it clear that Caasah had a different clientele now and that it came directly from the public health care system. "AIDS keeps spreading in the streets. But it is difficult for a homeless person to be confined here and to adhere to our norms. We created all these norms so that the institution itself would not be marginalized. Before, patients at the hospitals said they did not want to come here, that this was a place for *marginais*. This is no longer the case. We also couldn't let people just stay here, doing nothing, not producing."

As I heard Celeste, I thought that the form of patient-citizenship previously nurtured at Caasah had run its course. These socially and medically regenerated persons were a welfare population of sorts through which a new form of self-governance—an auto-bioadministration—was gaining form. As disability recipients, they were doomed to be "old" in the eyes of new institutions emerging around ARV distribution that emphasized a capitalistic drive for productivity. Yet for Evilásio and others, this identity was a means of claiming actuality.

Naiara was still Caasah's vice president. Two of the psychologists had left, and two had stayed—most of their work involved reestablishing the family ties of transient patients. The medical care was now under the control of an infectious disease specialist and a nurse, both paid by a local private hospital. "We had to develop partnerships with the private sector," stated Celeste. The psychologists were also in charge of training the forty volunteers who kept the place working "like a hospital," said Naiara, as she later joined the conversation: "We did not envision Caasah to be like this. We wanted to be able to deal with social aspects of AIDS, but the general health care infrastructure for AIDS patients is still in the same precarious situation. Public officers and doctors are happy that therapies are available but, for the most part, they don't give a damn about treatment adherence. So we have to do this work. . . . Street people only look for help when they are ill, but they don't want to stay."

"To tell you the truth," intervened Celeste, "when the health secretary came here to inaugurate the building, I told him that I wanted to help these people living with AIDS in the streets, but that I needed specialized staff and additional funds. Do you know what his answer was? 'My daughter, only put your hands where something can be done.'" Such discouragement, she said, was more than frustrating. It is clear, I replied, that these people have no political value. "Indeed."

Yes, distribution programs make antiretroviral therapies accessible, but they are one element in the full treatment of a disease that, as the health secretary made clear in his statement, remains a matter of a regional politics of *nonintervention*. Beyond the mere distribution of drugs, there is a need for clinics and treatment centers, for active approaches to providing care. Here, grassroots initiatives and patients themselves are overburdened with this task, and as AIDS spreads amid misery, entrance into treatment programs is restricted, under the name of providing quality care.

Both women then proudly told me that they were taking classes in hospital administration at a local college. "I never wanted our work to get so professionalized, for I think that it is love that makes us do this work. But we had no choice. In all levels of government, nobody respected us," said Naiara. "They either ignored what we said or called us incompetent. We have to play their technical game so that we can help people better. The state has no sentiments." Naiara added that she herself underwent psychotherapy at Caasah: "I was too nervous and angry. I was getting a Caasah personality, full of revolt and yelling; I had to cool down, I no longer recognized myself."

Celeste, in turn, confided that she had been receiving physical threats from Naum Alves, a former resident, who was now running his own

AIDS project among transvestites in the Pelourinho District and resented Caasah's success. "He accuses me of corruption. He asked the Public Ministry to investigate me. This is character assassination. I am not prepared for this. Other NGOs have lawyers, but I only have myself and my honesty. I get my salary as a coordinator and a food basket like any other volunteer here, nothing else. I am not afraid of the investigation. Let them come." She was afraid for her safety: "He calls me and leaves threatening messages, that he will cut me when he sees me in the streets. I have already contacted the police and put a restraining order on him."

At any rate, the institution now had a team that worked directly with local hospitals, said Celeste: "A patient comes here with a hospital referral. Before that, the hospital's social worker contacts us, and we send our social worker, head nurse, and psychologist to evaluate the potential patient, to see whether he fits into the institution and its norms. Most come here to consolidate their health; we help them adhere to their treatment. They are allowed to stay for three to six months. We also have patients from the interior who stay here while they undergo exams or pick up medication."

Caasah's operations now also reflected the interiorization of the epidemic and the fact that there were no specialized services in the countryside. The interpersonal networks of which Caasah was part compensated for the lack of a treatment planning framework. And in their precariousness, these networks decisively impacted clinical continuity, which is necessary for successful treatment. Yet, when I asked Celeste what differentiated AIDS patients who adhered to their treatments from those who did not, she was adamant that "in the end, they themselves make the difference, it has to do with who they are."

From Epidemic to Personalized Disease

Later that week, I visited the AIDS unit of State Hospital Luis Souto where, unlike Caasah, there had been no physical reconstruction. The unit still had only sixteen beds and was now mostly populated by "homeless people and by patients who have been rejected by other medical services," said Dr. Nanci, the energetic and outspoken infectious disease specialist who was endearingly called "*Māezona*" (Big Mama) by many of her patients, among them several former Caasah residents. "We now get more women and young people . . . people from the interior, rustic people who don't even know how to speak right. Hospitals there do not treat AIDS." The availability of ARVs had not been met by improvements in the care infrastructure. Patient discharge and continuity of treatment remained the service's main problem.

"Did you see this young woman who was leaving with her mother?" Dr. Nanci asked me as we sat in the room where she had just finished teaching a group of second- and third-year medical students. I nodded "yes," and before I could say anything else, she condensed her critique of what was happening to AIDS and what, for her, the newly discharged patient embodied:

> Unfortunately, this is what I see: the poor patients return to the hospital. They no longer come here simply to die, for we can physically rescue them. The patient leaves here walking, in an overall good state, but without a change in value systems. This girl will most likely go back to the world of drugs, to that circus. It's a shame what is happening to AIDS. The socially well-off patients benefit from the antiretroviral therapies, but the poorest are consumed by AIDS. They have no work, the family abandons them, they lose their values, and there is no way to socially reintegrate these persons . . . no medication can do this. This girl needs psychiatric help. She does not know how to take care of herself. She would need a total rescue operation. And what do we do for her? Nothing. Do you understand? She is a social problem. The mother is overburdened, she cannot take it anymore. She has to feed her daughter, but where is there money to sustain this family? Where do they find jobs? To live is too complex and, in comparison, the world of drugs is an easy way out for all parties. So she will come back . . . and die, not of AIDS, but of drug abuse.

Against this grim reality is the fact that "antiretrovirals are available, and this is great. Even though the results are not those we would like, many poor patients are using the drugs and benefiting from them. Very few patients who come regularly to the ambulatory are hospitalized," Dr. Nanci emphasized.

She was particularly proud of her "Caasah team," as she put it. "Edimilson, Luis, Rose . . . These are people who had a history of self-abuse, illiterate, but they were taught the benefits of medication. I find this fantastic. The patient remains poor but rescues himself and teaches others to do the same. Luis is strong and working. See Rose, how conscientious she is. She was using a medication that was not recommended for pregnant women and even before she saw me she stopped taking it. Few people would do that. She was wrong in getting pregnant, but was responsible with her new life."

"I regret to say this," continued Dr. Nanci, "but for me Caasah lost its role." With Caasah's transformation into a "house of passage," a crucial institutional nexus of AIDS care was now missing: "Patients who have nowhere else to go should be allowed to live there. I have a seventeen-year-old man with AIDS. He has all kinds of neurological sequelae and no family. He was raped by his uncle, who died of AIDS. Caasah does not accept him, for it is now a house of passage. The hospital ends up becoming a deposit of patients. This is wrong. We need more, and not fewer, shelters."

Unlike Celeste, Dr. Nanci had a more negative view about the fate of those evicted from Caasah and spoke of her frustration in not being able to reverse systemic health care failures: "Marta Damião, for example, fell back into alcoholism. She feels abandoned and cheated by Caasah. The other day, she came to the ambulatory in bad shape. I almost had to hospitalize her. Some people say that I endorse paternalism—so be it. If you want treatment adherence for this miserable population, you must keep an eye on them, bring them by the hand to schedule appointments, monitor their medical exams, so that they don't get lost in the system."

The unit's new social worker, Leila Andrade, also regretted the limits of Caasah's current role, especially in light of the recent closure of two state-run homeless shelters in the city: "All of a sudden we have all these medical technologies, but we have no means to help a person to reorganize his or her life. What is a life chance for someone who has nowhere to go?" And sharply, she added: "I think that this is one of our gravest public health issues: in spite of all these advancements, people come in from the streets dying of AIDS, without knowing it."

Of the eight hundred patients listed as taking ARVs, some two hundred were not picking up medication, Leila told me. "Some might have migrated; others probably died or simply stopped the treatment. Many complain of side effects and stop, others take the medication until they gain weight and then relapse. Patients who are employed are afraid people will learn they have AIDS, and they have to find all kinds of hidden ways to take the pills. It is difficult to be a lifelong patient. But when they see the results, it is great: their emotional side begins to improve as well." There was no way of tracking the missing AIDS patients, she said. The service was not computerized and "because we have to protect the patient's privacy we cannot contact them." She also regretted that there were no studies underway to track local cultures of adherence. "It wouldn't be that difficult to identify what works and what does not . . . but our scientists study therapy failure rather than the patient."

Clearly, a concept of spontaneous service demand guided this local ARV rollout. Given the latest developments around AIDS, I thought, the medico-political question shifted from controlling the epidemic to controlling individualized disease. Whereas places like Caasah had initially helped to consolidate a differentiated AIDS population among the diseased urban poor, the now democratized ARV seemed to have built in a mechanism of self-inclusion or exclusion, I told the social worker. AIDS was here further individualized. To which she replied, "This is real. The human misery is too massive, and we have no structure to help them. Moreover, the medical teams by and large do not help much. They see the pathology and forget the world." Earlier, Dr. Nanci had also conveyed her frustration with the lack of a multidisciplinary team in the unit and her impatience with the medical ethics displayed by her students: "They only see risk and stupidity in the AIDS patient. They do a linear reading of the patient. They don't analyze the subject's complexity and withhold the compassion the patient needs."

The social worker said that as far as she knew, there were only two street patients regularly coming to ambulatory checkups (every three months). "Yes, the homeless AIDS patient remains outside the system." She then conveyed a difficult truth: the existence of a valuation system by which doctors, in spite of universal ARV availability, decided whether to put patients on lifesaving treatments. "How can doctors make sure these patients will take the medication? . . . So, it is better not to start them on ARVs, they say, for there is no guarantee that they will continue the treatment. Doctors are concerned about the creation of viral resistance to medication."

Dr. Nanci confirmed this: "It is very difficult, but at times I make the choice not to begin treatment. It does not help to have the medication available if the individual cannot use it adequately. Who will teach him? How to achieve adherence when there is no self-esteem left? As I told you, drug users are difficult to deal with. They are the most rebellious patients we have. They no longer adapt to anything, and we don't have a structure to help them. They have no sense of guilt, of being responsible for the life of the Other."

The patient's lifestyle and supposed lack of responsibility justifies the withdrawal of antiretrovirals. This decision stands for medical competence: "It is worthless to give out therapies for those who will not take them . . . to do a pretend medicine." Yet a change in values, a conversion of sorts, expressed in routine medical appearances, might make medical technology worthwhile: "In theory, obviously, the doctor cannot condemn a drug user patient not to take medication . . . but the fact is that the homeless patient does not return for routine ambulatory checkups. So what I do is to tell the patient that he has to come back. If he returns and demonstrates a strong will, we begin treatment. But they never, or rarely, come back. For the street patients, we are like firemen; they call on us for an emergency rescue."

Physically Well, Economically Dead

At Caasah, institutional maintenance was a daily struggle. "This is a beautiful building, but that's all the state gave us. We owe more than $1,000 to local pharmacies. Our patients come from the hospital with their ARVs but nothing else. No vitamins, no painkillers, no Bactrin to treat opportunistic diseases." As AIDS became more chronic than fatal, local programs were not necessarily readjusting themselves to meet the new needs of patients. The national ARV rollout was supposed to be matched by the regional government's provision of treatments for opportunistic infections. But it was clearly up to proxy health services such as Caasah or to the patients themselves to regiment treatments beyond ARVs.

The city administration kept providing Caasah with security guards and health technicians. Caasah's project with the Health Ministry did not include aid for treatment, only for legal and volunteer work and job training workshops. "After we moved to this building, community donations declined. People think we have a continuous flow of money, but that is not true. The government now has other priorities. Officers speak of shelters for battered women—that's what makes the headlines these days." Celeste also mentioned that a new AIDS outpatient clinic was under construction, with the caveat that "I only believe in things related to the state when I see them working."

What came up next in the conversation was a shock to me, revealing both the makeshift quality of Caasah's earlier form of care and the murkiness of its present-day operations. What happened to Professor Carlos? "It was a bombshell," Celeste replied. "Some time ago officers of the Bahian Nursing Association came to my office. They had been investigating Carlos and had found out that he had never graduated from nursing school. He had falsified all his diplomas and credentials. I did not know what to say. Thank God they could not indict us, as I had kept the dossier he had given us before we hired him."

Unbelievable, I said. Carlos was a quack nurse . . . all the people he had trained and all the patients he had treated, so many he had helped but so many had also died under his care. . . . Celeste did not want to go there, she told me. And to my surprise, she added that Carlos was still working at Caasah. "Not with patients though. He is our administrative right arm." What a remarkable complicity and second chance. Celeste asked me not to bring up the fraud in my conversation with Carlos later that morning.

"We now have this paradox that the poor AIDS patient is physically well, yet he is still economically dead," Celeste continued. "They don't survive by medication alone. The person with AIDS continues to face discrimination in the family, cannot easily get a job, and the Ministry is making it harder and harder for newly diagnosed patients to get disability pensions. In order to get it, you must really be close to death." In other words, "the NGOs [nongovernmental organizations] don't help patients, they only work for themselves—you know that. The state needs us. We have the people. We are the only institution that delivers systematic care to the disenfranchised. We are the transition from the hospital to society. But to do it well, unfortunately, we have to reduce the number of people we take in. Our patient only leaves here after all his exams show that he is well, and after we find a place for him, either back with his family or renting a shack with someone else."

No longer addressing those considered antisocial, Caasah had recast itself as a temporary nexus between the public hospital and the patient's family or family-substitution unit. "We don't handle the person who has just learned that he is HIV-positive or the terminal patient," as Celeste put it. Caasah now took in the AIDS patient who was (according to an intricate valuation system not yet clear to me) potentially able to adhere to his ARV regimen. ARV adherence was not in the state's purview, so to speak, and the family or its absence was the ostensible social problem. Caasah filled in this medical and social void and realized adherence, in a demonstrative way:

> We are the ones who really teach the patients how to manage their drugs. Many don't know how to read or write, so we teach them to read the drug by its shape and color. . . . We give them the initial material support needed to make the treatment work. They leave the house with the knowledge that they cannot stop taking the medication. We know the difficulties people will face outside and we support them so that these difficulties might be alleviated. But in the end, it is the patients themselves who make the difference.

Drug Resistance and Rescue Treatments

For Carlos, with whom I met later that day, Caasah's fundamental task is "to create a new psychology" that could counter what he calls the "auto-biologia," or the *self-biology*, of AIDS patients. Carlos had physically changed. As a result of hormones, I suspected, he now had noticeable breasts and a curvaceous body and dressed in feminine fashion. In spite of being Caasah's operations manager, he spoke as if he were still a nurse:

> When a patient who has no symptoms sees someone else taking ARVs and falling ill, quite often he reasons that "as long as I am okay, I will not take the medication." Many stop after regaining health. Moreover,

patients know that ARVs alter the body, that a new set of symptoms is developed—hair loss, dry skin, urinary problems, fat accumulation, and so on. . . . The person fears this. Not so much because others will know that he has AIDS, but because the drugs change his self-biology. So we have to train them well on the effects of each drug, so that they are able to identify the origin of what they are experiencing.

The villains, according to Carlos, were the drugs' side effects linked to an "I can't adjust" mentality. But this, I thought, covered up people's concrete relations to AIDS, the meaning it had for them in the quotidian and the difficulties they had in re-envisioning themselves as lifelong patients.

Carlos had no qualms about saying that former residents kept contacting him over the phone, telling him of their symptoms: "And I always tell them not to stop the medication." He had also remained a vocal critic of the ways doctors used their work in the public health care system to advance their financial interests: "What we see today is the evisceration of health as a public good. Doctors decide what and who to treat according to how much money they will get from SUS [Sistema Único de Saúde]. . . . If it is too little, they don't bother. Doctors also gain extra incentives from drug companies to shift patients' treatment regimens, and this is not always in the best interest of the patient, who unfortunately does not know what is going on."

•••

Nurse Elida later told me that indeed not all medication side effects were patient-induced:

> One of our patients, Raimundo Nonato, is now at the University Hospital with a strange fever. He was not our patient at first. He was part of the University patient cohort. He first got the fever there, then he had a setback at home, and he was brought here because he had volunteered at Caasah before. We knew him. He improved, but then the fever came back. He told me that all began after a change in medication. The doctors told him that they did it because his CD4 count was low. But he is taking a drug that is not yet registered in Brazil. I saw the bottle. There is a note saying that this is a specially released medication for him. Our doctor thinks that he is part of a clinical trial.

Carlos, 2001

Elida said that, unfortunately, at Caasah they had no access to the patients' medical records, and that all they knew about the treatment came from the patients themselves. She also found it strange that most patients had medical follow-up every three months, and a few were seen once a month, which for her signaled that they were part of some form of clinical trial, "most likely of rescue treatments."

Pharmacist Francisco Moraes confirmed the expansion of ARV-related trials in the city: "Yet the information that is available to both doctors and patients is quite limited. Many times patients do not know the specifics of their participation in the trials." Bahia's reputation as a leader in infectious disease research, combined with the weakness of regulatory institutions, he argued, made it easier for companies to carry out trials on their own terms. "And add to the mix the generalized political irrelevance of human life in Bahia."

At the end of my conversation with Elida, she recalled that at a local meeting for AIDS NGOs, patients had discussed possibly forming a group of those who wanted to be guinea pigs for new drugs. "I remember people saying, 'I have nothing to lose.'"

•••

There is no uniformity among ARV distribution programs in the country. Regional politics, local patterns of medical sovereignty, and inconsistent patient mobilization consolidate variation in both AIDS treatment and life chances. Within an unplanned and a highly unregulated medical system, clinical trials have emerged as one more site in which public health care infrastructural insufficiencies are palliated and through which patients negotiate and experiment with their idiosyncratic genre of AIDS. Pharmaceuticals thus calibrate people's sense of belonging (or lack thereof) to the state and the economy.

"Medication is me"

From AIDS indigence to Caasah's refuge. From all kinds of medical and inner crises and foretold death to an unexpected future made possible by adherence to antiretroviral therapies. "This was then. . . . Today is another

Luis, 2001

world," Luis Cardoso told me as he looked at the portrait Torben had made of him in March 1997, the time he was beginning to take the AIDS cocktail.

"Times have changed for the HIV carrier. There is no cure, but with medication we can have an extended life and be healthy. I came to Caasah in bad shape, sad, ill. I thought that the world had come to an end for me, that I was here and the world was there. That I wouldn't survive. But I recovered. And I got the courage to face the world." Luis was happy to show me the photograph of Davi, the three-year-old boy he had informally adopted. "His parents died of AIDS. He was living here in Caasah. I now pay his grandma to care for him. All expenses are on me. I love him."

First diagnosed with AIDS in 1993, Luis lived in Caasah from 1995 to 1999. "Now Caasah is like a hospital. It is no longer a house for people to eat, fight, sleep, and wait for death to come. Now you come here, recover, and leave. Life is to struggle." Disability pension and the salary he earned as Caasah's office assistant allowed Luis to pay for his bills and save for the home he dreamed of owning. "I am in good health now. I get a few colds, sometimes, but that's normal." Open about his homosexuality, Luis implied that he was dating: "Today I can do everything, but I know that I have to play it safe."

Yes, ARVs opened up a space that did not exist before. But alone they don't explain everything that has happened to Luis. Caasah's collective arrangement replaced the family that was not there in the first place and enabled him to access medical and psychological care. "Here I stopped blaming others for what was happening to me. Self-pity and hatred keep you imprisoned, don't let you live the present. One has to think differently, forget the past. . . . I have nothing to say against the antiretrovirals. I am under Dr. Nanci's care. Celeste, the psychologists, and the volunteers motivated me a lot. But I don't live here anymore, and I must take care of myself. I got used to the medication. *Medication is me.*" For Celeste, "Luis is like a son." He represents Caasah and the state of Bahia in national meetings of people living with HIV, and he runs HIV/AIDS prevention workshops in the interior. Even his doctor calls Luis "my teacher."

Luis's new life is intimately linked to having Caasah as a recent past, to being the object of medical competence, and to developing the thoughtful capacity to monitor bodily effects. When the time of treatment came, a different subjectivity had to be produced. "I learned that the AIDS patient has the right to struggle for a future, like any other human being. I always look first at what is up to me to do. I must have this conscience: to have to take food and medication and to sleep at the right time, to always be alert to my symptoms and to the weather, to schedule my medical appointments. *I am the effect of this responsibility.*" Treatment regularity occasioned a previously unknown gratification. Luis says, "I gained another body. But it is a mistake to think that one can put the guards down."

This novel engagement with technology is lifelong. It constantly challenges and interrogates the self. It requires personal virtue, Luis insists, a new way of claiming actuality and existing beyond immediacy: "If one wants to have a real life one must make medication a routine and to know what is good for oneself, beyond the moment. The world presents all possible paths. It is a school in getting lost. But it is up to me to take life forward. I know that beyond the immediate pleasure of a drug or of alcohol, there are grave consequences. So I choose to do what is best for me, and I don't counsel anyone to take the easiest route."

In this work, I am not concerned with finding psychological origins for the will to live that AIDS survivors voice and enact. Their sense of interiority is, rather, ethnological, that is, the variable whole of their desires and actions in relation to the outside world of AIDS—those highly competitive social, medical, and political spaces that must be won, over and over again. "Nature is not a form, but rather the process of establishing relations" (Deleuze 1997, p. 59).

The ethnography of AIDS after the introduction of ARVs thus illuminates processes of individual and group becoming, as Luis and others suggest, taking place through medicines and multiple sites, relations, and intensities—fields of immanence (Deleuze 2006, p. 130). It is within this circuitry, as it unequally determines life chances, that AIDS survivors articulate their "plastic power" (instead of a given truth or life form) and invent a domesticity and health to live in and by (Biehl 2005, pp. 14–18).[1]

I asked Luis whether religion had helped him adhere to the ARV regimen. "I think religion is within us," he said. "I always believed in God, and now more than ever before. But religious talk does not help if you don't have the will to live inside you." For Luis, organized religion is actually a fraudulent practice of medicine when it comes to AIDS: "How many times I saw people having religious conversions in Caasah. Priests coming and asking patients to surrender to God, saying that their crises were henceforth over, that they would be cured. But cure never came. It is when you go to the doctor, do the tests, that you see whether your CD4 count increased or not."

God, the individual's will to live, and access to ARVs, combined, make for a vital force, a mental compass of sorts, that helps Luis "not to embrace the world" and decide "step by step" what is good for him, that is, the value of health. Subjectification is thus a way to move beyond oneself and to protect life unfolding: "I know that alcohol interferes with the medication. . . . So why would I drink? When a person has the will to live he adheres to medication. Many however say, 'I will die anyway,' and return to their vices. I don't see things this way. Life is not an adventure. I know that one day death will come, but I don't search for it."

Luis is an amazing person, hardworking, wise, witty, *and* a master of moral discourse. He speaks of a new economy of life instincts organized around AIDS therapies. And he himself is the dominant human form that emerges from this economy: "I face my problem. I take advantage of the help I get. I struggle to live." He is indeed the representative of a new medical collective, and his discourse conveys present-day forms and limits of society and state: "I have nothing to do with society," he says. "From my perspective, society is a set of masters deciding what risk is and what is bad for them. I have never participated in that. As for the government, I must say that I am thankful for the medication. People struggled a lot, and we are now a country in which the AIDS cocktail has not been missing. This is the good aspect of the state. The rest is for me to do."

He has harsh words for those who throw medication away—"it's a crime." A crime against the state and the person himself, Luis reasons: "The government is paying a lot for medication. To keep the person hospitalized would be even more expensive. People don't sell the antiretrovirals, for no one would buy them. Many people with AIDS already gave me their drugs, and I brought them to Caasah. Tell me: what is the point of dying in a hospital bed? I must follow the medication rightly."

Luis made treatment adherence seem too easy. As much as I admired his resilience and uplifting presence, I also found his righteousness quite disturbing. For him, individual conscience was the a priori of a healthy existence, and mourning a loss, any kind of loss, was a defect to be overcome. Moreover, the overemphasis on individual responsibility was self-serving. It clearly reflected Caasah's house-of-passage modus operandi and, more broadly, the hegemonic neoliberal discourse that one has to be ever more self-conscious, lord of oneself, upbeat and upward. The institutional and interpersonal forces that have thrown Luis into action in the first place were absent from his life-extending account, particularly as he spoke of noncompliant *marginais*. It was evident from his recent past recollection that without his belonging to Caasah, ARVs wouldn't have had the same kind of efficacy they had for him, and that he kept drawing strength from being the object of regular public attention (as "good" AIDS patient, volunteer, and educator). In Luis's technocratic moral discourse, social abandonment exists in a vicious circle with self-destruction and self-created risks. His pharmaceutical subjectification has indeed led to salutary effects, yet it remains built on the exclusion of those who cannot conform:

> It is not a matter of getting them [homeless AIDS patients] help. For they already have it [in the form of medication]. They use their social condition as an excuse to keep their habits. . . . It is a question of self-destruction. As I see it, these people are more for death than for life. They have no love for their own life. I know guys who are infected and their wives keep getting pregnant. It is evil. They are putting an innocent life in the world that has nothing to do with their problem. But I also know many people who struggle to live and to earn their money honestly and don't surrender. . . . See Rose and Evangivaldo. . . . It is your mind that makes the difference.

"I am mother and father"

I rented a car and Luis took us to Rose's house. I would have never found it on my own. There are no traffic signs to guide you through these labyrinthine *favelas* turned commercial districts. Houses and informal businesses blur in an unplanned and overcrowded way. After missing a few exits and having to make some dangerous maneuvers, we finally reached the bumpy highway that connects southern and northern Brazil. Five miles north, we turned right onto an unpaved road and into the heart of Salvador's misery: the Cajazeiras District.

Rose was waiting for us. "It's a poor shack, but it is mine." She introduced us to Jessica, her one-year-old daughter, and Ricardo, her eleven-year-old son who had previously been under the care of Professor Carlos. The kids kept watching television as we sat to talk.

She immediately made fun of her new hairdo: "It's modern. I straighten it and put some color in it. Men will be knocking at my door. . . . For someone who passed through hell, I am in good shape. I am strong and conscious, right, Luis?"

Rose wanted to know what I had been up to. I said that besides having taken up a teaching position and working on another book project, Adriana and I were expecting our first child in a month. No, we did not know if it was a boy or a girl. "I also did not want to know," she added: "I only wanted the baby to be healthy."

Tearful, she recollected the death of her partner Jorge from AIDS-related diseases, a few months before Jessica was born. "He wanted this child so much." Rose knew that Jessica's HIV-positive status could still change. "She has never been ill and we hope for the best."

Rose had given birth to another daughter in 1993 in Caasah. Rose referred to her as "Naiara's girl" and did not speak of the two other teenage children that I knew she had given up for adoption. I had also learned from Dona Conceição that Ricardo's father lived in the streets and also had AIDS—a word Rose did not utter.

"Life continues. When one has the 'girl' [*menina*] one must have the responsibility. I can do everything, but within my limits. I am not a fanatic patient. There are people who think about *it* twenty-four hours a day and who live in hospitals. I live a normal life. I know how to have fun too." As Rose used the word *menina* and the pronoun *it* to refer to AIDS, she gesticulated that she was talking in disguise because of the kids and the neighbors: "They don't know I have the *menina*. It's tough to go about life without *it*, and if you have *it*, Oh Lord, no one wants anything to do with you":

The first time we met in Caasah—it was 1995, right?—we were all losers there. If I look back, I consider myself victorious.

I am thankful to the government for the medication and to Dr. Nanci, she is a blessing, a mother to me. I never had another doctor, and I never had this problem of having to change treatment. People

Rose, 2001

fight to have Dr. Nanci as their doctor. We are siblings in Christ and in doctor, right, Luis? When the antiretrovirals first came in, we were among those she gave priority.

I live from aid. We were lucky to get disability pension before the medication. We are the founders of the old institution. Caasah was my salvation. The majority who now apply for disability do not get it.

We had everything done for us, food, medical appointments. . . . That routine helped. We did not know the price of anything though. And once you are on your own, my brother, it is a whole new ballgame.

This is how I lead my life. . . . always struggling to pay the bills, raising my children, for I am mother and father. I must thank God for existing. I don't live better for I lack material conditions.

Like Luis, Rose spoke of being medically conscious and responsible. Yet, in her account new life was anchored in the experience of being saved *by* Caasah. To adhere to ARVs, to make a child, and to take on maternity/paternity in such vulnerable and precarious conditions—"life is on a hanging thread"—was already the effect of something, evidence that a change had taken place. Like most poor AIDS patients, Rose had experienced cruel forms of abandonment and self-abuse and had no family, no one to count on. "If I thought then what I think now I would not have done one-quarter of the things I did. It was craziness. I had no support. I was lost in the world. The world was the only school I had, and it ended in what it did." Death by AIDS, she meant, a destiny she escaped.

Caasah made it possible for Rose and other socially dead subjects to reconstruct themselves from ashes. In her account, "salvation" had to do with distancing oneself from and elaborating upon losses and failures, with finding a place for death other than in one's body. "One was the mirror of the Other. This proximity and constant interaction made one reflect on what had passed." In this interval grassroots space—which redefined what family was—a discourse of morbidity was channeled into agency and labor vis-à-vis the newly available medical technologies.

Caasah remained a foundational reference (material and symbolic) to the AIDS- and class-defying breadwinning mother Rose had become. "I know that they have financial difficulties there. It is a big operation. And they have the children now. But I am glad I can count on them. When Celeste and Naiara see that I have a serious problem they say, 'Come here and we will see what we can do.'" Rose confided that Professor Carlos kept playing nurse to her: "He helps me a great deal. I don't care what people say about that creature . . . [she imitated the way he talked]. Whenever I need a drug I cannot buy or I have a wound that won't heal, I can count on him."

She complained bitterly about the lack of medicines for opportunistic diseases in the hospital and local health post, and told me how she herself had become a proxy-physician/pharmacist, recycling available medication through trial and error: "The basic pharmacy program and health post is just façade. A few weeks ago Jessica got this horrible scabies. I couldn't get a specialist in this end of the world. So I had to tell Dr. Nanci about it. She gave me a prescription but had no samples. The medication was too expensive. I couldn't buy it. Carlos had given me another medication. So I used it in Jessica. Thank God, it worked."

Rose's powerful discourse of salvation from the world, from herself, and from social AIDS—"I am a new creature, you can write my story"—was matched by an equally powerful account of the ways she was able to establish and sustain transference with people standing for state and medical institutions and thus realize her new life. The actions she microscopically alluded to reflected long-lasting social and subjective forms based on *apadrinhamento*—the practice of godfathering as revealed by the canonical works of Gilberto Freyre (1987) and Sergio Buarque de Holanda (1956). Pragmatically, to have one's child adopted by a powerful someone greatly helps one to traverse institutions and to access goods. This engineered kinship stands for and does the welfare work of community/neighborhood, and, ultimately, it also makes the vision of social mobility possible.

Rose went to the kitchen to brew coffee. Quite poignantly, she served us *cafezinho* on a tray. She was proud of her well-taken-care-of house, of her urbane manners, and of her children—"the boy is my right arm"—and she was always captioning her moves: "Who would have imagined that Rose, from Pelourinho, would be giving interviews, telling the professor and the gringo [Torben] what life is all about? I tell you, I want to be alive to see a cure. In the name of Jesus, I want to be a guinea pig when they test the vaccine."

Rose also made it evident that she was expecting some financial aid at the end of our session: "Any aid is welcome, my brother. You know that

I limp, from those mad times [intravenous drug use]. I need help to buy Voltaren [a painkiller]. The other day the pharmacist sold me five tablets. I didn't have enough money to buy the whole box. And I couldn't buy meat that week. It's not easy."

"It is the financial part of life that tortures me"

Caasah's former residents are the new people of AIDS. After experiencing family dissolution and social abandonment, they have come into contact with the foundational experiences of sustained care and biotechnology. Back in the world, these patient-citizens refuse the position of leftovers and break open new pathways. Life is in transit. They humanize technology and redo themselves in familiar terms. Rather than an initial scenario, here families are epiphenomena. The unwanted transform themselves into objects of desire and punctuate the world with medical signifiers. They have little to share and keep to themselves what is working. There are no institutions or policies actively looking after them or tracking the forms AIDS, the disease, is taking. Those alive keep returning to the "old institution"—Caasah. All the time and at every turn, they have to consider the next step to be taken to guarantee survival.

Luckily, we happened to be back at Caasah for a final visit when Evangivaldo showed up looking for help. "When it was just Fátima and I we could improvise things, but now that we have a child it is another responsibility. Her name is Juliana, she is two years old. I can go hungry, but she cannot. I love my daughter. God give me this force to live because of her. My little girl, she keeps me going."

We were happy to help him out so that he could provide for his family. And in a low voice he added: "Don't do it in front of the coordinators though. I don't trust them anymore." He suggested that the institution favored some individuals and that there was a generalized unaccountability as far as donations were concerned: "I am also a founding member of Caasah. We all know that much money enters here. . . . I have a family, I have to pay rent. . . . The other day we had to take the girl to a specialist. She had worms. I now owe 75 reais [$30] at the pharmacy."

Evangivaldo, 2001

After a few shots, Torben told Evangivaldo, "You are a natural." He smiled and said that he would show the old portrait to his doctor and nurses at the hospital where he picks up ARVs: "This work was important to me, it marked my history. I felt calm and a curiosity. I know that this is a kind of scientific work for all to see what we go through in Brazil. . . . I did not expect that my body would be clean as it is today. I want the doctor to see how I changed. Now I only have hope."

"Before being kicked out of here," Evangivaldo recalled, "I told Fátima, 'You have no family and I am alone. I have nobody for me. Let's live together.' She said 'yes,' the important thing is to be happy." Fátima had been a mother before, "but the girl is being raised by a judge in Minas Gerais. Juliana is my first child. It was not planned. The condom broke. But now that she is here I see that this is what I wanted the most in life. I thought I would die without being a father, but now I have a fruit of the earth."

Evangivaldo lived for his family. He was grabbed around the throat by all kinds of real impossibilities. Yet he found ways to transcend his sense of being choked:

> I am not finding work these days. I feel defenseless without a job. I cannot stay still, I want to work, I pick up bottles, cans, paper to sell, I clean toilets, wash cars. Last month I got pneumonia. I don't have a family to help me, only the government. But the disability pension goes for rent and some food. Sometimes I help Seu Marcos to sell sodas, but he did not need me today. I then went to a lady who sometimes helps us too. Fátima was her maid. But she was not at home. The government, the big ones, the politicians . . . they should not deny work to those who need it and want to work. I can stay without food, but medication I always take, even if I only have coffee to drink. I put my medicines in a little container when I go looking for jobs. I always hear the forecast on the radio. I don't want to be surprised by the weather and get a cold. It is the financial part of life that tortures me.

Evangivaldo said that he planned to find to a better shack—"now we live in a zone of bandits." He wanted us to then visit him and meet Fátima and Juliana. Neighbors were presently treating them well: "It helps that we have the child. But they don't know of AIDS. I tell them that I get benefits because I have mental problems." In fact, now and then Evangivaldo looked for psychiatric help at a local health post, he said, "when I feel an anguish that does not go away. It is hard, I hear voices, and when I look around nobody is calling me." He carried with him a prescription for the antipsychotic and sedative levomepromazine and the hypnotic diazepam. "I have the prescription, but I have no money to buy the drugs. Sometimes I get free samples, other times I buy a few pills."

Evangivaldo then showed us a piece of paper on which he had listed how his income was allocated and the debts he had to pay. I will never forget his calculation, what he goes through in life:

> I owe 75 *reais* to the pharmacist. I get 200 *reais* from disability. I pay 120 *reais* for rent. Electricity takes 20 *reais*. I also now owe 30 *reais* at the grocery store. We are trying to get Fátima a disability pension too. I am a Brazilian citizen. I walk with my head up in my country. I am not ashamed of putting my hands into garbage to pick up what I need. Those who rob should be shackled, not us.
>
> When Fátima cannot do the work, I am the man and woman of the house. Sometimes I wake up at 4:00 a.m., leave everything ready, and ride my bike for two hours, to get downtown. I go door to door, asking for a job.
>
> There are days when I cannot get the money we need and I panic. My head spins, and I fall down. I hide in a corner and cry. Then I don't know where I am. You cannot imagine how painful this is. But I tell myself, "Focus, Evangivaldo, focus, you will find your bike and your way home."
>
> And do you know why I manage to do this?
>
> It is because my daughter is waiting for me at home.

Global Public Health

Large-Scale Medical Change

AIDS emerged and thrived alongside globalization. Amid denial, stigma, and inaction, it became the world's largest epidemic. By the end of the 1990s, about 20 million people had died of AIDS. But HIV infection is no longer a death sentence. In the past ten years—due to combination regimens of antiretroviral drugs (ARVs)—we have seen the transformation of AIDS from a fatal to a lifelong disease. These therapies have an overwhelmingly positive effect on prognosis. Nonetheless, 3 million AIDS-related deaths took place globally in 2004 alone, most of them in poor countries, where life expectancy has been dropping dramatically.[1]

HIV/AIDS ravages bodies, families, populations, countries. Yet, at the same time, the virus and the syndrome are also a driving force of personal and political possibilities across the globe. AIDS has given rise to extensive creation—of solidarity and organizations, of science and policy, of new modes of thought and life. Brazil's bold, multi-actor, and large-scale therapeutic response to AIDS has made history. In this book, I have explored the broad economic and political effects that treating AIDS had on health services, both national and local, and how this life-extending policy influenced international efforts to reverse the pandemic's course. I have also illuminated individual and communal modes of life that have emerged around ARVs among the country's most vulnerable urban populations. In highlighting the successes, failures, and complexities of the Brazilian response to AIDS, I have revealed significant structural, logistical, and conceptual changes in governance and citizenship—groundbreaking in their own right.

The Brazilian AIDS policy is emblematic of novel forms of state action on and toward public health. Pressured by activists, the democratic government was able to negotiate with the global pharmaceutical industry, making ARVs universally available to its citizens and also opening up new market possibilities for that industry. The sustainability of the policy has to be constantly negotiated in the marketplace, and one of the unintended consequences of AIDS treatment scale-up has been the consolidation of a model of public health centered on pharmaceutical distribution. This intervention gains social and medical significance by being incorporated into infrastructures of care that are themselves being reshaped by state and market restructuring.

ARV rollouts are matters of intense negotiation; their local realizations are shaped by contingency and uncertainty. Such realizations encode diverse economic and political interests, as well as the anxieties and desires of both groups and individuals. They also involve shifts from one form of bodily and medical knowledge to another. Amid a striking decrease in AIDS mortality, this pharmaceuticalization of public health also promotes discourses that define segments of the population as disposable. That is, in their poverty and deviancy, members of these subpopulations are stigmatized as noncompliant, making their deaths inconsequential. As my ethnography shows, local AIDS services triage quality treatment, and wider rights for the poor and sick to housing, employment, and security remain largely unavailable.

This book has been concerned with a series of interconnected problems linking AIDS, pharmaceuticals, global health initiatives, the state, inequality, social experience, and subjectivity. Readers have been introduced to changing institutions and to individuals at the intersections of science, medicine, activism, public health, charity, and homelessness. In this concluding chapter, I return one more time to Brazil and look for the book's main characters—the activists and policy makers who decisively shaped the country's response to the epidemic as well as the more anonymous Others living with AIDS and those who care for them. Refusing to quickly generalize, in dialogue with these various subjects and perspectives, I appraise the AIDS policy's institutional, discursive, and human outcomes.

What has happened to the social mobilization that made this therapeutic policy possible in the first place? What are the present economic, technological, and political challenges the model policy faces? How has ARV rollout impacted the overall health infrastructure and the institutions of care that serve the poorest? What is the lasting power of the mechanisms of self-governance that AIDS patients developed in order to adhere to treatments and extend their lives? And more broadly, what does this longitudinal study reveal about contemporary Brazil, global approaches to AIDS, and the impact of economic globalization on local social realities?

The book's conclusion thus exposes the logics of dead ends and articulates the paths to large-scale medical change, as well as their points of entry. At both the macro- and micro-levels, we see a state of triage and a politics of survival crystallizing.

"A little more reverence for life"

Many public- and private-sector treatment initiatives are being launched worldwide, and today the AIDS crisis in the developing world is finally on the radar of transnational organizations, governments, and citizens alike. According to activist groups, the Global Fund to Fight AIDS, TB, and Malaria "represents the globalization of Brazil's model of harnessing the forces of government and civil society to confront the AIDS challenge."[2] More than one hundred countries have together committed a total of $3 billion to the Global Fund—an international health financing institution—with the United States pledging to donate the most, $2 billion.

Here, governments and civic organizations focus on funding rather than implementation. The development of aid projects (mostly aimed at helping women and vulnerable children) is left to local groups. When UNAIDS, the United Nations' AIDS program, was founded in 1996, it had $300 million available for loans to middle- and low-income countries. This budget increased to $4.7 billion by 2003. The World Bank, which has supported the development of the Brazilian AIDS program, has played the largest role in financing UNAIDS.[3]

This increase in AIDS funding in recent years "is largely a fruit of the well-coordinated activism of the international community," stated Dr. Paulo Teixeira, Brazil's former AIDS coordinator, at the Global Health Governance Workshop in São Paulo in June 2005. Organized by sociologist Gilberto Calcagnotto and the German Overseas Institute,[4] the workshop aimed to facilitate a discussion over the ways in which global, national, and local institutions interacted and affected the control of infectious diseases among the poor. Dr. Teixeira was there to discuss the work of the national AIDS program, and three other research teams were presenting the results of studies carried out in São Paulo, Belo Horizonte, and Recife. I was invited as a commentator. In my remarks, I pointed to the new geographies of access and marginalization that have emerged alongside pharmaceutical globalization and to the circuits of care that shape the environment in which poor AIDS patients engage in their daily struggles to survive.

"With respect to actual socioeconomic analyses, I am of course not unaware that without models, paradigms, ideal types, and similar abstractions we cannot even start to think," writes Albert O. Hirschman, whose words I take to heart. "But cognitive style, that is, the kind of paradigms we search out, the way we put them together, and the ambitions we nurture for their powers—all this can make a great deal of difference" (1970, p. 338). Yes, "model policies," such as the Brazilian response to AIDS, can disturb and redirect some of the metaphysical assumptions of policy design. But our thinking should not stop at the paradigmatic power of a given intervention. Following Hirschman, ethnography can help us to tease out the many components of a policy, the empirical ways in which it is assembled, and how it takes institutional hold and fits into ongoing and largely unequal social relations.

For Hirschman, as for the ethnographer, people come first. This respect for people, this attention to the manufacturing of political discourses and to the sheer materiality of life's necessities, makes a great deal of difference in the knowledge we produce. Large-scale processes are not abstract machines overdetermining the whole social field. Social mobilization and personalized interventions have made a world of difference in expanding treatment to the poorest afflicted by AIDS, for example. Neither can the micro-arrangements of individual and collective existence be described solely in terms of power. The overconfidence in power arrangements and rational choice is itself a cultural product to be scrutinized. As Hirschman writes, "In all these matters I would suggest a little more reverence for life, a little less straitjacketing of the future, a little more allowance for the unexpected—and a little less wishful thinking" (1970, p. 338).

The ethnographer upholds the rights of a micro-analysis and thus brings into view the immanent fields people invent to live in and by. In doing so, he still allows for some veritable principles of unification, or dispersion for that matter, to surface. Far beyond gross dualisms, ethnography's unique theoretical force lies in recording competing rationalities and vital experimentations, in conceptualizing fine articulations of worlds, differentiated, in flux, and impending. Amid much mindlessness and misunderstanding, there is also the potential of art for the social scientific venture—as the late Gilles Deleuze put it, "There is no work of art that does not call on a people who does not yet exist" (2006, p. 324).

The Future of Treatment Rollouts

"We have changed the discourse and paradigm of intervention," Dr. Teixeira told me at the Global Health Governance Workshop. "It has become politically costly for development agencies and governments not to engage AIDS." Yet, the operations of global AIDS initiatives and their interface with governments and civic organizations "reflect and extend existing power relations, and this synergy can be quite negative," he added. "The negotiating power of developing countries is simply too low, be it at the United Nations or at the World Trade Organization [WTO]. . . . AIDS gave poorer countries a small window of opportunity to intervene in global governance and to try to recast an uneven correlation of forces." Brazil has done so by challenging the patents and pricing structures of global pharmaceutical companies at the WTO and by spearheading alternative south-south cooperation programs.

The Lula administration has only recently begun to embrace some of these political options. As Dr. Teixeira recollected, "I always voted for the Workers' Party, and I had high hopes in this government. But for reasons that have not been made public, the current government has been reluctant to make bold moves as far as generics, patents, and international relations are concerned." By early 2004, for example, the national AIDS program had taken the technical and juridical measures that were needed for the government to issue compulsory licenses for the production of two patented drugs that took up almost 60 percent of the country's AIDS treatment budget. "We had preliminary agreements with Indian companies to provide us the necessary chemical materials, and I was at the WHO [World Health Organization] to provide international support," Dr. Teixeira stated. "It was just a matter of the health minister appearing on national television and announcing it, but he did not."

Other public health scholars at the workshop told me that the national AIDS program had actually lost some of its political currency, as it was taken as a "success story of the previous administration." The current administration wants to construct "its own success stories." As is always the case in Brazil's political culture, electoral motives take priority over policy continuity. Besides political factors, "there is also confusion and administrative incompetence," pointed out Michel Lotrowska, a workshop participant and an economist working for Doctors Without Borders' research program on neglected diseases in Rio de Janeiro. Given new budgets and bureaucracies, for the first time in 2005 there were shortages of ARVs in the health care system, Lotrowska stated.

"The preparedness that was in place is being compromised," Dr. Teixeira added. "We are lagging in technology." The ARV reverse engineering program at Farmanguinhos (the state's main laboratory) has been partially dismantled and generic drug development is not keeping pace with the market. Lotrowska gave the example of Tenefovir, an important rescue drug:

> Brazil is one of the only emerging markets in which companies make money with ARVs. So they isolated Brazil in terms of pricing. It is a very expensive drug, it takes a lot of the AIDS budget, and there is nothing to replace it. India never got interested in producing it, and Brazil did not think prospectively. The government cannot issue a compulsory license for it. . . . Things are disorganized, and people at various levels of government are fighting each other. The country's machinery of AIDS drugs development is stalled. Of course, all this is good for big pharma.

Brazil is now experiencing what other countries treating AIDS will soon face. It has a very inexpensive first line of ARVs, but a growing number of people are going into new drug regimes (either because earlier combinations did not work or because patients and doctors are demanding access to more sophisticated drugs, with fewer side effects) that are entering the market. With patients taking advantage of new treatments Brazil's ARV budget has doubled to nearly $500 million in 2005. In spite of the country's generic production, some 80 percent of the medication in the budget is patented. "We are moving toward absolute drug monopoly," concluded Lotrowska:

At Doctors Without Borders, we estimated that in a few years, the price of antiretroviral therapy will increase significantly. Given patent restrictions and all kinds of bilateral agreements in place, we have less and less generic competition: we have to find a mechanism that can lead to price reduction without this competition. Without such a mechanism, medics will soon have to tell patients, "I will only give you first-line treatment, and if you get drug resistance, then you will die for I don't have the means to pay an expensive treatment as Brazil is doing now." And we don't know for how long Brazil can sustain this budget. Doctors Without Borders treats some 20,000 AIDS patients. Even though the price of Tenefovir has been lowered to $400 in Africa, we would not be able to afford putting these patients on it.

Dr. Teixeira is an insider to these emergent forms of transnational (pharmaceutical) governance. Alongside Dr. Jim Yong Kim, a co-founder of Partners In Health, he helped coordinate the joint WHO and UNAIDS "3 by 5" campaign, aimed at providing ARVs to 3 million people living with HIV/AIDS in low- and middle-income countries by the end of 2005. Announced in December 2003, this initiative began with a good deal of momentum, and the Indian generic triple fixed dose capsule (FDC) provided its pharmaceutical backbone. Supporters of the initiative hailed "3 by 5" as opening up "a new era in international public health" (WHO 2005, p. 5),[5] while critics claimed that it was irresponsibly unrealistic, with "little vision, no credible framework, and insufficient technical assistance."[6]

In June 2005, the WHO reported that approximately 1 million people were on ARVs in low- and middle-income countries, in contrast to 400,000 in December 2003. Dr. Kim reflected on falling short of the desired "3 by 5" target: "We didn't do enough, and we began to deal with the problem too late." Yet, "before '3 by 5' there was no emphasis on saving lives," he said. "Many world leaders thought that we had to forget this generation of HIV-infected people and to think only about the next generation. We did something to change this."[7] Indeed, increased availability of ARVs averted an estimated 250,000 to 350,000 premature deaths in the developing world in 2005 alone. Several countries have taken up the challenge of treating AIDS, and the international debate has now shifted to how this can be most effectively done in contexts of limited resources.[8]

Yet funding bottlenecks, personnel shortages, and continuing debates on drug pricing and patents have limited this and many other AIDS initiatives. As Dr. Teixeira put it, "In the name of their own interests, private foundations, rich governments, and pharmaceutical companies keep putting all kinds of obstacles to a more rapid scale-up of AIDS treatments. Interventions of the pharmaceutical companies are paralyzing the WHO." Lotrowska vehemently criticized the "shameful" ways in which large-scale drug distribution programs have rejected generics: "The WHO has a mandate to pre-qualify the medication that countries benefiting from the Global Fund can use. It is supposed to do the work that poor countries cannot do by themselves. But there is now a long list of medications waiting to be qualified, and no adequate measures are being taken to expedite this process."

In the meantime, as I have argued throughout this book, a pharmaceutically centered model of public health is being consolidated worldwide, and medicines have become increasingly equated with health care for afflicted populations. As with other disease entities, pharmaceutical companies have operated astutely within legal and regulatory windows of opportunity in the case of AIDS, redirecting activist and political gains to their own advantage—be it as public relations gains through corporate philanthropy, as financial profits from global treatment projects, or as market expansion via developing states that have made AIDS "the country's disease" (as it is with Brazil, now a captive buyer of ARVs).

Consider Roche's recently introduced drug, T-20 (Fuzeon). This drug is the first of a new class of drugs—called fusion inhibitors, which keep HIV particles from fusing with lymphocytes—that will undoubtedly have great impact in preventing or coping with drug resistance. In Brazil, some 1,200 patients were prescribed T-20 immediately after the drug's debut, with a yearly cost of $20,000 per patient. "When the starting price of a drug is as T-20's, it is evident that after some time you will get a 30 to 50 percent price reduction," Lotrowska told me. "But even with this reduction, what will happen to the country's AIDS budget when thousands more will need it or want it?"

While back in Salvador for a final research trip in June 2005, I learned that Roche was training local infectious disease experts to make T-20 a first-line treatment rather than simply a rescue drug. This is a common practice, according to Bart Kroger, a Dutch medical researcher now living and working in Salvador. "These opinion-makers are extremely well paid, and they present the drug and treatment options in local congresses," he said, astounded by the translocal state of medical science and ethics. "The specialists take on a 'neutral' position, generally presenting positive aspects of the drug in question but also criticizing less important aspects of the drug. They don't want to sound as if they had been bought by the company. This is important for them not to lose credibility among peers and also to keep open the possibility of working for other companies in the future."

I also heard of cases where doctors began prescribing the rescue drug Kaletra at the time of its 2002 launch in the United States, before its registration in Brazil. These doctors referred patients to a local nongovernmental organization (NGO) and to successful public-interest lawyers, forcing the state to provide medication not yet approved by ANVISA, the country's National Health Surveillance Agency. For better or worse, such developments compromise the sovereignty of the state in the fields of biological and pharmaceutical governance. In the face of pervasive pharmaceutical marketing enmeshed with patient mobilization, regulatory incoherence thrives. And these local medical sovereigns are now also market operators. They mediate the introduction of new drugs in the public health care system and, in the name of adherence and concern over drug resistance, triage away patients who could benefit from the system's caregiving capacity, dismal as it is.

Meanwhile, policy makers have to ceaselessly invent new political strategies to keep the country's pharmaceutical policy in place. Brazil crossed a new threshold when for the first time it issued a compulsory license for an AIDS drug in May 2007. The government stopped price negotiations with Merck over Efavirenz which is used by 75,000 Brazilians, and decided to import a generic version from India. Officials claim that this will save the country some $236.8 million by 2012. Activists praise this move as an important advance in the widening of access to the newest and most expensive therapies.

Pharmaceutical Philanthropy and Equity

In early October 2005, I talked to Dr. Jane Walker, the executive vice president of a large pharmaceutical corporation. For her, the Brazilian AIDS treatment program worked "not so much because of politics, but because of a good allocation of resources." As for treating AIDS in poorer regions, Dr. Walker insisted that "drug price is not the problem; the problem is infrastructure." Dr. Walker was now leading her corporation's efforts to "not just" bring antiretrovirals to women and children in hard-hit places in southern and West Africa, "but to build up local treatment capacity."

This medical care and research endeavor was carried out in partnership with global AIDS initiatives, local health care groups, and NGOs. For the executive, it seemed matter of fact that public-private partnerships did better infrastructural work than state institutions alone. This discourse of state replacement, I thought, added an activist and morally urgent spin to a central tenet of neoclassical economics: the idea of a self-regulating market. The challenge, Walker told me, "is to find treatment models that can be inexpensively scaled up. Every one of the estimated 40 million people living with HIV is a person. We must do something as a world. We must save every one of these lives. The solution is not medicine as we practice and as we know it. We must save every one of these lives."

Here, one saves lives by finding new technical tools and cost-effective means to deliver care: that is, medicines and testing kits. The civil and political violations that precede disease are apparently lost sight of in this pharmaceutical humanitarianism, and the economic injustices reflected in barely functioning health care systems are depoliticized (Epstein 2005a; Farmer 2003; Gruskin et al. 2005). In the end, governments function on the business side, merely purchasing and distributing medicines, while nurture—now a technological endeavor—is left to communities and patients.

The U.S. president's $15 billion Emergency Plan for AIDS Relief (PEPFAR) reflects this global pharmaceutical frame of assistance. Announced in early 2003, PEPFAR aims to bring therapy to 2 million people and to prevent 7 million new infections by 2008 in fifteen of the neediest countries in Africa and the Caribbean.[9] However, there is a catch: rather than subscribing to the WHO's drug-approval process, PEPFAR requires separate FDA approval. Officials claim that this is to protect the safety and quality of drugs. But critics have accused the Bush administration of delays and of actually reserving money for expensive brand-name drugs, thus reducing the number of potential recipients.[10] Defying these and other criticisms, in May 2004 PEPFAR began buying generics, and in July 2006 the FDA approved a generic 3-in-1 combination ARV drug made by the Indian manufacturer Aurobindo Pharma. According to Dr. Mark R. Dybul, acting U.S. global AIDS coordinator, it is unclear if the generic drug will significantly cut costs, but by requiring patients to only take one pill two times a day the combination drug "should facilitate better therapies and better adherence" (*New York Times*, July 6, 2006).[11]

The methodological designs of AIDS treatment programs (pilot and otherwise), as well as the models they employ, have to be scrutinized and politicized. PEPFAR, for example, has an expeditionary quality, implemented from without, and is designed to save lives. It favors large-scale drug distribution but does not adequately address the issue of public health care infrastructure improvements, or, for that matter, prophylaxis and treatment of opportunistic diseases. It is very much in line with previously developed programs such as "the excellent program Merck has in Botswana," in Dr. Walker's words.

The Botswanan program, the African Comprehensive HIV/AIDS Partnership (ACHAP), was created in 2000 as a joint venture of Merck, the Bill and Melinda Gates Foundation, and the government of Botswana, with a financial commitment of $100 million—half each from Merck and Gates—for HIV/AIDS prevention and care (Ramiah and Reich 2005; Buerki 2005). ACHAP was committed to building institutional capacity within the government; one of its major activities was a national treatment program in communication with global networks. Planned by an international management consulting firm, the program was not initially based on knowledge of the local pathways of the disease and did not draw from lay resources. Rather, it was based on clear-cut calculations—of the number of potential ARV doses and of the availability of clinical personnel to dispense them and administer follow-up CD4 tests.

However, even this ostensibly scientific reliance on numbers does not necessarily evade politics and deceit. My analysis of the epidemiological surveillance system in Bahia (chapter 3), for example, revealed striking gaps in the government's ability to track the trajectory of AIDS. This casts doubt on the reliability of data produced to evaluate the impact of programs. In Brazil and elsewhere, such distortion of the epidemiological reality is not new, nor is it specific to AIDS. For example, a recent report in *The Lancet* suggests that the World Bank has overwhelmingly exaggerated the impact of its anti-malarial efforts in both Brazil and India (Attaran et al. 2006). Numbers and statistics are intensely political.

Regardless of the accuracy of the Botswanan numbers, the fact is that following the program's inception, both attention and funding were diverted from prevention efforts, and rates of HIV prevalence remain among the highest in the world, estimated—for what it is worth—at 24 percent among the general adult population. With the routinization of testing for HIV/AIDS and the growing awareness of the positive effects of ARVs, the demand for treatment rose sharply. Amid rising costs and limited financing and capacity, a "need" emerged to curtail further growth, in order to continue treatment of those patients already under care. As AIDS became more chronic than fatal (for some), ARVs remained the axis of program adjustment. The partnership will run through 2009, and the uncertainty over sustainability persists.

Global ARV rollouts rightly open the door to drug access, but they also exemplify the inadequacies of a magic bullet approach to health care. Drugs are ancillary to the full treatment of the disease. Alone, neither money nor drugs nor even a sophisticated pilot model guarantee success. Healing, after all, is a multifaceted concept, and "healing" is no more synonymous with "treatment" than "treatment" is with "drugs." Statistical strategies and corporate profit motives hover above, by and large missing the interpersonal networks that link patients, doctors, and governments, which are especially important in resource-poor settings, where clinical infrastructures are not improving. This displacement of the local from the planning framework leaves unaddressed the clinical continuity necessary for successful AIDS treatment. As a result, extremely well-endowed efforts—facing the humanitarian paradox of "life-saving drugs versus caregiving infrastructure"—are by and large falling short of the mark, without effecting the changes hoped for.

As Susan Reynolds Whyte and colleagues note in the context of ARV access in Uganda: "In principle, affordable treatment will change the meaning of AIDS (and of life!). . . . But the process is a rough and inequitable one. As drugs for AIDS become more common, they expose the nature of healthcare . . . its dynamism, its unevenness, and the order in its disorder" (2006, p. 241). With treatments available and structural violence ongoing, politically motivated and deceiving discourses have surfaced to rationalize in a perverse fashion the survival dilemmas the most vulnerable now face in the absence of improved living conditions—poor HIV-positive mothers in sub-Saharan Africa, for example, who used medicines to prevent vertical transmission are now left "to choose" to breastfeed their newborns (as the least lethal option) because lack of clean water makes formula feeding a riskier practice. Pre-modern and modern ways to access resources and convert risk into life possibilities routinely overlap to redistribute technology and care unequally.

The work of Paul Farmer and Partners In Health provides an opposing community-based model for AIDS treatment. The HIV Equity Initiative in Haiti does not operate like a traditional NGO, that is, removed from people. A pragmatic solidarity with the poor is its starting point. It uses the local clinic as the nexus of care within integrated prevention activities and ARV administration (Walton et al. 2004). "Improving clinical services can improve the quality of prevention efforts, boost staff morale, and reduce AIDS-related stigma," writes Farmer (2005, p. 1). In this holistic approach, accounting for individual trajectories and staying with patients through the progression of the disease (the work of *accompagnateurs*) is considered as important as tackling the economic and social factors that impact their families and mitigating the decay of clinical infrastructures.

Yet, funding for this and other community-based projects has been scarce. With the exception of a $100,000 start-up grant from a foundation based in Haiti, this initiative has never received support from a major foundation or international body charged with responding to AIDS, writes Farmer (2005, p. 22). Instead of mathematical modeling, a "biosocial framework" of analysis is needed. This experientially grounded and social justice–minded approach could help to redirect transnational funding and thus "remediate the obscene disparities of risk and access that characterize the global AIDS pandemic" (p. 23).

While Farmer's project is still by no means accepted as a gold standard, its presence has created "dents" in the prevailing rationalities that guide the treatment of AIDS in resource-poor settings. In challenging the view that comprehensive care of this sort is unsustainable, the project has gained a kind of iconic function/value, expressing unforeseen possibilities and articulating a new human rights imperative. However, its expansion also begets an array of questions concerning the ethical grounds for prioritizing AIDS over other afflictions of poverty (malaria or diarrhea, for example—see Das 2006), as well as political questions regarding its operationalization and sustainability over time. The WHO's difficulties in pushing forward with the "3 by 5" campaign leave no doubt that even the noblest of efforts must be politicized and understood in relation to the strategies of both national governments and global initiatives. Nonetheless, Partners In Health has opened up new spaces and redefined the perceived boundaries of feasibility.

Where Is the State?

An ethnographic analysis of the Brazilian AIDS policy shows how empowering pharmaceutical access can be, but also how much additional effort is required to transform drugs that are "accessible" into drugs that are both present and effective in the everyday lives of poverty-stricken patients. Access is an essential beginning, but it means neither "delivery" nor "adherence." While acknowledging Brazil's success in treating AIDS, throughout this book I have also charted the complexities involved in large-scale ARV rollout, depicting the institutions and the social and cultural phenomena that emerged on the ground in response to the novel concept of accessibility.

Although drug distribution has been the focus of the national AIDS program, caregiving has become the responsibility of regional and municipal governments, as well as community initiatives. "AIDS remediation is about pact-making between various levels of government and society," Dr. Teixeira told me, referring to the state's overall decentralization policy and the Health Ministry's guidelines that assigned provinces and municipalities specific responsibilities in drug assistance.

HIV/AIDS prevention and assistance have been progressively but inconsistently incorporated in the country's ailing universal health care system (Sistema Único de Saúde, or SUS). In reality, administrative discontinuities abound. Not even the national AIDS and TB programs work in tandem. Different provinces allocate resources differently according to the pressure of interest groups. Overall, local governments basically continue to operate as providers of complex medical services, and poor citizens must either wait indeterminately in line to access these services or find means to buy them in the medical market.

At the Global Health Governance Workshop, researchers working in the city of São Paulo showed that the quality of AIDS care varies considerably across adjacent towns. Divergences in administrative capability and civic engagement occasion different levels of policy implementation. "The intensities and fragilities of local medical contexts have to be attended to. Governments have not found ways to work the data we produce on the barriers patients face to access care," these politically frustrated researchers told us. "Intelligence remains centralized." That is, national and regional registries and the medical establishment have created their own self-reinforcing truths.

Dr. Teixeira had harsh words for what he saw as "typical public health research only focusing on administrative deficiencies" and for a retrograde desire for "normativity" that punctuated the studies presented at the workshop: "SUS is a political principle. Stop thinking that there must be a model, a design, something normative coming from the federal government. SUS is a principle of right and responsibility. The rest are pacts between government and society. . . . Each place is a different situation. We must evaluate things from the patient's point of view, from his lab reports. Then we will see the real impact of the policy."

One of the public health experts then stated, "Yes, SUS is a principle and social construction. Yet it is also a concrete health care system. And if we presuppose 5,000 systems [the approximate number of Brazilian municipalities], we can no longer analyze anything. . . . Then when medication is lacking, for example, we can blame it on the town's isolation or corrupt mayor. We will never be able to say whether a policy is good or bad." Dr. Teixeira ironically replied, "You must analyze what is happening with the epidemic and with the patients. If AIDS mortality is falling, patients are living longer and better, why bother if there is a problem in the little town of 'xyz'?"

Dr. Teixeira was rightly emphasizing the dynamic character of the AIDS policy—"we need to put things in temporal perspective and assess, step by step, what is working and what is not"—but in working with patients, one also sees the policy's limitations, both in its technologically centered design and in its implementation. The AIDS policy relies on organizations that are both concerned and indifferent to the full scope of the epidemic, and its impact is indeed socially differentiated. On the ground, the policy reproduces the fault lines of race and poverty.

A recent survey on mortality in the state of São Paulo, for example, revealed that AIDS is two times more fatal among black patients than it is among white patients. According to researcher Luís Eduardo Batista, "The majority of Blacks have less formal education, lower income and live in the peripheries" (in Brazil, the census uses terms such as *Black*, *White*, and *Brown*).[12] On average, a white person in São Paulo earns almost double what a black person earns. From Batista's perspective, "racism impacts health" because blacks receive substandard care and go unaddressed in prevention campaigns. The violence of daily life is reinforced in this case by interlocking and discriminatory organizational contexts that overdetermine AIDS as a medical failure.

Moreover, the NGOs that were supposed to have taken over assistance "have long lost idealism and passion. They keep selecting their clientele and find all kinds of ways to pretend that they do the work of their projects," activist Gerson Winkler told me during our last meeting in September 2005. "It's all a make-believe game: I pretend that I do a project, you pretend that you see it, the government pretends that it is monitoring and evaluating the project. And we all pretend that the results are positive." A national conference of AIDS NGOs was taking place in the city of Curitiba later that month. "The agenda is quite progressive: drug patents and vaccine development. But when one brings up the question of prevention and assistance, the major axis of the epidemic, then the AIDS movement seems to be at a loss."

For Winkler, it was the word *chaos* rather than *control* that better described AIDS on the ground. "AIDS incidence keeps growing among the miserable. For them AIDS is one tragedy among many others. And there is very little understanding of what in fact they have to do in order to adhere to treatments. HIV vertical transmission has not been zeroed—and it could have been. . . . If you address the chaos you will have to conclude that programs are actually not working, and this is too costly."

•••

We met at a café in Porto Alegre. Winkler had introduced me to the world of AIDS in Brazil in 1992 (see chapter 2). He had survived AIDS for almost two decades now, and I was happy to see that he was doing well, as critical and politically minded as ever. The previous day he had had silicon implanted in his cheeks to counter the disfiguring effects of ARVs. He was proud to say that he and Junior Batista, his partner, had recently adopted Wesllen, a five-year-old HIV-positive orphan:

When Wesllen was two years old his mother exchanged him for drugs. The police found him during a raid. He had become a commodity in drug traffic. He was placed in a child welfare institution, where he witnessed and suffered all kinds of abuse. He recalls a lot. When we got him, he spoke very little. He is in school now, a very creative and smart child . . . loves to swim and play soccer, but has many sequelae of maternal abandonment. Institutional marks are not easily overcome. We must set limits all the time, and he has a tough time with affection. A psychiatrist is helping us out. We love the kid. He will have to find ways to deal with his history . . . at least he has a chance to live now.

Both Junior and Wesllen are black, and the couple decided that Junior would be the "father" and Winkler the "uncle" in the new family unit. "The other day Wesllen said that his mother was dead . . . but she is in prison. . . . He also asked me if I was gay. I was caught off guard. We think it is too early for him to have this information and to learn of AIDS. It is not easy. In the morning we all sit the table and take our set of antiretroviral drugs. He has never asked why we do that. He clearly understands that we have something in common through these drugs and that this something is veiled."

Winkler was also a recent cancer survivor: "I had a cancer last year. It is under control now. But every six months, before checkup, I panic. To have a cancer in my life is worse than having HIV. Cancer makes the idea of death imminent again, and I now have much more than my own life to lose."

Always attuned to the present configurations of AIDS, Winkler was now running an income generation project for sixty people living with HIV. With the defeat of the Workers' Party in the 2002 regional election, Winkler lost his job at a governmental witness protection program. AIDS activist organizations did not welcome him back. "You know that your ideas have been surpassed when nobody calls you any longer for meetings to discuss the destiny of the world," he jokingly remarked. Using his personal savings, he set up a bakery at home, applying for funding from UNESCO and the Health Ministry to help people to enter the bread and pastries market:

I have been framed as standing for the *hay gobierno soy contra* ["if there is a government I am against it"] position, but in fact I am interested in bringing the state back into the picture. AIDS is a big market now. The composition of NGOs, government, and international agencies has become an intermediary thing, a kind of parallel state, with the usual corruption, conflicts of interest, and lack of self-reflexivity over projects and trajectories. Projects are generally short-lived, and there is no systematic auditing going on. Moreover, this hyped image of Brazil being the best AIDS program in the world validates people's self-interested sense that what they are doing is actually good. It is certainly not solidarity that fuels NGOs' activities today.

State institutions of care have continued to deteriorate in the Lula administration, and social policies remain "disconnected and charity-based," Winkler fiercely pointed out. "Any kind of work or results coming from NGOs, as limited as they are, are welcomed by the government, for they are better than nothing. In reality, there are too many people searching for help in vain. There are too few patient home visitation programs. Their everyday conditions of survival have to be composed and understood—this is fundamental for adherence programs." Medical technologies have come to stand in for the non-work of state institutions: "When one goes and visits patients they ask for help as if one were the state, but where can one objectively refer them to? I am not the state. I am an NGO. Where is the state?"

A 2005 study suggests that the poor have become increasingly excluded from the Brazilian job market over the past decade. The average income of the poorest 10 percent has declined 39.6 percent, and their dependence on governmental assistance programs has swelled. In 1995, 89 percent of this population's income came from formal and informal sector work. By 2002, the share of income due to work had fallen to 48 percent; more than half of their income now came from government assistance.[13]

Winkler spoke at length about the difficulties he faced in helping poor AIDS patients develop a work mentality and in placing them in the market. "AIDS stigma is a constant reality. Poor patients are highly dependent on NGO or church-based assistance. Most of them come from disaggregated families and have never been employed or worked before. The logic of the job market never existed for them. During our training program they learned skills and to deal with money. There is agency here. But as all projects, we are only funded for one year. There will be no continuity, and they will, most likely, fall back into scavenging for assistance."

Active—not just activist—state institutions must come into play, argues Winkler: "The state cannot reduce itself to allocating money for projects. We trained these people and helped them to seriously consider treatment adherence. But only four out of sixty participants found jobs. Government-sponsored programs for job placement are needed. It's a farce to think that NGOs can be the executor of state services. Our task is not to put medicines on the shelves or place people in the market."

The most important thing AIDS activism can do "is to create space," concludes Winkler, "not a technical but a civil space, a space for people to occupy and from which to engage in politics."

A Vanishing Civil Society

Throughout this book, I have considered the multiple realities that make patients' attachments to the AIDS policy and to a new life substantially uneven. Ethnography addresses the world emerging in between policy and indicators. It maps the fractured and intense ways in which this particular pharmaceutical initiative takes institutional grip and personal adherence. Such a cartographic account of local trajectories and intensities unsettles the technological determinism that pervades the discourses making the AIDS policy iconic. It also exposes how politics has changed in Brazil's neoliberal turn.

The idea of a "state-society interface" does not have much currency these days. One could argue that the 1990s discourse of a mobilized civil society was by and large a reaction of certain sectors to a neoliberalizing government. The Cardoso administration, eager to reduce and circumscribe the scope of its operations, willingly considered the demands of organized groups. As the government increasingly outsourced assistance to nongovernmental and philanthropic organizations, it reconceptualized itself as an activist state.

Today, there is a clear dispute over who represents the people, as Michel Lotrowska told me in June 2005: "Workers' Party politicians and the government both understand themselves as being the people. I don't see civil society welcome in this administration. It is welcomed by specific programs, such as the AIDS one." Lotrowska mentioned that the national AIDS program had invited him and several other NGO representatives to Brasília in order to discuss possible wordings for a future ARV compulsory license: "The lawyers from the Health Ministry reacted very strongly to our presence. They said that they couldn't discuss politically sensitive issues in front of us." Moreover, to speak of an interface between state and society assumes that both are coherent; in reality, neither are.

When I began this work in the mid-1990s, there was much debate over the co-optation of AIDS activism (read: civil society) by the state. Now, however, activists and the anthropologist are left to reckon with a different paradigm: that of a *vanishing civil society*. Politics has been increasingly individualized—a matter of survival. Against the background of budgetary constraints and the industrialization of the nongovernmental sector, a multitude of interpersonal networks and variations in AIDS care have emerged locally, creating different levels of quality of life for the patient—the underside of the pharmaceuticalization of public health.

To bridge the gap between the potential of pharmaceuticals, governmental distance, and the lack of care planning among the poorest, certain sectors of Brazilian society constructed *casas de apoio* (houses of support)—places like Caasah. From an abandoned building, governmental and philanthropic support, and a few individuals' uncommon will to live, in a world where therapies were promised but not initially present, Caasah arose amid misery to ensure and to orchestrate the fulfillment of that promise. From the perspective of an outsourcing state, such grassroots entities are important, if transient, social instruments of remediation. Yet for AIDS patients, they become proxy-families and communities, providing the ground necessary to make ARVs work. Patients are active in the pursuit of health, although few are activists.

Caasah's institutionalization legitimized and collectivized the voices of its poor residents, encapsulating them in the type of environment necessary to promote treatment adherence and consequent renewal of life chances. Here, people did not have to worry about the stigma that came with having AIDS "on the outside," and a daily routine made it easier, at least for some, to follow drug regimens. An intense process of individuation—"salvation from my previous life," as some put it—and a spirit of competition with fellow residents motivated adherence as well. The restructuring of the patient's local world was an integral part of this process, a part that

FAILED LOVE

Broken Heart. Great Art.

You might not be looking for a relationship right now, but everyone can commit to chocolate, cynicism, and art. The Art Museum's Student Advisory Board invites you to mend your broken heart with our great art.

Drown your pain in dark and twisty refreshments.

February 11, 2010, 7 to 10 p.m.
Art Museum Galleries

PRINCETON UNIVERSITY
ART MUSEUM

present large-scale interventions such as PEPFAR miss altogether and that national measures of cost-effectiveness (trends in drugs dispensation and overall mortality) alone do not capture.

•••

I collected the life stories of twenty-two residents during my 1996–97 work in Caasah. When I returned in December 2001, I learned that twelve of them were dead. Among the dead were Edileusa, Lazaro, Marilda, Jorge Leal, Alcida, and Valmir—all died in the "old Caasah," that is, before the institution moved to its new state-built headquarters and transformed itself into a house of passage and AIDS orphanage. Their lives had been marked by poverty, family abandonment, drug abuse, homelessness, and uncertainties of all kinds. Four of the six women worked as prostitutes and two worked as maids. Marilda and Alcida were socially recognized as "mentally retarded." Two of the six men were transvestites, three were intravenous drug users, and one worked as a seasonal truck driver. They all learned of AIDS after falling sick. After repeated hospitalizations, they were sent to Caasah.

Nerivaldo and Maria Madalena did not want to submit to Caasah's norms and were asked to leave the institution before the transition. They died as indigents in the makeshift hospice of Mãe Preta. Soraia and Altemar began ARV treatment in 1997 and left after improving their health. They were never again seen in any medical service and are now also presumed to be dead. Nadia and Jorge Ramos initially did well on ARVs, but both died a few months after leaving Caasah.

Celeste, Caasah's president, gave me two explanations for these patients' fates: either they were in "too advanced a stage of disease" to benefit from therapies, or they had failed to adhere to therapies because of "their culture and ignorance." In this administrative account, culture is *méconnaissance* (misunderstanding). It is equated with the patient's lack of good judgment as far as health is concerned. Many of the AIDS patients I met in the streets of Salvador and in Caasah indeed trivialized AIDS and its treatment—"I will die anyway." But if they did so, I thought, it was because AIDS was just one among the many tragedies they faced; on their own, they could only do so much to deal with it.

Moreover, residents like Nerivaldo and Maria Madalena strongly reacted against the moral imperatives of being a certain kind of "good person" in order to be attended to as an "AIDS patient." They knew that *marginais* like themselves were unwanted there, so they decided to return to the streets. They died refusing an AIDS identity, without a life project of their own. Nadia and Jorge Ramos, however, found partners at Caasah, and this, according to Celeste, helped them as they moved back into the world. "They had someone to live for. But as soon as they regained a body, they relapsed and stopped taking medication. *They didn't continue Caasah's work.*" It was up to each person to become the institution of care.

The ten residents who survived the four-and-a-half-year gap between my visits to Caasah were Tiquinho, Edimilson, Evilásio, Marta Damião, Rose, Dulce, Luis, Evangivaldo, Valquirene, and Fátima. Before falling sick and moving into Caasah, six of these patients had participated in the informal economy in some way (as a maid, a waiter, or a carpenter, for example), and three of them had been in stable marriages. Two women and one man had found themselves involved in sex work and the drug trade. By charting the trajectories of these AIDS survivors—those who lived pre- and post-ARV rollout—we can identify some of the elements that, along with medical technology, trigger and sustain a patient's return to a common being, to a normality of sorts: that is, the everyday mechanisms that, revolving, make AIDS a chronic condition.

Tiquinho, the child who grew up in Caasah, was the only one allowed to live in the new facility. The institution had helped Marta Damião, Edimilson, Fátima, and Rose get shacks from the government and was providing rent aid to Evangivaldo, Evilásio, and Luis, who was employed at Caasah. Dulce had moved in with her sister. All of the adult survivors created new family units. They lived with other AIDS patients, had children, or reunited with estranged relatives. Valquirene had moved with her new partner to the interior. All of them had disability pensions (today, with the availability of ARVs, pensions are rarely granted to newly diagnosed patients) and were entitled to a monthly food basket at Caasah.

Understanding the Nexus of AIDS, Poverty, and Politics

I returned to Salvador in June 2005 for final fieldwork, before heading to the Global Health Governance Workshop in São Paulo. "If you look carefully, nothing has changed. Things are the same as you saw last time," a tired and doubtful Celeste told me. She was struggling with vision loss in her left eye (from a viral infection acquired at Caasah) and seemed quite depressed. Celeste no longer spoke of herself and Caasah as agents of change. To a certain extent she had taken up a reactionary rhetoric that Caasah's work was futile, yet she insisted, "I can't give up this work."

Caasah was still the only place in Salvador that provided systematic care to poor AIDS patients who had been discharged from public hospitals. "They recover here, but medication for opportunistic diseases is difficult for us to get. Some patients return to their families. Others go back to the streets. I would say that half of people living in the streets are HIV-infected. The situation remains the same: disease keeps spreading, and the government pretends not to know of it, so that it does not have to intervene."

Celeste was outraged with the recent news that, in 2004, the Ministry of Health had allocated $1.2 million to the state of Bahia; $350,000 went unused and had to be returned. And the money that had been spent, Celeste added, had basically gone to Bahia's new state-of-the-art AIDS outpatient care and research center: "Doctors and administrators are always in international conferences. They publish glossy reports that have nothing to do with what is actually happening. Where is the evidence of their work with the poor patient? Whenever they organize an event, they call us and ask us to bring our children. It is all for photos. They don't have the AIDS public; Caasah has it."

At the Luis Souto AIDS unit, Dr. Nanci, who cared for several of Caasah's survivors, also told me that "things here have not changed." The new AIDS center was an important ambulatory alternative, but it was too centralized and limited in reach. "The reality of our unit is the same as it was at the beginning of the epidemic: full of pauper and wasted patients. The difference is that they now come from the interior, where no new services have been created." She also regretted the absence of a Caasah-like institution to help the service with patient discharge and adherence: "We know what is needed, what could be done, but we have no political force. There is so much corruption and money diversion on top. Access to therapies has been democratized, but health has not."

Caasah still received some aid from the Health Ministry and the city's Health Division, but basically, "It is society that sustains us." Most of

Caasah's maintenance money now came from partnerships with the private sector and from community donations, Celeste told me: "We now have a call-center, with eighteen volunteers constantly calling people." Becoming a member of the Municipal Health Council has helped Celeste to better understand "how all this decentralization works, how the mafia operates. You wouldn't imagine all the bureaucratic hindrances we have to overcome to qualify for projects. At least now I can put direct pressure on the health secretary." She struggled for institutional survival.

The fact is that, with the plateau that epidemiologists projected long ago still unattained, AIDS continues to grow amid poverty and marginality in all regions of the country. Officials have attributed the absence of a plateau to improvements in the notification system, but with few politically unattached analysts, such claims cannot be taken at face value. Regardless, official statistics report that more than 610,000 people are currently infected with HIV; 200,000 people are being treated; and about 11,000 people die of AIDS per year. Given the epidemiological service's sluggishness in keeping up with the epidemic, the figures on morbidity and mortality—which likely represent lower bounds—are alarming. Beyond national prevention campaigns and the public demystification of the disease, there are no coordinated programs to identify HIV infection among highly vulnerable groups. What is more, the Ministry of Health is actually operating under the assumption that the "epidemic is under control."

In November 2005, the Health Ministry issued a technical brief stating that the government was not going to issue a compulsory license for the drug Kaletra—the country's "low prevalence and control of HIV infection" did not amount to a "national emergency."[14] NGOs and the National Health Council had demanded this patent break in order to cut growing treatment costs and ensure the sustainability of the ARV policy. The Ministry's note contradicted the country's most recent epidemiological bulletin, which affirmed the growth of the epidemic, particularly among women and blacks, as well as the overall high seroprevalence at the national level—18.4 cases per 100,000 people, up from 15.7 in 2001.

This "official trivialization of AIDS" is part and parcel of an electoral-minded politics of disease. In light of the forthcoming presidential election, the Lula government was choosing to address other infectious diseases that affect larger segments of the population, such as dengue.[15] A new methodology was helping the government to identify the regions at higher risk of a dengue epidemic and devise punctual prevention campaigns rapidly. Moreover, estimates suggest that some 4 million people were affected by dengue in 2002, during the presidential race between Lula and José Serra, the former minister of health. This time Lula's main adversary was Geraldo Alckmin, also from PSDB (Partido da Social Democracia Brasileira), the party previously in power. It does not help the AIDS cause that Serra and former president Cardoso have portrayed the AIDS policy as evidence of their activist and technologically sophisticated state.

Indeed, rhetoric can only go so far in maintaining low HIV infection rates. Real life has to be put onto the map of the AIDS policy. This requires going to where people are and recounting their experiences and the daily world they face. Pauper patients are not the problem in themselves. With no political voice, they have been both disregarded and made invisible. This is not due to governmental inability or ignorance. Where there has been an active HIV search, testing, and care—in maternity wards, for example—infection has been curtailed. If this is ethically acceptable and technologically possible, why not reach out to other vulnerable groups and organize alternative forms of on-site testing, side-by-side with medical care?

To ensure quality care, policy makers would need to discuss interventions with particular vulnerable groups and make adequate medical information and technology available to them, along with sustained assistance. A deliberate engagement by AIDS NGOs in local politics might break open some new ground on this front.

Local Economies of Salvation

I asked Celeste for news about the patients I had followed over the years. Out of the initial group of twenty-two patients with whom I had worked in 1997, seven were still alive in 2005: Luis, Rose, Evangivaldo, Marta Damião, Dulce, Valquirene, and Fátima. How wonderful. This life extension is obviously a result of technological advancements, argued Celeste, "but it would not have happened if they had not learned to care for themselves." In the end, treatment adherence "is relative to each person. It requires a lot of will."

Subjectivity—a person's manufactured will to live—had become a fundamental cog in the ARV adherence machine. Yet, as I would soon learn, all of the former residents who were still alive also possessed a place they called home, a steady if meager income, and a social network. And, in case of an emergency, they could still resort to Caasah. This tie to Caasah, as momentary and uncertain as it now was, remained vital to them.

Consider the way Tiquinho died. The young hemophiliac man was healthy when he visited his family in the interior in June 2003, Celeste sadly recounted. "It was ignorance on the part of the family." Aside from the perfunctory "call us if there is a problem," Tiquinho had received no specific therapeutic plan, and the family (poor and illiterate) was unable to help him adequately. "They only called us when it was already too late. The town's ambulance brought him and we took him directly to the state hospital, where he died a few days later. It was a shock for us. This was the first time he had fallen ill."

Luis was still working at Caasah. He was in charge of the institution's fund-raising activities. "I am not concerned with HIV. What I want is to live. If there is medication, let's take life forward. Life is to fight for." In the previous year, Luis had experienced kidney failure and had been hospitalized for two weeks. "Work keeps my mind occupied and one needs to have projects and objectives to meet—if not, life has no meaning." Becoming a father, he said, "is the best thing that ever happened to me." Davi, his adopted son, was now a healthy seven year old—"He is a prankster. He is my passion. He makes it all worthwhile."

Luis told me that Edimilson, one of Caasah's founders, had died, living by himself and disgruntled with what he saw as the failure of the institution to provide steady help. He died desiring some form of institutional belonging, I thought, needing it to live. Evilásio "also died alone." Luis said that he had split with his girlfriend and had gone back to smoking crack. The week before, Luis had seen Marta Damião, "old and wasted" in a bus.

Professor Carlos remained in charge of Caasah's outpatient food basket program. He confirmed that Marta was having a difficult time staying away from alcohol. "Her children come and pick up her basket." He had also recently seen Valquirene in the streets, "lost in time and space, and very skinny." "After a while, she recognized me," said Carlos, "but refused to come with me to Caasah. She said that she had a mission to fulfill." Her health had begun to deteriorate after her husband's death from AIDS. "Her son was already under her mother's care, and she lost her reason to live," Carlos argued.

Dulce, however, was doing much better. "She converted, and now goes to a Pentecostal church. She sells beauty products door to door, and comes here every month to pick up her basket. She had a boyfriend but broke up with him. She lives with her sister and children and this greatly motivates her." Fátima was back to the Pelourinho. A former prostitute, she now lived with her adult daughters in a renovated apartment that the government had made available to her, living off her AIDS pension and her meager income from cleaning houses.

These AIDS survivors extend life through and beyond the clinic and Direct Observed Therapy (DOT). They live in flux. Moving between institutions, they look for usable resources. At every turn, they must consider the next step to be taken to guarantee survival. And this interstitial domain to which I refer as *local economies of salvation* is now a foreseeable reality for millions of poor AIDS patients worldwide (at least as far as first-line antiretroviral regimens are concerned).

●●●

"What joy you give me by coming back," beamed Evangivaldo, the kindest and most resilient man I have ever met. For him, human companionship—caring for the Other—is foundational to medical responsibility: "God gives me this force to live because of my daughter." To have someone to live for and to be desired by seemed to be a constant element in the account of the long-term AIDS patients with whom I worked.

"I don't have the aid of a father and a mother, and I can only count on the tenderness of Fátima and Juliana," Evangivaldo continued. "When I see them with no food, it makes me ill. But when I find a job or get a donation, and there is nothing lacking at home, and all is normal, then for me it is another life, and it is all good."

Poor AIDS patients like Evangivaldo also interact and trade with AIDS NGOs and civic groups that channel assistance, albeit minimal, from regional and national programs. The NGOs, which depend on their clientele to back up reports and authorize new projects (now mostly related to treatment adherence and income generation), become venues for some patients to access food, rent aid, and specialized medical consultations, among other things. Overwhelmed with assistance demands and concern for their own institutional survival, NGOs rarely succeed in placing the person in the market, but they do successfully differentiate politicized patients who defend their rights from those who passively circulate in the medical service system.

Only a few, like thirty-year-old Sonara, manage to become "AIDS workers." She was Caasah's new poster-person. Professor Carlos spoke highly of her: "She was a drug user and came here in bad shape. But she now takes the medication, eats well, and takes care of her daughter, who is also HIV-positive. She is our representative in the AIDS NGO Forum, gives lectures, sells jewelry, and is also a teacher here." He introduced me to Sonara as she was running a candle-making workshop for a group of twelve patients. She was the only white person there. Her style of dress, manners, and speech were characteristic of the Brazilian middle class. As much as I admired her transformation, I could not have been more disturbed by her moral reasoning: "Today, people only die of AIDS if they want to."

The AIDS survivors I interviewed acted coldly toward fellow patients. For many, I thought, health corresponded to a measure of moral uprightness. Mutual empathy was rare. I will never understand why, for example, Luis did not let us take Rose's food basket to her as we were heading back to her shack in the Cajazeiras District in early June 2005. The previous day, over the phone, Rose had asked me to do just that. She would avoid a long trip and transportation expenses, I told Luis. But my request met a series of obstacles, both external and internal: "The baskets are not ready. Professor Carlos is not here to release them. I don't have much time. I must be back no later than 11:00 a.m. We have to go."

Rose was euphoric to see us. She was doing great. I was particularly happy to learn that her daughter had turned HIV-negative. Ricardo, her fifteen-year-old-son, was helping two workers to finish the house's second floor: "It is my skyscraper. Water was infiltrating, and in the long run I plan to rent it out." She was disappointed that we had not brought her basket. I offered Rose a ride back to Caasah, but she said that she couldn't leave the construction unattended: "That's life. Each one is on her own."

Rose intelligently navigated the local circuits of AIDS care. She had garnered the support of other NGOs and opened up a little business she called "Rose tem de tudo" (Rose has it all) and had also devised a construction fundraising campaign among religious philanthropists. She was proud of having been able to enroll her son in project Teenage Citizen (Adolescente Cidadão), which Dona Conceição was running with World Bank funds.

•••

Later that week, I met with Dona Conceição. She had accomplished much and now headed IBCM (the Conceição Macedo Assistance Institute). With the help of a local sociologist, she had designed a project to employ 120 children of AIDS patients in local industries. "We were asked to go to Washington, D.C., to present it to the Bank, and we won. I almost fainted when they called my name. . . . After school, a bus picks the kids up at home and they spend the afternoon learning a trade." All teenagers receive $50 (then half of a monthly minimum wage) and food vouchers. In order to carry out the project, the Bank also helped Dona Conceição develop her own institution. She kept working with homeless and poor AIDS patients. "In the morning I am at IBCM, and in the afternoon I am in the streets."

Dona Conceição aided a total of two hundred families, she said: "Once a month, I also hold a general meeting for these AIDS patients to share experiences. I offer breakfast, and they get their food baskets." Rose and Evangivaldo told me that they also participated in these meetings. Dona Conceição regretted that she remained the only institution to address AIDS in the streets; her funds from the World Bank would only last a year: "We cannot meet all the demand for help. There are many, many, many with AIDS in the streets. I have never seen one single homeless patient who is tested getting an HIV-negative result. It's a disgrace! He leaves the hospital with medication and it is up to me to help. How in the world will they give up on drugs? It is the only way for them to escape the reality they are locked in."

Medics, professional activists, and model patients often fail to address the complexities of life with AIDS and pharmaceuticals among the poorest. During our last conversation, I asked Evangivaldo whether he had told his doctor all he has to go through in life. "Yes," he had once mentioned to his doctor that he routinely rode his bike for two hours "with only coffee and medication in the body" to get to downtown Salvador in search of job.

"Dr. Jackson said that he did not believe it. That my HIV was almost undetectable and that I acted as if I did not have AIDS. I told him that my bike was parked outside the hospital, that I would show it to him. He was amazed. He then called his superior and some residents and asked me to tell them my story."

After the spectacle Evangivaldo had become, "the doctors said that they were proud of me, and that if all HIV-positive people had the same will to live that I have then no one would have to be hospitalized. They said that I was an example for other patients."

Evangivaldo took the opportunity to ask the doctors for advice on where to go to actually find a job. To which Dr. Jackson replied: "I feel bad for not being able to help, but I am sure that God will show a path for you to get where you want to." Meanwhile, Evangivaldo had to take twelve pills a day, and his doctor never considered putting him on a newer medication already made available by the government (fewer pills and fewer side effects).

AIDS survivors are divided by the development of two distinct identities, one to successfully navigate the AIDS institutional landscape and one to guarantee social recognition and participation in neighborly exchanges. In the language of economics, informational asymmetries abound, and those on ARVs must balance their social and biological needs in deciding how to signal their disease status to the world surrounding them.

Rose, for example, hides her condition from neighbors—"to protect my children." Evangivaldo tells his landlord that he gets disability aid because he is "mentally ill." Luis, on the other side, says that "I wear my AIDS T-shirt all the time. Once I stopped discriminating against myself, I had no problem confronting the world." His neighbors know that he works at Caasah, and he has even brought some of them to tour the institution: "I wanted them to move beyond their prejudices, to see that AIDS affects not just *marginais*."

The Unexpected and the Possible

Luis, Rose, and Evangivaldo have come a long way from facing social and biological death in Caasah to now facing the immanence of life. Refusing to be overpowered, they plunged into new environments and became agents in other people's routes. They have by all standards exceeded their destinies. And throughout these medical, economic, and affective trajectories—a second nature of sorts—they began to actualize the dignity and the desires that had hitherto been virtual to them. Ethnography captures this human force that is capable of acquiring sufficient consistency for turning a situation around—call it a language of hope—and transforms it into a map of the present world: a broken world, full of rifts that deepen, yet also a world of previously unimaginable possibilities.

Anthropologists have long faced criticism for avoiding normativity and evading policy prescription; *Will to Live* may also be subjected to such charges. However, anthropology's task, as I have tried to show, is to produce different kinds of evidence. We are motivated by bold questions—how, for example, neoliberalism and globalization recast contemporary institutions and power arrangements. Yet, we approach them with a deep and dynamic sense of the local: those practices and knowledges people wage and articulate as they claim a chance at life.

This collaborative study has illuminated profound institutional and medical changes that have occurred in Brazil and in the field of AIDS.[16] There are times when political and market institutions cannot so easily resist demands for change. Social mobilization and novel rational-technical interventions have universalized AIDS treatment in Brazil. In charting the unfolding of this policy of biotechnology for the people, this book has also located specific points of entry for re-envisioning prevention and care as far the urban poor are concerned. One of the positive and perhaps unintended effects of therapy access has been to "denaturalize" unequal laws of reality and reveal them as amenable to human action. Therapy access reveals the urgency of improving people's basic living conditions. Moreover, damaging side effects should not be diverted to people themselves but should be guarded against by more and not less preventive policy making.

Public institutions are indeed co-functions of successful AIDS treatment. This calls for ongoing self-examination by those who implement policies to their own effects on events. It also involves a rethinking of how to reach the afflicted in their own terms, acknowledging self-destructiveness and human struggles for recognition in a largely hostile world. Likewise, at issue is a reconsideration of the systemic relation of pharmaceutical research, commerce, and public health care and a search for ways to break open the widespread societal deafness to those most vulnerable, people who remain unheard despite all they have to say.

Rather than specify a route, the anthropologist demarcates uncharted territories and tracks people moving through them. In the field, unexpected events happen all the time and new relations of causality are created. An openness to the unexpected and the deployment of categories that are important in human experience make science "more realistic, better," writes Albert O. Hirschman: "I like to understand how things happen, how change actually takes place" (1998, p. 67). And the map the anthropologist produces allows the navigator—the interpreter—to consider the territories explored and their life force. The interpreter's beliefs regarding justice, the reach of the state, and the capabilities and responsibilities of citizens inform how the ethnographic evidence will be mobilized. Like every human map, the evidence here is made of impasses, thresholds, breakthroughs, and enclosures on the ground—lives and social fields in transit. By evaluating these displacements, we get an actual understanding of the ways politics matter and also broaden our sense of what is possible and desirable.

•••

"Fátima had a stroke," Evangivaldo told me: "She has difficulty walking. She already fell twice trying to leave our shack. She does not want to stay still. But I tell her that the important thing is that she is alive, that I do not mind doing everything. She hurts inside because she cannot help. And she says that I am a father to her. I tell her that God knows the gift of everyone. Today I take care of you. Tomorrow it could be your turn to help me. The one who is strong now has to help the weaker. The important thing is to have life to care for our daughter and our home, to have a dignified life and to be healthy to see Juliana grow. That's what I say."

Acknowledgments

Many people in Brazil and in the United States have helped make this book possible. I am deeply indebted to Gerson Winkler for introducing me to the reality of AIDS in Brazil, as well as to the people of Caasah and to Dona Conceição and her homeless patients for letting me document their existence. Their day-to-day thinking has been the compass of my research. I hope that they and those who follow in their footsteps might have some use, however small, for this book in their work, lives, and projects.

I am forever thankful to Torben Eskerod for collaborating with me on this project and for his friendship. Torben's beautiful and haunting artwork captures the singularity of the lives that compose the book and lends a much-needed depth to social science.

I have also been most fortunate in having the wonderful help of Tom Vogl. His scrupulous, caring, and always illuminating reading and editing are wholeheartedly appreciated.

I began the study leading to this book in the mid-1990s while a graduate student in the Department of Anthropology of the University of California at Berkeley. I am grateful to Paul Rabinow for his advice and encouragement throughout the years. I also want to thank Nancy Scheper-Hughes, Lawrence Cohen, and Stefania Pandolfo for their teaching and support. A Chancellor's Fellowship helped me to complete, in 1999, the dissertation that this book expands. Milton Quintino and Jane Galvão at ISER (Instituto de Estudos da Religião), in Rio de Janeiro, supported the early stages of research. I am grateful to them and to Jessica Blatt for their invaluable aid and insight.

I also want to acknowledge the support of colleagues from the Instituto de Saúde Coletiva of UFBA (Universidade Federal da Bahia), where I was affiliated in 1996–97. I am greatly inspired by the work of Naomar de Almeida-Filho, Denise Coutinho, and Sérgio Cunha, and I thank them for their help. I am grateful to Ana Luzia Outeiro, Anamélia Franco, Vicente D. Moreira, Roselene de Alencar, and Felix Drexler for their assistance. Esdras Cabus Moreira has been extremely helpful at various stages of this study. Other institutions on whose help I have relied include GAPA Porto Alegre, the public hospitals and HIV/AIDS services of Salvador, and the Programa Nacional de DST e AIDS of the Brazilian Health Ministry. My gratitude goes to the AIDS activists, health professionals, and policy makers who shared their work and vision with me. In the interest of anonymity, the names of some institutions and people interviewed have been changed.

In doing the research from which this book draws, I had generous support from the Program on Global Security and Sustainability of the John D. and Catherine T. MacArthur Foundation, the Wenner-Gren Foundation, and the Brazilian Council of Technological and Scientific Development (CNPq). Torben Eskerod was supported by a research grant from the Erna and Victor Hasselblad Foundation. Princeton University, where I have been teaching since 2001, has been wonderfully supportive of my work. I completed this book while holding a Harold Willis Dodds Presidential University Preceptorship. Princeton's University Committee on Research in the Humanities and the Social Sciences has sponsored research and the book's production. I have also drawn on support from the Program in Latin American Studies.

It has been a great pleasure to be part of Princeton's Department of Anthropology. The department's commitment to excellence in research and teaching has been a sustaining force. I am deeply grateful to James Boon for his brilliant insights and support. I also want to thank Carol Greenhouse, John Borneman, Isabelle Clark-Deces, Alan Mann, Abdellah Hammoudi, Rena Lederman, Janet Monge, Mekhala Natavar, Lawrence Rosen, and Carolyn Rouse for helpful conversations. Carol Zanca, Mo Lin Yee, and Gabriela Drinovan make it all work in the best possible way—thank you.

Over the years, I have also greatly benefited from intellectual exchanges and the assistance of a superb group of undergraduate and graduate students. In particular, I am thankful to Leo C. Coleman for his editorial help. I also thank Elizabeth Courtney Crane-Sherman, Alexander Edmonds, Christopher Garces, William Garriott, Matthew Goldberg, Adrienne Gropper, Caroline Lee, Peter Locke, Kavita Misra, Rachel O. Okunibi, Michael Oldani, Sarah Pinto, Steven L. Porter, Eugene Raikhel, Amy Saltzman, Ari Samsky, Ian Whitmarsh, and Jessica Zuchowski. Students in my medical anthropology courses have engaged with the book's materials and I am thankful for their comments and insights.

At Princeton, I have also fruitfully interacted with Fernando Acosta-Rodriguez, Elizabeth Armstrong, Anne Case, Marcia Castro, Angus Deaton, Helen Epstein, Peter T. Johnson, Stanley Katz, Patricia Fernandez-Kelly, Evan Lieberman, Donald Light, Stephen Macedo, Adel Mahmoud, Christina Paxson, Rosalia Rivera, Burt Singer, Peter Singer, Kate Somers, Michael Stone, Trisha Thorme, Robert Wuthnow, Lisa Wynn, Deborah Yashar, and Viviana Zelizer. I am very grateful to Nancy Watterson for reading the manuscript and for editorial help.

The unique environment of the Institute for Advanced Study has significantly shaped my scholarship. In 2002–03, while a member of the School of Social Science, I had the privilege to meet regularly with Albert O. Hirschman, and this experience has expanded my horizons and reframed this book in terms of hope. I also thank Sarah Hirschman, the late Clifford Geertz, Joan W. Scott, Michael Waltzer, Eric Maskin, Adam Ashforth, Claude Rosental, Brenda Chaffin, and José Serra for very helpful conversations during that time. I brought this work to conclusion while a member of the School of Historical Studies during 2005–06. I am deeply thankful to Heinrich von Staden for his support and to the staff for all their help. Thank you also to Caroline Walker Bynum, Nicola Di Cosmo, Michael Braddick, Warwick Anderson, Marianne Constable, Jeffrey Prager, Wilhelm Schmidt-Biggemann, and Carol Gluck for stimulating conversations.

I am grateful to Arthur Kleinman, Kay Warren, and Paul Farmer for their insightful comments and for generously supporting this work. I also thank Richard Parker for his critical reading of the manuscript. I have received helpful comments on the basic ideas and on various chapters of the book from a wonderful group of scholars: Joe Amon, Alide Marina Biehl Ferraes, Philippe Bourgois, Gilberto Calcagnotto, Maurice Cassier, Arachu Castro, Patricia Clough, Jean Comaroff, Veena Das, Joseph Dumit, Didier Fassin, Ilana Feldman, Michael M. J. Fischer, Robson de Freitas Pereira, Lucia Serrano Pereira, John Gershman, Byron Good, Mary-Jo DelVecchio Good, Clara Han, Jennifer Hirsch, Marcia Inhorn, Craig Janes, Frédéric Le Marcis, Claudio Lomnitz, Michel Lotrowska, Paulo Picon, Deborah Posel, Kaushik S. Rajan, Steven Robins, Aslihan Sanal, Ann L. Stoler, Luis Guilherme Streb, Kok-Chor Tan, Miriam Ticktin, Greg Urban, Susann Wilkinson, and the late Iris Young. Thank you always to Robert Kimball.

I profited greatly from discussions as I presented my work-in-progress to the Department of Anthropology at Brown University; the Department of Anthropology at Columbia University; the Departments of Social Medicine and of Anthropology at Harvard University; the Department of Anthropology at the New School for Social Research; the Department of Anthropology at the University of Chicago; L'Ecole des Hautes Etudes en Sciences Sociales in Paris; the Ethnohistory Program at the University of Pennsylvania; the Wits Institute for Social and Economic Research (WISER) in Johannesburg; and sessions of the American Anthropological Association.

Fred Appel has been an enthusiastic supporter of *Will to Live*. Torben and I greatly appreciate his editorial guidance and help at every step. Also at the Press we are thankful to Dimitri Karetnikov, Maria Lindenfeldar, and Brigitte Pelner for their fine work. Thank you also to Sylvia Coates and Linda Forman for editorial help. Sections of the introduction appeared in a 2006 article in *Public Culture* 18(3):457–72; sections of chapter 1 appeared in a 2004 article in *Social Text* 22(3):105–32; and sections of chapter 4 appeared in a 2001 article in *Culture, Medicine and Psychiatry* 25(1):87–129.

Finally, my wife, Adriana Petryna, has helped me during fieldwork and has carefully read many drafts of this work. Her comments and advice have always been tremendously valuable, and her own anthropological work has been a rich source of ideas. Thank you, Adriana, for everything. Our five-year-old Andre came with me to Brazil several times as I finished fieldwork. My little research assistant is the life force of this book—he animates us for what is to come.

Will to Live is dedicated to my mother, Noemia Kirschner Biehl, and to the memory of my father, Fernando Oscar Biehl—for all they did in life.

Notes

Note to Epigraph: Melo Neto, João Cabral de. (2005). New York: Archipelago
Books, pp. 81–82. I retranslated a few verses.

Introduction: A New World of Health

1. There is a significant activist and social scientific literature on the evolution of the Brazilian response to HIV/AIDS, particularly vis-à-vis political forces and cultural influences, as well as a vast documentation of the unfolding of the treatment policy that the national program itself has made available (Parker and Daniel 1991; Parker 1994; Parker et al. 1994; Galvão 2000, 2002b; Castilho and Cherquer 1996; Camargo Jr. 1994; Bastos and Barcellos 1995, 1996; Bastos 1999; Levi and Vitória 2002; Serra 2004). My work has unfolded in dialogue with this highly relevant literature.

2. See Deleuze and Guattari's discussion of "What Is an Assemblage?" in *Kafka: Toward a Minor Literature* (1986, pp. 81–88). See Ong and Collier (2005) and Rabinow (2003) for anthropological elaborations on contemporary technological and political assemblages.

3. Deleuze 1997, p. 3.

4. The epidemiological information in the paragraph was retrieved from http://www.cdc.gov/nchstp/od/gap/countries/brazil.htm and http://www.usaid.gov/our_work/global_health/aids/Countries/lac/brazil.html. See also Berquó 2005; Hacker et al. 2006; Pechansky et al. 2006; Trevisol and Silva 2005.

5. See Farmer et al. 2001; Galvão 2000; Rosenberg 2001; Wogart and Calcagnotto 2006; *New York Times*, June 23, 2005.

6. See http://www.worldbank.org/aids-econ/arv/.

7. See Barnett and Whiteside 2002; Farmer 2003; Walton et al. 2004; D'Adesky 2004; Nguyen 2005; Fassin 2007.

8. See Das and Poole 2004; Geertz 2004; Greenhouse, Mertz, and Warren 2002; Hansen and Stepputat 2006; Ong 2006; Paley 2001; Trouillot 2003.

9. My discussion of patient- and pharmaceutical-citizenship is informed by Adriana Petryna's anthropological work on "biological citizenship." She developed this concept in the context of people's struggle for care and accountability in the Chernobyl aftermath (2002). See also Rabinow 1999; Scheper-Hughes 2004; Knorr-Cetina 2001; Rapp 1999; Fassin 2001; Biehl 2005.

10. See Das and Das's discussion of "local ecologies of care" and the deployment of pharmaceuticals among the urban poor in India (2006).

11. The distribution of wealth in Bahia is one of the most unequal in Brazil, already the world's second most unequal country.

12. Criticism, I believe, performs its practical calling by participating in the reemergence of subjugated knowledges, "namely, a whole set of knowledges that have been disqualified as inadequate to their task or insufficiently elaborated: naive knowledges located low down on the hierarchy, beneath the required level of cognition or scientificity," in Michel Foucault's words (1980b, p. 82).

13. Anthropologist Marc Abélès has been studying the dual displacement in govermentality and resistance that accompanies the work of nongovernmental organizations as "life and survival" are put at the heart of political action (2006a, p. 493). As he argues, "A sense of powerlessness has become the backdrop for political action. It is as though the citizen's capacity for initiative were going through a more or less explicit reassertion of this admission of powerlessness, tied to the awareness of a radical reappraisal of our terms of belonging. The other side of this position is a projection towards a vaguer collective interest relating more to survival (*survivance*) than to the art of harmonious living together (*conviv-ance*)" (p. 494). See also Abélès 2006b.

Chapter 1: Pharmaceutical Governance

1. See Cardoso's essay on the reform of state (1998). See also Cardoso's account of his two administrations (2005).

2. On inflation and currency devaluations, see Terra and Bonomo (1999). See Font (2003) for a review of Fernando Henrique Cardoso's presidency. See also Sousa (1999) and Sands (2004).

3. For a discussion of social movements and the politics of citizenship in Latin America, see Alvarez et al. (1998); Escobar (1995); Warren (1998); Yashar (2005). See Doimo (1995) for a recent history of social movements in Brazil and Caldeira (2000) for a discussion of democratization and human rights in the country. See Paley (2001) for a discussion of health movements and democratization in Chile. See Das (1999) for a critique of the measures, practices, and values related to international health interventions and Appadurai (2002) for a discussion of the urban poor and new forms of activism and governmentality in India. See Coma-roff and Comaroff (2003) for a discussion of the ethnography of local worlds vis-à-vis economic globalization.

4. The Brazilian AIDS policy is a contemporary *form/event,* using Paul Rabi-now's terminology, through which novel political rationalities and infrastruc-tures of care are actualized. Forms/events are not straightforward realities with predetermined outcomes. "Analytic attention to forms/events," writes Rabinow, "brings us closer to the shifting practices, discursive and otherwise, as well as to the shifting configurations that both shape and are shaped by such practices" (1999, p. 179; see also Rabinow 2003).

5. See Slaughter (2004) for a discussion of sovereignty in a networked world order.

6. See Appadurai 2002; Biehl 2005; Caldeira 2000; Das 1999; Fassin 2001; Ferguson and Gupta 2002; Fischer 2003; Ong 1999; Ong and Collier 2005; Petryna 2002; Rabinow 1999, 2003; Riles 2000; Roitman 2005; Sassen 1998.

7. See the Comaroffs' discussion of the contemporary "judicialization of politics" (2006, pp. 26–31). Class struggles, they argue, "seem to have metamorphosed into class actions. Citizens, subjects, governments, and corporations litigate against one another, often at the intersections of tort law, human rights law, and the criminal law, in an ever mutating kaleidoscope of coalitions and cleavages" (p. 26).

8. See *Jornal NH*, November 7, 1994, "Conferência de saúde possibilita intercâmbio." Just as capital market liberalization had been pushed full force despite any evidence that it spurred economic growth, the outcome of Brazil's experimental public-private health system was not necessarily one of equality or social justice. Brazilian social scientists and activists alike are pointing to the effects of the country's health remanagement as an "excluding universalism." That is, today the elite and the reduced middle class are excluded from the constitutionally mandated universal access to public health care. Formerly state-fostered, the private health sector is now well established and can comfortably draw from its own pool of resources and clientele. In the meantime, the public health services remain in a constant state of collapse and without new funds allotted to them. In their precariousness and mishandling, all over the country, these services remain universally accessible to the majority of poor populations. See http://www.country-studies.com/brazil/the-health-care-system.html.

9. In 1985, transvestite Brenda Lee founded the country's first *casa de apoio* in São Paulo (see chapter 2).

10. According to Richard Parker, the phenomenon of "sexual fluidity," that is, men who are sexually active with both men and women, is a referent "which must be considered so that AIDS epidemiology or its effective control might acquire some meaning" (in Parker and Daniel 1991, p. 71). Sexual fluidity has created a situation in which, as Nancy Scheper-Hughes puts it (1991), "a large proportion of men (especially in the popular classes) who are sexually active with both men and women, but who define themselves publicly as exclusively heterosexual, . . . thus put a great many unsuspecting women at risk of HIV infection." On one level, it is politically problematic to speak of risk groups, which are perceived as stigmatizing and as part of an epidemiological practice that creates difference and locates blame. At the same time, the social location of disease is crucial, especially in the context of limited health resources and of gender inequalities and asymmetry in what Scheper-Hughes calls "sexual citizenship." See also Goldstein (1994).

11. See report in *Isto É/Senhor* (1991, p. 52). For more information on the loan agreement with the World Bank, see Galvão (2000); Levi and Vitória (2002).

12. Here, we have moved a long way from the kind of activism that Steven Epstein chronicled in his book *Impure Science* (1996). Epstein showed how organized groups of laypeople invaded the domain of HIV/AIDS scientific fact

making in the United States. Activists appropriated the languages and cultures of the biomedical sciences, enrolled allies with moral and political argumentation, and successfully established themselves as a population of research subjects. According to Epstein, these politics over credibility and knowledge have transformed the procedures by which drugs are tested, the ways in which test results are interpreted, and the processes by which those interpretations are then used in the licensing of drugs for sale. Absent the AIDS activists, CD4 counts would not have been accepted as a surrogate marker of treatment efficacy in 1991; and without the adoption of the CD4 marker, the AZT/DDC combination would not have been licensed in 1992. Epstein understands this as a democratization of science and technology. In the Brazil case, representatives of people with HIV/AIDS took up the latest HIV/AIDS scientific developments and, operating at the level of social representation and mobilization as well as lawmaking, they demanded that the state fulfill its constitutionally mandated biopolitical obligations, and this would take all parties into new directions.

13. See Diário Oficial 1994.

14. See Ong's discussion of graduated sovereignty (2006) and Ferguson and Gupta's discussion of new forms of neoliberal governmentality (2002). Scheper-Hughes discusses changes in the concepts of bodily integrity, sociality, and human values in the context of the global market in human organs for transplantation (2003). Paley (2001) elaborates on social/health movements in Chile's neoliberal democratization.

15. See Rabinow's discussion (2003) of the concepts of progress and motion in Canguilhem. See also Rabinow's general introduction to Canguilhem's work (1994).

16. See the press release by IMS Health (March 21, 2006): "IMS Health Reports Global Pharmaceutical Market Grew 7 Percent in 2005, to $602 Billion," http://www.imshealth.com/ims/portal/front/articleC/0,2777,6599_3665_77491316,00.html.

17. See "Pharmaceutical Growth Opportunities in Brazil, Russia, India and China: Healthcare Reform, Market Dynamics and Key Players," *Business Insights*, August 1, 2006, http://www.marketresearch.com/product/display.asp?productid=1331044. Note that Mexico has recently displaced Brazil, becoming the largest pharmaceutical market in Latin America. Argentina and Venezuela rank third and fourth, respectively.

18. On the anthropology of pharmaceuticals, see Petryna, Lakoff, and Kleinman (2006); Geest, Whyte, and Hardon (1996); Whyte, Geest, and Hardon (2003); and Nichter and Vuckovic (1994). For a discussion of the marketing practices of pharmaceutical companies, see Oldani (2004). On access to ARVs and human rights, see Farmer (2003).

19. See Applbaum (2006) and Lakoff (2005) for a discussion of how pharmaceutical companies inform public health practices. See Chawla, Diwan, and Joshi (2004).

20. See Lee and Goodman 2002; Yamey 2001; WHO 2002; Aylward et al. 2003.

21. In *Global Responses to AIDS: Science in Emergency*, Cristiana Bastos argues that without state incentive and money, and without the technical know-how to develop original protocols, Brazil's complex AIDS clinical practice "could not be converted into scientific knowledge that would be accepted by the international system" (1999, p. 150). Global pharmaceuticals have recast the workings of this local AIDS science.

22. See Aron (1951).

23. See "The Brazilian Experience in Universal Access to Antiretroviral Therapy," WHO-WTO Secretariat Workshop on Differential Pricing and Financing Drugs, Høsbjør, Norway, April 8–11, 2001, by Paulo R. Teixeira (slide show; http://www.aids.gov.br/politica/exp_univ_therapy.htm). See Galvão (2000b, p. 1863) for information on the logistics of ARV distribution in Brazil.

24. According to Veena Das and Deborah Poole, one of the first tasks of political anthropology is "sighting instances of the state as it exists on the local level and then analyzing those local manifestations of bureaucracy and law as culturally informed interpretations or appropriations of the practices and forms that constitute the modern liberal state. These parochial sightings of the state lead, in turn, to a more spatially and conceptually dispersed picture of what the state is, albeit one that is still basically identifiable through the state's affiliations with particular institutional forms" (2004, p. 5). See also Das and Das (2006).

25. See Petryna 2005, 2006.

Chapter 2: Circuits of Care

1. On policy analysis and pragmatism and democracy, see Ostrom (2002) and Dryzek (2002, 2004). See also Wogart and Calcagnotto (2006).

2. Jessica Blatt, a colleague at the University of California at Berkeley, and Danish photographer Torben Eskerod agreed to collaborate. See Biehl (1995).

3. See Drauzio Varela's (1999) account of life and AIDS in the Carandiru prison in São Paulo.

4. See Petryna's discussion of "illness as work" in the context of the Chernobyl aftermath (2002).

5. *New York Times*, July 16, 2005, p. C2. On inequality in Latin America and the Caribbean, see De Ferranti et al. (2004).

6. See Freire Costa 1992, 1995; Fry 1982; Parker 1990.

7. See Deleuze and Guattari's discussion of "minor literature": "its cramped space forces each individual intrigue to connect immediately to politics" (1986, p. 17).

8. See Teixeira 1997, pp. 64, 65.

9. Amid the lack of any kind of comprehensive social contract and an almost limitless consumption of illegal substances among all sectors of the population, the country is experiencing an unprecedented "moral crisis," argues Jurandir

Freire Costa, one of Brazil's most acclaimed ethicists, and this translates into rampant criminality and violence. "The miserable and the elite want the same thing. Both search for unrestricted enjoyment and are irresponsible in the face of life. They have no commitment to their children and ironize any kind of concern with the Other's wellbeing. There is no civil war, but an agreement of mutual killing" (see interview at http://www.terra.com.br/istoe/1836/1836_vermelhas_01.htm).

10. The 1998 report of the national AIDS program showed that 53 percent of the World Bank funds were going into prevention, 21 percent were going into institutional development, 19 percent were going into assistance, and 7 percent were going into epidemiological surveillance. The majority of the NGOs' projects (90 percent) were aimed at prevention, the majority of federal projects (70 percent) were aimed at institutional development, and the majority of municipal projects (73 percent) were aimed at assistance.

11. See Misra's discussion of how AIDS "performs institutionality" in India (2003, 2006).

12. See Foucault's statement on "Confronting Governments: Human Rights" (2000, pp. 474–75).

Chapter 3: A Hidden Epidemic

1. See Briggs (2003) for a discussion of the control of a cholera epidemic in Venezuela in the early 1990s. Briggs and Mantini-Briggs show that public health intervention led to discrimination on the basis of race and class.

2. Castel suggests that neoliberal societies are developing new strategies and forms of managing individuals and populations through the assignment of "different social destinies . . . in line with their varying capacity to live up to the requirements of competitiveness and profitability" (1991, p. 294). See also Foucault's 1976 lectures at the College de France (1992). See also Agamben's discussion of the "politicization of death" (1998).

3. Psychologist Denise Coutinho produced a study in 1976 on the prevalence of mental illness among 742 poor adults (15 and older) living in Pelourinho (Coutinho 1976; see also Almeida-Filho 1998). She found that mental disorders were extremely prevalent, at 49 percent. Alcoholism and neurosis-like disorders were the two diagnoses most frequently found (22.6 percent). See Bourgois's discussion of the history of the underground economy in inner-city New York (2002, pp. 48–76). See also Bourgois and Schonberg (forthcoming).

4. As John Collins (2003, p. 81) states, "A 1984 federal landmarking of ten central neighborhoods that contained approximately 7,000 residents and 3,000 buildings and monuments dating mainly from the seventeenth through the nineteenth centuries—Santo Antonio Além do Carmo, the Passo, the Carmo, the Pelourinho Square, the Maciel, the Taboão, the Misericórdia, the Terreiro de Jesus and the Praça da Sé, and the 28 of September Street—established today's borders of the Centro Histórico, subsequently recognized in 1985 by UNESCO

as number 309 in the World Heritage List maintained by the International Commission of Monuments and Historic Sites (ICOMOS)."

5. This information was gathered at the website of the Coordenação Nacional de DST e AIDS, Brasília, Ministério da Saúde: http://www.aids.gov.br.

6. After our preliminary analyses revealed flaws in the Bahian epidemiological surveillance system and the state's AIDS program in general, one of my leading collaborators restricted my access to the data. Thus the analyses presented here are limited, based on just a few early cross-tabulations.

7. Much of the medical literature relies on odds ratios for this type of prospective analysis, perhaps for their ease of computation using logistic regression. However, odds ratios are necessary only for retrospective studies, and they are often misinterpreted as rate ratios, which can lead to erroneous conclusions, especially if the outcome of interest occurs more than 10 percent of the time (Zhang and Yu 1998), as is the case here. Following Zou (2004), Figures 3.3–3.5 report ratios based on Poisson regression with robust standard errors.

8. Naomar de Almeida-Filho, personal communication, October 2000.

9. The categories traditionally used by epidemiologists and social scientists to map and interpret the impact of social realities on health (age, gender, ethnicity, sexual identity, for example) cannot account for the technical and institutional dynamics at work here. Insights from the sociology of science are helpful, to a point (Latour 1988, 1990; Shapin and Schaffer 1985). Latour, for example, highlights the "cascading" power of scientific representations to socially "draw things together," thus allowing "harder facts" to be produced (1990, pp. 40, 41). He explicates the "paradox" that "by working on papers alone, on fragile inscriptions that are immensely less than the things from which they are extracted, it is still possible to dominate all things and all people" (p. 60). Thus, ideas and representations become social technologies in that they function for "accumulating time and space" (p. 32), because they enable one to present (and therefore control) "absent things" (p. 27) in a persuasive and efficient way. His caveat: "To take the existence of macro-actors for granted without studying the material that makes them 'macro,' is to make both science and society mysterious" (p. 56).

Ethnography can illuminate how historically specific and contingent these transactions are—that is, what happens in the meantime as science and technology are integrated into policies and bureaucracies. This is particularly relevant as science and technology (in the form of epidemiological modeling, clinical trials, and life-extending drugs, for example) inform the idea and organization of health care and patients' demands.

10. I am here rethinking one of Foucault's maxims that biopower dominated mortality rather than death: "power does not know death anymore and therefore must abandon it" (1992, p. 177; 1980; see also Agamben 1998; Biehl 2005).

11. For examples of local medical research, see Brites et al. (1997); Pedral-Sampaio et al. (1997); Brites et al. (1996).

Chapter 4: Experimental Subjects

1. As of December 1996, there was only one CTA service in the state of Bahia. Given the administrative success of this initiative and also the availability of federal funds, other CTA units have been created since.

2. Epi-Info version 6.3 was used for data entry, data checking, and statistical analysis of the questionnaires.

3. "We are embedded, ethically as well as existentially and materially, in technologies and technological prostheses," notes Michael M. J. Fischer. "[Our] technological prostheses are also taking us into models of ethics with which our older moral traditions have little experience or guidance to offer . . . we are again thrown . . . to ungrounded ways of acting, to new forms of social life" (1999, p. 467).

4. Note that the following biases are potentially at work here (see Rugg et al. 1991): (1) recall bias, when a respondent has difficulty remembering past events accurately; (2) social desirability bias, when a respondent provides answers he or she believes to be socially appropriate; (3) situational demand bias, when the respondent provides responses based on the nature of the situation; and (4) selection bias, owing to volunteerism or motivational differences, such as differences in individual motivation to alter risk. However, these biases are unavoidable, and in the absence of better data, the evidence presented here can shed some light on the local dynamics of HIV/AIDS.

5. On a critique of fantasy regulating social reality, see Zizek (1997).

6. Gurtler (1996, p. 303) notes: "Whether the seroconversion lag differs between groups and how much of the interindividual variation in seroconversion lag can be accounted for by measurable factors such as age, size of inoculum, immune status of the host at the time of infection, or route of acquisition of infection remains unknown."

7. On research aimed at shortening the window period, see Hashida and colleagues (1996). On research about window-period blood donations and on the potential effect of new screening tests, see Kitayaporn and colleagues (1996) and Lackritz and colleagues (1995). On the ability of the polymerase chain reaction (PCR) to detect the HIV virus during the window period, see Barbara and Garson (1993).

8. See Prakash (1999) for a discussion of the politics of health in colonial contexts.

9. See anthropologist Nicolas Sheon's discussion of how HIV counseling and testing aims at sexual normalization and at a renewal of confessional modes of self-control. The confessional quality of the HIV counseling and testing interaction leads, according to Sheon, to a misreckoning of risk behaviors and produces resistance to prevention messages. Moreover, "repeat testers and recidivists are addicted to the cathartic effect of confessional ritual, consciously indulging in risks and using testing as a periodic status check" (1996, pp. 13, 19).

10. Robert Castel (1991) argues that in our neoliberal societies (with an increasingly reduced social welfare apparatus), the administration of population

and individual life has gone beyond the problematic of discipline, treatment, and normalization. "We are situated in a perspective of automated management of populations conducted on the basis of differential profiles of those populations established by means of medico-psychological diagnoses which function as pure expertise" (p. 291). Multifarious sciences, expertise, and apparatuses of risk assessment and prevention are invested in the pragmatic mutation of dangerous and morbid acts into predictable risk facts (Barata 1996; Beck 1992). Castel concludes that this scientific evaluation and management of life is, in its extreme form, "a myth whose logic is already at work" (p. 296).

11. According to Almeida-Filho (1992), clinical epidemiology replaces the clinic's technical role in the constitution of this cyborg-subject. For a critique of clinical epidemiology, see Almeida-Filho (1992, pp. 78–89). Beyond a clinical epidemiology, Almeida-Filho suggests the workings of an "ethno-epidemiology" (p. 111).

12. Lacan created the term *surplus jouissance* as an analogy to Marx's concept of *surplus value* (Marx 1983, pp. 407–9; Zizek 1997, pp. 325–27).

13. This expression was coined by Lacan in the essay "Science and Truth" (1989, p. 17).

14. The federal government took several initiatives to lower the price of condoms (including state production) and to make them widely available. Nongovernmental organizations played a key role in making condoms available to specific risk groups, and local health posts were enlisted to distribute them to the general population.

15. See Petryna's work on globalized clinical trials (2005, 2006). She identifies an ethical variability in the ways these trials are rationalized and run in zones of crisis, with troublesome medical effects. From this ethical variability emerges unregulated risk variability and dubious scientific evidence.

16. See http://www.fiocruz.br/sinitox.

Chapter 5: Patient-Citizenship

1. *Isto É/Senhor*, September 9, 1992.

2. Foucault's insights on pastoral power (1983) are helpful to understand these post-welfare state operations. Pastoral power ensures the salvation of the individual and the community, and it takes care of them for the whole life. By the eighteenth century, argues Foucault, pastoral functions had left ecclesiastic organizations and had spread throughout the social body/political institutions (p. 215). In fact, says Foucault, we can consider the state as a modern matrix of individualization or a new form of pastoral power. He notes: "It was no longer a question of leading people to their salvation into the next world, but rather insuring it in this world" (p. 215) through statistics, clinical biomedicine, psychiatry, police, families, and humanitarianism, for example. See Abélès (2006a).

3. The basic basket includes cleaning materials (one small box of laundry soap, 2 bath soaps, one bottle of detergent, one bottle of disinfectant, 4 rolls of toilet paper, 2 tubes of toothpaste) and food (4 kg beans, 6 kg rice, 3 kg sugar,

one package of noodles, 2 kg flour, one can of powdered milk, cookies, crackers, oil, 1 kg dried meat, 1 kg chicken, one sausage, one package of coffee, one box of milk, margarine, baking products, and eggs).

4. Byron Good's insight on salvation as a key component to medicine as a social technology is helpful in thinking through the health-conversion experiences of Caasah's inhabitants. Good (1994, p. 84) argues that "the juxtaposition of the rational-technical or physiological with the existential or soteriological is essential to our understanding medicine as a symbolic formation."

5. Brazilian scholars have been using some of Arendt's insights to problematize the operational logics of Brazil's crumbling welfare state and the inability of market-oriented policies to work in terms of the poor. Escorel (1993), for example, argues that fragmented and stratified concepts of citizenship legitimate a political order in which social policies are unequally distributed according to the citizen's participation in or exclusion from production processes. Escorel identifies the continuous social exclusion of the poorest masses as a trait of a totalitarian state. For the excluded, she says, there are no social policies; "the only social policy is the police" (p. 36).

6. Drawing from Foucault's work on biopower, anthropological studies have paid attention to the centrality of error to (modern) life, charting the emergence of "mutant ecologies" (Masco 2004) and "biological forms of citizenship" (Petryna 2002) in the wake of industrial disasters, for example. Studies of media and medical technologies have shown the truly prosthetic quality such technologies possess as people deploy them to refigure capacities and value systems (Fischer 2003; Young 1995; Cohen 1998; Rapp 1999; Rabinow 1999; Rajan 2006).

7. Deleuze 1995, p. 118.

Chapter 6: Will to Live

1. Philosopher Gilles Deleuze unremittingly resists closure in his conceptual work, challenging us to bring movement and incompleteness into view (1997, 2005, 2006). In his *Essays: Critical and Clinical* (1997), Deleuze takes the act of writing as a starting point and explores language as the very gate through which limits of all kinds are crossed and the energy of the "delirium" is unleashed (p. 1). The "delirium" suggests alternative visions of existence and of a future that clinical definitions tend to foreclose. This radical work of literature moves away from "truths" and "forms" (since truth is a form in itself) and toward intermediate, processual stages that could even be virtual. Writing is inseparable from becoming, says Deleuze, and becoming "always has an element of flight that escapes its own formalization" (ibid).

To become is not to attain a form through imitation, identification, or mimesis, but rather to find a zone of proximity where one can no longer be distinguished from a man, a woman, or an animal (as in Kafka's work, for example)—"neither imprecise, nor general, but unforeseen and nonpreexistent, singularized out of a population rather than determined in a form." One can institute such a zone of indifferentiation with anything "on the condition that one creates the literary means for doing so" (ibid).

For Deleuze, the real and the imaginary are always coexisting and complementing each other. They are like two juxtaposable or superimposable parts of a single trajectory, two faces that ceaselessly interchange with one another—"a mobile mirror"—"bearing witness until the end to a new vision whose passage it remained open to" (p. 63). Deleuze seems to be drawing here from Lacan's idea that some realities have to be fictionalized before they can be apprehended (1980)—an idea that Das has put to test in her luminous essay on "Language and the Body" (1997) where, against violent world-historical trajectories, people develop the capacity to imagine life beyond.

Actualized by literature, writes Deleuze, this mobile mirror reveals beneath apparent persons the power of an impersonal—"which is not a generality but singularity at the highest point: a man, a woman, a beast, a child. . . . It is not the first two persons that function as the condition for literary enunciation; literature begins only when a third person is born in us that strips us of the power to say 'I'" (1997, p. 3). The shift to the indefinite—from "I" to "a"—leads to the ultimate existential stage where life is simply "immanent," a transcendental field where man and woman and other men and women/animals/landscapes can achieve the web of variable relations and situated connectedness called "camaraderie." "Camaraderie is the variability that implies an encounter with the Outside, a march of souls in the open air, on the 'Open Road'" (p. 60). Inventing one's relation to the Other, sampling fragments, turning existence into a way, into an art, is also an ethnographic endeavor—"To write for this people who are missing . . . ('for' means less 'in the place of' than 'for the benefit of')" (p. 4).

Conclusion: Global Public Health

1. "Veja principais pontos do relatório da ONU sobre Aids e virus HIV," Folha Online, November 21, 2005.

2. Langevin, South Asian Research Institute for Policy and Development at http://www.sarid.net/health/healthdocs/050701-hiv.htm.

The Global Fund is meant to be an "innovative approach to international health financing" between governments, civil societies, afflicted populations, and the private sector. See http://www.theglobalfund.org/en. The Fund operates on a results-based system where Aidspan, an independent "watchdog," oversees the performance of the participating organizations. Almost two-thirds of the funds collected in rounds 1 and 2 were devoted to AIDS in sub-Saharan Africa, with 17 percent and 14 percent for TB and malaria. The presence of an independent review process for evaluating proposals, a technical review panel with experts tracking the progress of grant-funded programs, supposedly ensures that programs regularly report successes. Lastly, a simplified, rapid, innovative grant-making process operates transparently, posting all information on the website, and establishes accountability through local fund agents.

3. Though proposals of funding are always very generous, the amount actually disbursed never seems to equal the proposed amount. For example, "By January 2004, US$822.3 million had been committed to 24 countries in the region; US$170.6 million had been disbursed" (2004 Report on Global AIDS from the Bangkok conference, http://www.unaids.org/bangkok2004/GAR2004_html/GAR2004_10_en.htm).

4. Project "Global Health Governance: Institutional Change and the Interfaces between Global and Local Politics in the Poverty-Oriented Fight of Diseases," Research Group "Globalization and Social Development," German Overseas Institute, Hamburg University. See Wogart and Calcagnotto (2006).

5. According to Fleshman, "In the view of many public health experts, the rapid increase vindicates what some critics derided as WHO's "irresponsible and unrealistic 3x5 pledge" (http://www.un.org/ecosocdev/geninfo/afrec/vol19no1/191 aids.htm).

6. See http://www.healthgap.org/press_releases/03/051903_HGAP_BP_WHO _TA_TX.html.

7. Kim cited in "OMS admite fracasso em meta de combate ao HIV," Folha Online, November 28, 2005 (my translation).

8. WHO 2006.
By February 2006, more than 30 countries had convened national consultations on universal ARV access, and nearly 100 other countries had initiated the planning process. Over the past year, the number of people receiving treatment increased by about 300,000 every 6 months. Scale-up in sub-Saharan Africa has been particularly dramatic, from 100,000 at the end of 2003, to 310,000 at the end of 2004, to 810,000 at the end of 2005.

9. See http://www.avert.org/pepfar.htm.
By May 2003, this declaration had been approved into law as the United States Leadership Against HIV/AIDS, TB, and Malaria Act of 2003 (PL108-25). Initially, the commitment was set at $15 billion over 5 years in addition to $17 billion for domestic HIV/AIDS for 2005. Like the Global Fund, women and children are also the target of PEPFAR. Forty-one percent of the funds committed for 2006–08 will be used to purchase and distribute ARVs, with one-third of prevention funds spent on promoting abstinence and two-thirds spent on other methods of prevention, such as condom dissemination. Though proposed, the $15 billion is not guaranteed since Congress must appropriate the amount each fiscal year.

10. See http://www.un.org/ecosocdev/geninfo/afrec/vol19no1/191aids.htm.

11. Activists say that by using generics, PEPFAR's policies can finally be brought closer to those of other organizations, such as the Global Fund, helping reverse the confusion that marks drug procurement and supply in many places.

12. "Mortalidade de negros é maior do que a de brancos," Folha Online, August 3, 2005.

13. These are the conclusions of a study carried out by sociologist Álvaro Comin (CEBRAP—Centro Brasileiro de Análise e Planejamento).

14. "Governo vê AIDS sob controle para negar quebra de patente," Folha Online, November 11, 2005.

15. "Lula planeja usar dengue contra Serra," Josias de Souza's blog, Folha Online, December 20, 2005.

16. These concluding paragraphs draw from Kay Warren's insightful review of the manuscript. Thank you.

References

Aaby, Peter, A. Babiker, and J. Darbyshire, et al. (1997). "Correspondence: Ethics of HIV Trials." *The Lancet* 350:1546.

Abadia-Barrero, C. E., and A. Castro. (2006). "Experiences of Stigma and Access to HAART in Children and Adolescents Living with HIV/AIDS in Brazil." *Social Science and Medicine* 62(5):1219–28.

Abélès, Marc. (2006a). "Globalization, Power and Survival: An Anthropological Perspective." *Anthropological Quarterly* 79(2):483–508.

———. (2006b). "Parliament, Politics and Ritual." In *Rituals in Parliaments*, edited by Emma Crewe and Marion G. Muller. Frankfurt am Main: Peter Lang.

Acurcio, F., M. Guimarães, and C. Drew. (1996). "Accessibilidade de Indivíduos Infectados pelo HIV aos Serviços de Saúde: Uma Revisão de Literatura." *Cadernos de Saúde Pública* 12(2).

Adelman, Jeremy. (1988). "Tequila Hangover: Latin America's Debt Crisis." *Studies in Political Economy* 55.

Adorno, Theodor. (1982). "Freudian Theory and the Pattern of Fascist Propaganda." In *The Essential Frankfurt School Reader*, edited by Andrew Arato and Eike Gebhardt. New York: Continuum.

Agamben, Giorgio. (1998). *Homo Sacer: Sovereignty and Bare Life*. Stanford: Stanford University Press.

Aggleton, P., and J. S. Pedrosa. (1994). "Community, Solidarity, and Action—Grupo Pela Vidda, Brazil." *AIDS Care* 6(3):343–48.

Alcabes P., A. Muñoz, et al. (1993). "Incubation Period of Human Immunodeficiency Virus." *Epidemiologic Reviews* 15(2):303–18.

Almeida-Filho, Naomar. (1992). *A Clínica e a Epidemiologia*. Salvador and Rio de Janeiro: APCE/ABRASCO.

———. (1998). "Becoming Modern After All These Years: Social Change and Mental Health in Latin America." *Culture, Medicine and Psychiatry* 22(3): 285–316.

Altman, Dennis. (1999). "Globalization, Political Economy, and HIV/AIDS." *Theory and Society* 28:559–84.

Alvarez, Sonia, Evelina Dagnino, and Arturo Escobar (eds.). (1998). *Cultures of Politics—Politics of Cultures: Re-visioning Latin American Social Movements*. Boulder: Westview.

Andrade, Tarcísio Matos de. (1996). "Condições Psicossociais e Exposição ao Risco de Infecção pelo HIV Entre Usuários de Drogas Injetáveis em uma Comunidade Marginalizada de Salvador-Bahia." Ph.D. dissertation, Universidade Federal da Bahia (UFBA), Salvador.

Angell, Marcia. (2004). *The Truth About Drug Companies: How They Deceive Us and What to Do about It.* New York: Random House.

Appadurai, Arjun. (1996). *Modernity at Large: Cultural Dimensions of Globalization.* Minneapolis: University of Minnesota Press.

———. (2002). "Deep Democracy: Urban Governmentality and the Horizon of Politics." *Public Culture* 14(1):21–47.

Applbaum, Kalman. (2006). "American Pharmaceutical Companies and the Adoption of SSRI's in Japan." In *Global Pharmaceuticals: Markets, Practices, Ethics*, edited by Adriana Petryna, Andrew Lakoff, and Arthur Kleinman. Durham, NC: Duke University Press.

Arendt, Hannah. (1958). *The Human Condition.* Chicago: University of Chicago Press.

———. (1973). *The Origins of Totalitarianism.* New York and London: A Harvest/HBK Book.

Aron, Raymond. (1951). *Les Guerres en Chaine.* Paris: Gallimard.

Arrais, P.S.D., H. L. Coelho, M. C. Batista, M. L. Carvalho, R. E. Righi, and J. M. Arnau. (1997). "Perfil da Automedicação no Brasil." *Revista de Saúde Pública* 31(1):71–77.

Attaran, A., K. I. Bates, R. Bate, F. Binka, U. d'Alessandro, C. I. Fanello, L. Garrett, T. K. Mutabingwa, D. Roberts, C. Hopkins Sibley, et al. (2006). "The World Bank: False Financial and Statistical Accounts and Medical Malpractice in Malaria Treatment." *The Lancet* 368:247–52.

Aylward, R. B., et al. (2003). "Global Health Goals: Lessons from the Worldwide Effort to Eradicate Poliomyelitis." *The Lancet* 362:909–14.

Bacelar, Jeferson. (1982). *A Família da Prostituta.* São Paulo: Ática.

Barata, Rita. (1996). "Epidemiologia Clínica: Nova Ideologia Médica?" *Cadernos de Saúde Pública* 12:555–60.

Barbara J. A., and J. A. Garson. (1993). "Polymerase Chain Reaction and Transfusion Microbiology." *Vox Sanguinis* 64(2):73–81.

Barbosa, Denis Borges. (n/d). "O Acordo TRIPS e o Prazo das Patentes." Manuscript.

Barnett, Tony, and Alan Whiteside. (2002). *AIDS in the Twenty-First Century: Disease and Globalization.* New York: Palgrave Macmillan.

Bastos, Cristiana. (1999). *Global Responses to AIDS: Science in Emergency.* Bloomington: Indiana University Press.

Bastos, Francisco Inácio, and Christovam Barcellos. (1995). "Geografia Social da AIDS no Brazil." *Revista de Saúde Pública* 29(1):52–62.

———. (1996). "Redes Sociais e Difusão da AIDS no Brasil." *Boletim da Oficina Sanitária Panamericana* 121(1):11–24.

Beaglehole, Robert, and Ruth Bonita. (2004). *Public Health at the Crossroads: Achievements and Prospects.* 2nd ed. Cambridge: Cambridge University Press.

Beck, Ulrich. (1992). *Risk Society.* London: Sage.

———. (1997). *The Reinvention of Politics: Rethinking Modernity in the Global Social Order.* Cambridge: Polity.

Behforouz, H. L., P. Farmer, and J. Mukherjee. (2004). "From Directly Observed Therapy to Accompagnateurs: Enhancing AIDS Treatment Outcomes in Haiti and in Boston." *Clinical Infectious Diseases* 38 (Supp. 5):429–36.

Bermudez, Jorge. (1992). *Remédios: Saúde ou Indústria.* Rio de Janeiro: Relume Dumará.

———. (1995). *Indústria Farmacêutica, Estado e Sociedade: Crítica da Política de Medicamentos no Brasil.* São Paulo: Editora Hucitec e Sociedade Brasileira de Vigilância de Medicamentos.

Bermudez, J.A.Z., R. Epsztein, M. A. Oliveira, and L. Hasenclever. (2000). "O Acordo Trips da OMC e a Proteção Patentária no Brasil: Mundanças Recentes e Implicações para a Produção Local e o Acesso da População aos Medicamentos." Rio de Janeiro: Escola Nacional de Saúde Pública, FIOCRUZ/OMS.

Berquó, E. (2005). *Comportamento Sexual e Percepções da População Brasileira sobre o HIV/AIDS* [apresentação]. Brasília: Programa Nacional de DST e AIDS.

Bhabha, Homi K. (1994). *The Location of Culture.* New York: Routledge.

Biagioli, Mario. (2000.) "From Difference to Blackboxing: French Theory vs. Science Studies' Metaphysics of Presence." In *French Theory in America,* edited by S. Cohen and S. Lotringer. New York: Routledge.

Biehl, João (with Jessica Blatt). (1995). *Life on Paper: A Trip through AIDS in Brazil.* Rio de Janeiro: ISER. Study Document.

Biehl, João (with Anamélia Franco and Roselene de Alencar). (1996). "Uma Avaliação Qualitativa das Práticas de Prevenção em AIDS na Cidade de Salvador." Salvador: Secretaria Municipal da Saúde da Cidade de Salvador. Report.

Biehl, João. (1999a). *Other Life: AIDS, Biopolitics, and Subjectivity in Brazil's Zones of Social Abandonment.* Ph.D. dissertation, University of California at Berkeley.

———. (1999b). "Prefácio." In *Antropologia da Razão,* by Paul Rabinow. Rio de Janeiro: Relume Dumará.

———. (2001a). "Vita: Life in a Zone of Social Abandonment." *Social Text* 19(3):131–49.

Biehl, João (with Denise Coutinho and Ana Luzia Outeiro). (2001b). "Technology and Affect: HIV/AIDS Testing in Brazil." *Culture, Medicine and Psychiatry* 25:87–129.

Biehl, João. (2004a). "Life of the Mind: The Interface of Psychopharmaceuticals, Domestic Economies, and Social Abandonment." *American Ethnologist* 31(4):475–96.

———. (2004b). "The Activist State: Global Pharmaceuticals, AIDS, and Citizenship in Brazil." *Social Text* 22(3):105–32.

———. (2005). *Vita: Life in a Zone of Social Abandonment.* Berkeley: University of California Press.

———. (2006a). "Pharmaceutical Governance." In *Global Pharmaceuticals: Ethics, Markets, Practices*, edited by Adriana Petryna, Andrew Lakoff, and Arthur Kleinman. Durham, NC: Duke University Press.

Biehl, João (with Torben Eskerod). (2006b). "Will to Live: AIDS Drugs and Local Economies of Salvation." *Public Culture* 18(3):457–72.

Blower, S., L. Ma, P. Farmer, and S. Koenig. (2003). "Predicting the Impact of Antiretrovirals in Resource-poor Settings: Preventing HIV Infections Whilst Controlling Drug Resistance." *Infectious Disorders and Drug Targets* 3(4):345.

Borneman, John. (2001). "Caring and Being Cared For: Displacing Marriage, Kinship, Gender, and Sexuality." In *The Ethics of Kinship: Ethnographic Inquiries*, edited by James Faubion, pp. 29–45. New York: Rowman and Littlefield.

Bourgois, Philippe. (2002). *In Search of Respect: Selling Crack in El Barrio.* 2nd ed. Cambridge: Cambridge University Press.

Bourgois, Philippe, and Jeffrey Schonberg. (Forthcoming). *Righteous Dopefiend: Homeless Heroin Addicts in Black and White.* Berkeley: University of California Press.

Briggs, Charles L. (with Clara Mantini-Briggs). (2003). *Stories in the Time of Cholera: Racial Profiling During a Medical Nightmare.* Berkeley: University of California Press.

Brigido, L.F.M., R. Rodrigues, J. Casseb, D. Oliveira, M. Rossetti, P. Menezes, and A.J.S. Duarte. (2001). "Impact of Adherence to Antiretroviral Therapy in HIV-1 Infected Patients at a University Public Service in Brazil." *AIDS Patient Care and STDs* 15(11):587–93.

Brites, C., C. Pedroso, N. Silva, W. D. Johnson Jr., and R. Badaró. (1996). "The Influence of CD4+ T Cells, HIV Disease Stage and Zidovudine on HIV Isolation in Bahia, Brazil." *Revista da Sociedade Brasileira de Medicina Tropical* 29(1):5–9.

Brites, C., W. Harrington Jr., C. Pedroso, E. M. Netto, and R. Badaró. (1997). "Epidemiological Characteristics of HTLV-I and II Coinfection in Brazilian Subjects Infected by HIV-1." *Brazilian Journal of Infectious Diseases* 1:42–47.

Brites, C., C. Pedroso, E. Netto, W. Harrington Jr., B. Galvão-Castro, J. C. Couto-Fernandez, D. Pedral-Sampaio, M. Morgado, R. Teixeira, and R. Badaró. (1998). "Co-Infection by HTLV-I/II Is Associated with Increased Viral Load in PBMC of HIV-1 Infected Patients in Bahia, Brazil." *Brazilian Journal of Infectious Diseases* 2(2):70–77.

Buerki, Robert. (2005). "Antiretroviral Pharmaceutical Distribution at Home and Abroad: Broken Promises or Misapplied Models?" Princeton University. Unpublished paper.

Butler, Judith. (1997). *The Psychic Life of Power: Theories in Subjection.* Stanford: Stanford University Press.

Caldeira, Teresa. (2000). *City of Walls: Crime, Segregation, and Citizenship in São Paulo.* Berkeley: University of California Press.

Camargo, Kenneth, Jr. (1994). *As Ciências da AIDS e a AIDS das Ciências: O Discurso Médico e a Construção da AIDS*. Rio de Janeiro: Relume Dumará.

Canguilhem, Georges. (1994). *A Vital Rationalist*. Cambridge, MA: Zone Books.

———. (1998). "The Decline of the Idea of Progress." *Economy and Society* 27(2/3):313–29.

Cardoso, Fernando Henrique. (1998). "Notas sobre a Reforma do Estado." *Novos Estudos do CEBRAP* 50:1–12.

———. (2005). *A Arte da Política: A História Que Vivi*. Rio de Janeiro: Record.

Cassano, Conceição, A. M. Frias Luiz, and Joaquim F. Valente. (2000). "Classificação por Ocupação dos Casos de AIDS no Brazil—1995." *Cadernos de Saúde Pública* 16(Supp. 1):53–64.

Cassier, Maurice, and Marilena Correa. (2003). "Patents, Innovation and Public Health: Brazilian Public-Sector Laboratories' Experience in Copying AIDS Drugs." In *Economics of AIDS and Access to HIV/AIDS Care in Developing Countries: Issues and Challenges*, pp. 89–107. Paris: ANRS.

Castel, Robert. (1991). "From Dangerousness to Risk." In *The Foucault Effect: Studies in Governmentality*, edited by Graham Burchell, Colin Gordon, and Peter Miller. Chicago: University of Chicago Press.

Castiel, Luis David. (1998). "The Next Millennium and Epidemiology: Searching for Information." *Cadernos de Saúde Pública* 14(4):765–78.

———. (1999). *A Medida do Possível: Saúde, Risco e Tecnobiociências*. Rio de Janeiro: Contra Capa Livraria e Editora Fiocruz.

Castilho, Euclides, and Pedro Cherquer. (1996). "Epidemiologia da AIDS no Brasil." In "A Epidemia de AIDS no Brasil: Resumo (Versão Preliminar)," by Programa Nacional de Doenças Sexualmente Transmissíveis e AIDS. Brasília: Ministério da Saúde.

Castro, Arachu. (2005). "Adherence to Antiretroviral Therapy: Merging the Clinical and Social Course of AIDS." *PLoS Medicine* 2(12), e338.

Castro, Arachu, and Merrill Singer (eds.). (2004). *Unhealthy Health Policy: A Critical Anthropological Examination*. Lanham, MD: AltaMira.

Centers for Disease Control and Prevention [CDC]. (1993). *Journal of the American Medical Association* 269(16):2072–75.

Cerqueira, Nelson (ed.). (1994). *Pelourinho, Historic Center of Salvador, Bahia: The Restored Grandeur*. Salvador: Fundação Cultural do Estado.

Chawla, H. P. S., N. Diwan, and K. Joshi. (2004). *Emerging Trends in the World Pharmaceutical Market: A Review*. Business Briefing: Pharmatech. (www.touchbriefings.com/cpds).

Chen, L., A. Kleinman, and N. C. Ware (eds.). (1994). *Health and Social Change in International Perspective*. Boston: Harvard School of Public Health.

Coetzee, D., K. Hildebrand, A. Boulle, et al. (2004). "Outcomes after Two Years of Providing Antiretroviral Treatment in Khayelitsha, South Africa." *AIDS* 18:887–95.

Cohen, Jillian Clare, and Patricia Illingworth. (2003). "The Dilemma of Intellectual Property Rights for Pharmaceuticals: The Tension between Ensuring Access of the Poor to Medicines and Committing to International Agreements." *Developing World Bioethics* 3(1):27–48.

Cohen, Lawrence. (1998). *No Aging in India: Alzheimer's, the Bad Family, and Other Modern Things.* Berkeley: University of California Press.

———. (1999). "Where It Hurts: Indian Material for an Ethics of Organ Transplantation." *Daedalus* 128(4):135–65.

Cohn, Amélia. (1997). "Considerações Acerca da Dimensão Social da Epidemia de HIV/AIDS no Brasil." In *Coordenação Nacional de DST e AIDS, A Epidemia da AIDS no Brasil: Situação e Tendências,* pp. 45–53. Brasília: Ministério da Saúde.

Collins, John F. (2003). *The Revolt of the Saints: Popular Memory, Urban Renewal and National Heritage in the Twilight of Brazilian "Racial Democracy."* Ph.D. dissertation, University of Michigan.

Comaroff, Jean, and John Comaroff. (2003). "Ethnography on an Awkward Scale: Postcolonial Anthropology and the Violence of Abstraction." *Ethnography* 4(2):147–79.

Comaroff, John L., and Jean Comaroff. (2006). "Law and Disorder in the Postcolony: An Introduction." In *Law and Disorder in the Postcolony,* edited by Jean Comaroff and John L. Comaroff, pp. 1–56. Chicago: University of Chicago Press.

Cooper, D. A., J. Gold, P. MacLean, et al. (1985). "Acute AIDS Retrovirus Infection: Definition of a Clinical Illness Associated with Seroconversion." *The Lancet* 1(8428):537–40.

Coordenação Nacional de DST e AIDS [CN]. (1996). "Simpósio Satélite: A Epidemia de AIDS no Brasil: Situação e Tendências." Document of the First Brazilian STD/AIDS Prevention Congress, Salvador.

———. (1997a). *A Epidemia da AIDS no Brasil: Situação e Tendências.* Brasília: Ministério da Saúde.

———. (1997b). *Catálogo de Organizações Não-Governamentais.* Brasília: Ministério da Saúde.

———. (1997c). *A Epidemia da AIDS no Brasil: Situação e Tendências.* Brasília: Ministério da Saúde.

———. (1997d). "Coquetel Contra AIDS Faz Cair Número de Mortes e Custos do Tratamento." Brasília, Ministério da Saúde, press release, October 31.

———. (1997e). *Boletim Epidemiológico de AIDS* 9(5).

———. (1998a). *AIDS no Brasil: Um Esforço Conjunto Governo—Sociedade.* Brasília: Ministério da Saúde.

———. (1998b). "Saúde Define Estatégias de Combate a AIDS entre População de Baixa Renda." Brasília, Ministério da Saúde, press release, June 6.

———. (2000). *The Brazilian Response to HIV/AIDS: Best Practices.* Brasília: Ministério da Saúde.

———. (2001a). *National AIDS Drug Policy*. Brasília: Ministério da Saúde.

———. (2001b). *Boletim Epidemiológico, Janeiro a Março de 2001*. Brasília: Ministério da Saúde.

———. (2001c). "WTO Panel Calls Brazilian Patent Laws into Question: Official Note." Brasília, Ministério da Saúde, press release, February.

Cosendey, M. A., J.A.Z. Bermudez, A.L.A. Reis, H. F. Silva, M. A. Oliveira, and V. L. Luiza. (2000). "Assistência Farmacêutica na Atenção Básica de Saúde: A Experiência de Três Estados Brasileiros." *Cadernos de Saúde Pública* 16(1):171–82.

Costa, Jurandir Freire. (1992). *A Inocência e O Vício : Estudos sobre o Homoerotismo*. Rio de Janeiro: Relume Dumará.

———. (1995). *A Face e o Verso*. São Paulo: Escuta.

Coutinho, Denise M. (1976). "Prevalência de Doenças Mentais em uma Comunidade Marginal." Master's thesis, Faculdade de Medicina, Universidade Federal da Bahia.

Crane, Johanna, Kathleen Quirk, and Ariane van der Straten. (2002). "Come Back When You're Dying: The Commodification of AIDS among California's Urban Poor." *Social Science and Medicine* 55:1115–27.

Cunha, Sérgio, et al. (1996). "Atividade Prática: Disciplina MED 100—Epidemiologia." Salvador: Faculdade de Medicina, Universidade Federal da Bahia. Manuscript.

Czeresnia, Dina. (1995). *AIDS*. Rio de Janeiro: Relume Dumará.

D'Adesky, Anne-Christine. (2004). *Moving Mountains: The Race to Treat Global AIDS*. New York: Verso.

Daniel, Herbert. (1991). "We Are All People Living with AIDS: Myths and Realities of AIDS in Brazil." *International Journal of Health Services* 21(3).

Das, Veena. (1996). *Critical Events*. New Delhi: Oxford University Press.

———. (1997). "Language and Body: Transactions in the Construction of Pain." In *Social Suffering*, edited by Arthur Kleinman, Veena Das, and Margaret Lock, pp. 67–91. Berkeley: University of California Press.

———. (1999). "Public Good, Ethics, and Everyday Life: Beyond the Boundaries of Bioethics." *Daedalus* 128(4):99–133.

———. (2006). "Power, Marginality, and Illness." *American Ethnologist* 33(1): 27–32.

Das, Veena, and Ranendra K. Das. (2006). "Pharmaceuticals in Urban Ecologies: The Register of the Local." In *Global Pharmaceuticals: Markets, Practices, Ethics*, edited by Adriana Petryna, Andrew Lakoff, and Arthur Kleinman. Durham, NC: Duke University Press.

———. (2007). "How the Body Speaks: Illness and the Life World among the Urban Poor." In *Subjectivity: Ethnographic Investigations*, edited by João Biehl, Byron Good, and Arthur Kleinman. Berkeley: University of California Press.

Das, Veena, and Deborah Poole (eds.). (2004). *Anthropology in the Margins of the State*. Santa Fe: School of American Research Press.

Dawson, Jill, Ray Fitzpatrick, et al. (1991). "The HIV Test and Sexual Behavior in a Sample of Homosexually Active Men." *Social Science and Medicine* 32(6):683–88.

De Ferranti, David, Guillermo E. Perry, Francisco H. G. Ferreira, and Michael Walton. (2004). *Inequality in Latin America and the Caribbean: Breaking with History?* Washington, DC: World Bank.

Deleuze, Gilles. (1995). *Negotiations, 1971–1990*. New York: Columbia University Press.

———. (1997). *Essays: Critical and Clinical*. Minneapolis: University of Minnesota Press.

———. (2005). *Pure Immanence: Essays on a Life*. Cambridge, MA: Zone Books.

———. (2006). *Two Regimes of Madness*. New York: Semiotext(e).

Deleuze, Gilles, and Felix Guattari. (1986). *Kafka: Toward a Minor Literature*. Minneapolis: University of Minnesota Press.

Desjarlais, Robert. (1997). *Shelter Blues: Sanity and Selfhood Among the Homeless*. Philadelphia: University of Pennsylvania Press.

Diário Oficial. (1994). "Acordo TRIPS ou Acordo ADPIC." Brasília, December 19, p. 1989, n. 239.

Doimo, Ana Maria. (1995). *A Vez e a Voz do Popular: Movimentos Sociais e Participação Política no Brasil Pós-70*. Rio de Janeiro: ANPOCS/Relume Dumará.

Dourado, M.I.C., M. Barreto, N. Almeida-Filho, J. Biehl, and S. Cunha. (1997a). "Região Nordeste." In *A Epidemia da AIDS no Brasil: Situação e Tendências*, by Coordenação Nacional de DST e AIDS, pp. 123–43. Brasília: Ministério da Saúde.

Dourado, M.I.C., C. Noronha, A. Barbosa, and R. Lago. (1997b). "Considerações sobre o Quadro da AIDS na Bahia." *Boletim Epidemiológico do SUS*. Brasília: Ministério da Saúde.

Dourado, M.I.C., M. A. Veras, D. Barreira, and A. M. de Brito. (2006). "AIDS Epidemic Trends after the Introduction of Antiretroviral Therapy in Brazil." *Revista de Saúde Pública* 40(Supp.):1–8.

Drahos, Peter, and John Braithwaite. (2004). "Who Owns the Knowledge Economy? Political Organising Behind TRIPS." *The Corner House*, Briefing 32.

Dryzek, John S. (2002). "A Post-Positivist Policy-Analytic Travelogue." *The Good Society* 11(1):32–36.

———. (2004). "Pragmatism and Democracy." *Journal of Speculative Philosophy* 18(1):72–79.

Dumit, Joseph. (1997). "A Digital Image of the Category of the Person." In *Cyborgs & Citadels: Anthropological Interventions in Emerging Sciences and Technologies*, edited by Gary Downey and Joseph Dumit. Santa Fe: School of American Research Advanced Seminar Series.

———. (2004). *Picturing Personhood: Brain Scans and Biomedical Identity*. Princeton: Princeton University Press.

Dye, C., G. P. Garnett, K. Sleeman, and B. G. Williams. (1998). "Prospects for Worldwide TB Control under the WHO DOTS Strategy." *The Lancet* 352: 1886–91.

Edmonds, Alexander. (2002). *New Bodies, New Markets: An Ethnography of Brazil's Beauty Industry*. Ph.D. dissertation, Princeton University.

Epstein, Helen. (2003). "AIDS in South Africa: The Invisible Cure." *New York Review of Books*, July 17.

———. (2005a). "Claiming the Right to Health." *The Lancet* 366:1155–56.

———. (2005b). "The Lost Children of AIDS." *New York Review of Books*, November 3.

Epstein, Steven. (1996). *Impure Science: AIDS, Activism, and the Politics of Knowledge*. Berkeley: University of California Press.

Eribon, Didier. (1996). *Michel Foucault e seus Contemporâneos*. Rio de Janeiro: Jorge Zahar Editor.

Escobar, Arturo. (1995). *Encountering Development: The Making and Unmaking of the Third World*. Princeton: Princeton University Press.

Escorel, Sarah. (1993). "Elementos para a Análise da Configuração do Padrão Brasileiro de Proteção Social: O Brasil Tem um *Welfare State?*" *Estudos— Política, Planejamento e Gestão em Saúde*, n. 1. Rio de Janeiro: FIOCRUZ/ ENSP/DAPS.

Faleiros, Vicente de Paula. (2003). "A Reforma do Estado no Período FHC e as Propostas do Governo Lula." In *A Era FHC e o Governo Lula: Transição?* edited by Denise Rocha and Maristela Bernardo. Brasília: Instituto de Estudos Sócio-Econômicos.

Farlow, Andrew. (2007). "The State of Global Health Funding Initiatives." Paper presented at Princeton University, April 24.

Farmer, Paul. (1992). *AIDS and Accusation: Haiti and the Geography of Blame*. Berkeley: University of California Press.

———. (1999). *Infections and Inequalities: The Modern Plagues*. Berkeley: University of California Press.

———. (2003). *Pathologies of Power: Health, Human Rights, and the New War on the Poor*. Berkeley: University of California Press.

———. (2005). "Introducing ARVs in Resource-poor Settings: Expected and Unexpected Challenges and Consequences" (manuscript at http://www.pih .org/library/essays).

Farmer, P., F. Léandre, J. Mukherjee, et al. (2001). "Community-based Approaches to HIV Treatment in Resource-poor Settings." *The Lancet* 358:404–9.

Farnham, P., R. Gorsky, D. Holtgrave, et al. (1996). "Counseling and Testing for HIV Prevention: Costs, Effects, and Cost-effectiveness of More Rapid Screening Tests." *Public Health Reports* 111(1):44–53.

Fassin, Didier. (2001). "The Biopolitics of Otherness: Undocumented Foreigners and Racial Discrimination in French Public Debate." *Anthropology Today* 17 (1):3–7.

———. (2005). "Compassion and Repression: The Moral Economy of Immigration Policies in France." *Cultural Anthropology* 20(3):362–87.

———. (2007). *When Bodies Remember: Experiences and Politics of AIDS in South Africa*. Berkeley: University of California Press.

Fassin, Didier, and Helen Schneider. (2003). "The Politics of AIDS in South Africa: Beyond the Controversies." *British Medical Journal* 326:495–97.

Fassin, Didier, and Paula Vasquez. (2005). "Humanitarian Exception as the Rule: The Political Theology of the 1999 Tragedia in Venezuela." *American Ethnologist* 32(3):389–405.

Ferguson, James, and Akhil Gupta. (2002). "Spatializing States: Toward an Ethnography of Neoliberal Governmentality." *American Ethnologist* 29(4):981–1002.

Fiori, José Luís. (2001). *Brasil no Espaço*. Petrópolis: Editora Vozes.

Fischer, Michael M. J. (1999). "Emergent Forms of Life: Anthropologies of Late or Postmodernities." *Annual Review of Anthropology* 28:455–78.

Fischer, Michael M. J. (2003). *Emergent Forms of Life and the Anthropological Voice*. Durham, NC: Duke University Press.

Fleury, Sonia. (1989). *Reforma Sanitária: Em Busca de uma Teoria*. São Paulo: Cortez/ABRASCO.

———. (1996). *Política de Saúde: O Público e o Privado*. Rio de Janeiro: Editora da Fiocruz.

———. (1994). *Estado Sem Cidadãos: Seguridade Social na América Latina*. Rio de Janeiro: Editora da Fiocruz.

Foege, William, Nils Daulaire, Robert Black, and Clarence Pearson. (2005). *Global Health Leadership and Management*. San Francisco: Jossey-Bass.

Fonseca, M. G., F. I. Bastos, M. Derrico, C. L. T. de Andrade, C. Travassos, and C. L. Szwarcwald. (2000). "AIDS e Grau de Escolaridade no Brasil: Evolução Temporal de 1986 a 1996." *Cadernos de Saúde Pública* 16(Supp. 1):77–87.

Fonseca, M. G., C. Travassos, F. I. Bastos, N. V. Silva, and C. L. Szwarcwald. (2003). "Distribuição Social da AIDS no Brasil, Segundo Participação no Mercado de Trabalho, Ocupação e Status Sócio-Econômico dos Casos de 1987 a 1998." *Cadernos de Saúde Pública* 19(5):1351–63.

Font, Mauricio A. (2003). *Transforming Brazil*. Oxford: Rowman and Littlefield.

Fortun, Kim. (2001). *Advocacy after Bhopal: Environmentalism, Disaster, New Global Orders*. Chicago: University of Chicago Press.

Foucault, Michel. (n/d). *O que é um autor?* Lisboa: Passagens.

———. (1972). *The Archaeology of Knowledge*. New York: Pantheon Books.

———. (1979). *Discipline and Punish: The Birth of the Prison*. New York: Vintage Books.

———. (1980a). *The History of Sexuality*. Vol. I. New York: Vintage Books.

———. (1980b). *Power/Knowledge: Selected Interviews and Other Writings 1972–1977*. Edited by Colin Gordon. New York: Pantheon Books.

———. (1983). "Subject and Power." In *Michel Foucault: Beyond Structuralism and Hermeneutics*, by Hubert Dreyfus and Paul Rabinow. Chicago: University of Chicago Press.

———. (1984). *The Foucault Reader.* Edited by Paul Rabinow. New York: Pantheon Books.

———. (1991). "Governmentality." In *The Foucault Effect: Studies in Governmentality*, edited by Graham Burchell, Colin Gordon, and Peter Miller, pp. 87–104. Chicago: University of Chicago Press.

———. (1992). "Del Poder de la Soberanía al Poder sobre la Vida." In *Genealogía del Racismo*. Buenos Aires: Editorial Altamira.

———. (1998). *Aesthetics, Method, and Epistemology.* Edited by James Faubion. New York: The New Press.

———. (2000). "Confronting Governments: Human Rights." In *Power*, edited by James Faubion, pp. 474–75. New York: The New Press.

Freud, Sigmund. (1955). "The Uncanny." In *The Standard Edition of the Complete Psychological Works of Sigmund Freud*. Vol. XVII. London: Hogarth.

———. (1959). *Group Psychology and the Analysis of the Ego.* New York: W. W. Norton.

———. (1961). *Beyond the Pleasure Principle.* New York: W. W. Norton.

———. (1963). "The Loss of Reality in Neurosis and Psychosis." In *General Psychological Theory: Papers on Metapsychology.* New York: Collier Books.

Freyre, Gilberto. (1987). *Masters and Slaves.* Berkeley: University of California Press.

Fry, Peter (1982). *Para Inglês Ver: Identidade e Política na Cultura Brasileira.* Rio de Janeiro: Zahar.

Galvão, Jane. (2000). *A AIDS no Brasil: A Agenda de Construção de uma Epidemia.* São Paulo: Editora 34.

———. (2001). "HIV/AIDS na América Latina: Desigualdades, Respostas e Desafios." Seminário: Violência Estrutural e Vulnerabilidade frente ao HIV/AIDS na América Latina: Práticas de Resistência.

———. (2002a). "A Política Brasileira de Distribuição e Produção de Medicamentos Anti-retrovirais: Privilégio Ou Um Direito?" *Cadernos de Saúde Pública* 18(1):213–19.

———. (2002b). "Access to Antiretroviral Drugs in Brazil." *The Lancet* 360(9384):1862–65.

GAPA Bahia. (1996). "Relatório do Programa AIDS e Mulheres na Periferia de Salvador." Salvador: GAPA Bahia. Report.

Garcia, R., R. Schooley, and R. Badaró. (2003). "An Adherence Trilogy is Essential for Long-term HAART Success." *Brazilian Journal of Infectious Diseases* 7(5):307–14.

Garrison, John, and Anabela Abreu. (2000). "Government and Civil Society in the Fight Against HIV and AIDS in Brazil." World Bank (Europe and the Americas Forum on Health Sector Reform, May 24–26, 2000). Study Document.

Gauri, Varun, and Evan Lieberman. (2006). "Boundary Politics and HIV/AIDS Policy in Brazil and South Africa." *Studies in Comparative International Development* 41(3):47–73.

Geertz, Clifford. (1973). "Religion as a Cultural System." In *The Interpretation of Cultures.* New York: Basic Books.

———. (2000). *Available Light: Anthropological Reflections on Philosophical Topics.* Princeton: Princeton University Press.

———. (2004). "What Is a State If It Is Not a Sovereign?" *Current Anthropology* 45:577–91.

———. (2005). "Shifting Aims, Moving Targets: On the Anthropology of Religion." *Journal of the Royal Anthropological Institute* 11:1–15.

Geest, Sjaak van der, Susan Reynolds Whyte, and Anita Hardon. (1996). "Anthropology of Pharmaceuticals: A Biographical Approach." *Annual Review of Anthropology* 25:153–78.

Gifford, A., J. Bormann, M. Shively, et al. (2000). "Predictors of Self-reported Adherence and Plasma HIV Concentrations in Patients on Multidrug Antiretroviral Regimens." *JAIDS* 23:386–95.

Goldstein, Donna. (1994). "AIDS and Women in Brazil: The Emerging Problem." *Social Science and Medicine* 39(7):919–24.

———. (2003). *Laughter Out of Place: Race, Class, Violence, and Sexuality in a Rio Shantytown.* Berkeley: University of California Press.

Good, Byron. (1994). *Medicine, Rationality, and Experience.* Cambridge: Cambridge University Press.

Good, Mary-Jo DelVecchio. (1995). "Cultural Studies of Biomedicine: An Agenda for Research." *Social Science and Medicine* 41(4):461–73.

Greenhouse, Carol, Elizabeth Mertz, and Kay B. Warren (eds). (2002). *Ethnography in Unstable Places: Everyday Lives in Contexts of Dramatic Political Change.* Durham, NC: Duke University Press.

Gruskin, Sofia. (2002). "The United Nations General Assembly Special Session on HIV/AIDS: A Landmark Event, But Were Some Lessons of the Last 20 Years Ignored?" *American Journal of Public Health* 92(3)337–38.

Gruskin, Sofia, Michael A. Grodin, George J. Annas, and Stephen P. Marks (eds.). (2005). *Perspectives on Health and Human Rights.* New York: Routledge.

Gurtler, L. (1996). "Difficulties and Strategies of HIV Diagnosis." *The Lancet* 348(9021):176–79.

Hacker, M. A., I. C. Leite, A. Renton, T. G. Torres, R. Gracie, and F. I. Bastos. (2006). "Reconstructing the AIDS Epidemic among Injection Drug Users in Brazil." *Cadernos de Saúde Pública* 22(4):751–60.

Hacking, Ian. (1990). *The Taming of Chance.* Cambridge: Cambridge University Press.

———. (1999). "Making Up People." In *The Sciences Studies Reader,* edited by Mario Biagioli. New York: Routledge.

———. (2002). *Historical Ontology.* Cambridge, MA: Harvard University Press.

Hansen, Thomas Blom, and Finn Steppunat. (2006). "Sovereignty Revisited." *Annual Review of Anthropology* 35:295–315.

Haraway, Donna. (1991). "The Biopolitics of Postmodern Bodies: Constitutions of Self in Immune System Discourse." In *Simians, Cyborgs, and Women: The Reinvention of Nature*. New York: Routledge.

Harvard AIDS Institute. (2000). *Consensus Statement on Antiretroviral Treatment for AIDS in Poor Countries*. Boston: Harvard School of Public Health.

Hashida, S., K. Hasinaka, I. Nishikata, et al. (1996). "Shortening of the Window Period in Diagnosis of HIV-1 Infection by Simultaneous Detection of p24 Antigen and Antibody IgG to p17 and Reverse Transcriptase in Serum with Ultra-sensitive Enzyme Immunoassay." *Journal of Virological Methods* 1(62):43–53.

Hecht, Tobias. (1998). *At Home in the Street: Street Children of Northeast Brazil*. Cambridge: Cambridge University Press.

Higgins, D.L., C. Galavotti, K. R. O'Reilly, D. J. Schnell, M. Moore, D. L. Rugg, and R. Johnson. (1991). "Evidence of the Effects of HIV Antibody Counseling and Testing on Risk Behaviors." *Journal of the American Medical Association* 266(17):2419–29.

Hirschman, Albert O. (1970). "The Search for Paradigms as a Hindrance to Understanding." *World Politics* 22(3):329–43.

———. (1971). *A Bias for Hope: Essays on Development and Latin America*. New Haven: Yale University Press.

———. (1991). *The Rhetoric of Reaction: Perversity, Futility, Jeopardy*. Cambridge, MA: Harvard University Press.

———. (1995). *A Propensity to Self-Subversion*. Cambridge, MA: Harvard University Press.

———. (1998). *Crossing Boundaries: Selected Writings*. New York: Zone Books.

Hochman, Gilberto. (1998). *A Era do Saneamento: As Bases da Política de Saúde Pública no Brasil*. São Paulo: Hucitec/ANPOCS.

Hofer, C., M. Schechter, and L. Harrison. (2004). "Effectiveness of Antiretroviral Therapy among Patients Who Attend Public HIV Clinics in Rio de Janeiro, Brazil." *JAIDS* 36(4):967–71.

Hoffman, Kelly, and Miguel Angel Centeno. (2003). "The Lopsided Continent: Inequality in Latin America." *Annual Review of Sociology* 29:363–90.

Holanda, Sérgio Buarque de. (1956). *Raízes do Brasil*. Rio de Janeiro: José Olympio.

Horsburgh, C. R. Jr., C. Y. Ou, J. Jason, et al. (1989). "Duration of Human Immunodeficiency Virus Infection before Detection of Antibody." *The Lancet* 2(8664):637–40.

Ickovics, J. R., A. C. Morrill, S. E. Beren, U. Walsh, and J. Rodin. (1994). "Limited Effects of HIV Counseling and Testing for Women." *Journal of the American Medical Association* 272(6):443–48.

International Conference on Primary Health Care. (1978). *Declaration of Alma-Ata*. Alma-Ata, USSR (www.who.int/hpr/NPH/docs/declaration_almaata.pdf).

Isto É/Senhor. (1991). "O Sexo Inseguro." November 20, p. 52.

Kaufman, Sharon R., and Lynn M. Morgan. (2005). "The Anthropology of the Beginnings and Ends of Life." *Annual Review of Anthropology* 34:317–41.

Keller, Evelyn Fox. (1992). *Secrets of Life, Secrets of Death: Essays on Language, Gender and Science.* New York: Routledge.

Kim, Jim Yong, and Charlie Gilks. (2005). "Scaling Up Treatment—Why We Can't Wait." *New England Journal of Medicine* 353(22):2392–94.

Kitayaporn, D., J. Kaewkungwal, S. Bejrachandra, et al. (1996). "Estimated Rate of HIV-1-Infectious But Seronegative Blood Donations in Bangkok, Thailand." *AIDS* 10(10):1157–62.

Kleinman, Arthur, and Anne Becker. (1998). "Editorial: 'Sociosomatics': The Contributions of Anthropology to Psychosomatic Medicine." *Psychosomatic Medicine* 60(4):389–93.

Kleinman, Arthur. (1995). *Writing at the Margins: Discourse Between Anthropology and Medicine.* Berkeley: University of California Press.

———. (1999). "Experience and Its Moral Modes: Culture, Human Conditions, and Disorder." In *The Tanner Lectures on Human Values.* Salt Lake City: University of Utah Press.

Kleinman, Arthur, and Joan Kleinman. (1997). "Cultural Appropriations of Suffering in Our Times." In *Social Suffering*, edited by Arthur Kleinman, Veena Das, and Margaret Lock. Berkeley: University of California Press.

Knorr-Cetina, Karin. (2001). "Postsocial Relations: Theorizing Sociality in a Postsocial Environment." In *Handbook of Social Theory*, edited by George Ritzer and Barry Smart, pp. 520–37. London: Sage.

Kober, K., and W. Van Damme. (2004). "Scaling Up Access to Antiretroviral Treatment in Southern Africa: Who Will Do the Job?" *The Lancet* 364:103–7.

Koenig, S., F. Léandre, and P. Farmer. (2004). "Scaling Up HIV Treatment Programmes in Resource-limited Settings: The Rural Haiti Experience." *AIDS* 18(Supp. 3):S21–S25.

Koenig, S. P., D. R. Kuritzkes, M. S. Hirsch, F. Leandre, J. S. Mukherjee, P. E. Farmer, and C. del Rio. (2006). "Monitoring HIV Treatment in Developing Countries." *British Medical Journal* 332:602–4.

Krakauer, Eric L. (1998). *The Disposition of the Subject.* Evanston: Northwestern University Press.

Krischke, Paulo. (1990). "Social Movements and Political Participation: Contribution of Grassroots Democracy in Brazil." *Canadian Journal of Development Studies* 11(1):173–84.

Kulick, Don. (1998). *Travesti: Sex, Gender and Culture among Brazilian Transgendered Prostitutes.* Chicago: University of Chicago Press.

Lacan, Jacques. (n/d). "Seminario XVI: De un Otro al Otro." Buenos Aires: Grupo Verbum (non-commercial edition).

———. (1977). "The Freudian Thing." In *Écrits: A Selection.* New York: W. W. Norton.

————. (1980). *O Mito Individual do Neurótico*. Lisboa: Assírio e Alvim.

————. (1989). "Science and Truth." *Newsletter of the Freudian Field* 3:4–29.

————. (1991). *The Seminar of Jacques Lacan: Book I—Freud's Papers on Technique, 1953–1954*. New York: W. W. Norton.

————. (1994). *O Seminário: O Avesso da Psicanálise, Livro 17*. Rio de Janeiro: Zahar.

Lackritz, E. M., G. A. Satten, and J. Aberle-Grasse, et al. (1995). "Estimated Risk of Transmission of the Human Immunodeficiency Virus by Screened Blood in the United States." *New England Journal of Medicine* 332(26):73–81.

Lakoff, Andrew. (2005). "The Private Life of Numbers." In *Global Assemblages: Technology, Politics, and Ethics as Anthropological Problems*, edited by Aihwa Ong and Stephen J. Collier, pp. 194–213. Malden, MA: Blackwell.

Lange, J., J. Perriens, D. Kuritzkes, and D. Zewdie. (2004). "What Policymakers Should Know about Drug Resistance and Adherence in the Context of Scaling Up Treatment of HIV Infection." *AIDS* 18(Supp. 3):S69–74.

Laplanche, J., and J. B. Pontalis. (1973). *The Language of Psycho-analysis*. New York: W. W. Norton.

Larvie, Patrick. (1997). "Personal Improvement, National Development: Theories of AIDS Prevention in Rio de Janeiro, Brazil." In *The Medical Anthropologies in Brazil*, edited by Annette Leibing. Berlin: Verlag für Wissenschaft und Bildung.

Latour, Bruno. (1988). *The Pasteurization of France*. Cambridge, MA: Harvard University Press.

————. (1990). "Drawing Things Together." In *Representation in Scientific Practice*, edited by Michael Lynch and Steve Woolgar, pp. 19–68. Cambridge, MA: MIT Press.

Le Marcis, Frédéric. (2004). "The Suffering Body of the City." *Public Culture* 16(3):453–57.

Lee, K., and H. Goodman. (2002). "Global Policy Networks: The Propagation of Health Care Financing Reforms since the 1980s." In *Health Policy in a Globalizing World*, edited by K. Lee, K. Buse, and S. Fustukian, pp. 97–119. Cambridge: Cambridge University Press.

Levi, Guido Carlos, and Marco Antonio A. Vitória. (2002). "Fighting against AIDS: The Brazilian Experience. *AIDS* 16:2373–83.

Lévi-Strauss, Claude. (1963). *Structural Anthropology*. New York: Basic Books.

Lock, Margaret. (2002). *Twice Dead: Organ Transplants and the Reinvention of Death*. Berkeley: University of California Press.

Longini, I. M., W. S. Clark, M. Haber, et al. (1989). "The States of HIV Infection: Waiting Times and Infection Transmission Probabilities." In *Mathematical and Statistical Approaches to AIDS Epidemiology, Lecture Notes in Biomathematics*, vol. 83, edited by C. Castillo-Chaves. New York: Springer-Verlag.

Lowy, Ilana. (2001). *Virus, Moustiques, et Modernité: La Fievre Jaune au Brésil*. Paris: Archives d'Histoire Contemporaine.

Luhrman, Tanya. (2000). *Of Two Minds: The Growing Disorder in American Psychiatry*. New York: Alfred A. Knopf.

Luiza, Vera Lucia. (1999). "Aquisição de Medicamentos no Setor Público: O Binômio Qualidade-Custo." *Cadernos de Saúde Pública* 15(4).

Lupton, Deborah, Sophie McCarthy, and Simon Chapman. (1995). "'Doing the Right Thing': The Symbolic Meanings and Experiences of Having an HIV Antibody Test." *Social Science and Medicine* 41(2):173–80.

Lurie, Peter, Percy Hintzen, and Robert Lowe. (1995). "Socioeconomic Obstacles to HIV Prevention and Treatment and Developing Countries: The Roles of the International Monetary Fund and the World Bank." *AIDS* 9:539–46.

Lush, L., G. Walt, and J. Ogden. (2003). "Transferring Policies for Treating Sexually Transmitted Infections: What's Wrong with Global Guidelines?" *Health Policy and Planning* 18(1):18–30.

Lyon, E., and P. Farmer. (2005). "Inequality, Infections, and Community-based Health Care." *Yale Journal of Health Policy, Law, and Ethics* 5(1):465–73.

MacRae, Edward. (1990). *A Construção da Igualdade: Identidade Sexual e Política no Brasil da Abertura*. Campinas: Editora UNICAMP.

Mainwaring, Scott. (1987). "Urban Popular Movements, Identity, and Democratization in Brazil." *Comparative Political Studies* 20(2):131–59.

Malta, M., C. Carneiro-da-Cunha, D. Kerrigan, A. S. Strathdee, M. Monteiro, and F. I. Bastos. (2003). "Case Management of Human Immunodeficiency Virus—Infected Injection Drug Users: A Case Study in Rio de Janeiro, Brazil." *Clinical Infectious Diseases* 37(Supp. 5):S386–91.

Martin, Emily. (1994). *Flexible Bodies: Tracking Immunity in American Culture from the Days of Polio to the Age of AIDS*. Boston: Beacon.

Marx, Karl. (1983). "Production of Surplus Value." In *The Portable Karl Marx*, edited by Eugene Kameka. New York: Penguin Books.

Masco, Joseph. (2004). "Mutant Ecologies: Radioactive Life in Post–Cold War New Mexico." *Cultural Anthropology* 19(4):517–50.

Mbembe, Achille. (2003). "Necropolitics." *Public Culture* 15(1):11–40.

Melo Neto, João Cabral de. (2005). *Education by Stone*. New York: Archipelago Books.

Menegoni, L. (1996). "Conceptions of Tuberculosis and Therapeutic Choices in Highland Chiapas, Mexico." *Medical Anthropology Quarterly* 10:381–401.

Miller, Elizabeth. (1998). "The Uses of Culture in the Making of AIDS Neurosis in Japan." *Psychosomatic Medicine* 60:402–9.

Ministério da Saúde [MS]. (1995). *Segundo Relatório Sobre o Processo de Organização da Gestão da Assistência à Saúde*. Brasília: Ministério da Saúde (third version).

———. (1997). *Farmácia Básica: Programa 1997/98*. Brasília: Ministério da Saúde.

———. (1999). *Política Nacional de Medicamentos*. Brasília: Ministério da Saúde.

———. (2002). *A Experiência do Programa Brasileiro de Aids*. Brasília: Ministério da Saúde.

Misra, Kavita. (2003). *A Safe Space: AIDS and New Sociality in an Indian Setting.* Ph.D. dissertation, Princeton University.

———. (2006). "Politico-moral Transactions in Indian AIDS Service: Confidentiality, Rights and New Modalities of Governance." *Anthropological Quarterly* 79(1):33–74.

Moatti, J. P., I. N'Doye, S. M. Hammer, P. Hale, and M. Kazatchkine. (2003). "Antiretroviral Treatment for HIV Infection in Developing Countries: An Attainable New Paradigm." *Nature Medicine* 9(12):1449–52.

Moreira, Edson D., N. Silva, C. Brites, E. M. Carvalho, J. C. Bina, R. Badaró, and W. D. Johnson Jr. (1993). "Characteristics of the Acquired Immunodeficiency Syndrome in Brazil." *American Journal of Tropical Medicine and Hygiene* 48(5):687–92.

Moreira, Esdras Cabus, and João Biehl. (2004). "Práticas Médicas de Aceitação da Morte na UTI de um Hospital Geral do Nordeste do Brasil." *Bioética* 12(1):19–30.

Mott, Luiz, and Marcelo Cerqueira. (1996). "Os Travestis da Bahia e a AIDS: Prostituição, Silicone e Drogas." Salvador: Grupo Gay da Bahia. Manuscript.

Mukherjee, J., M. Colas, P. Farmer, et al. (2003). *Access to Antiretroviral Treatment and Care: The Experience of the HIV Equity Initiative, Cange, Haiti: Case Study.* Geneva: World Health Organization.

Nattrass, Nicoli. (2004). *The Moral Economy of AIDS in South Africa.* Cambridge: Cambridge University Press.

———. (2006). "What Determines Cross-Country Access to Antiretroviral Treatment?" *Development Policy Review* 24(3):321–37.

Neisson-Vernant C., S. Arfi, S. Mathez, et al. (1986). "Needlestick HIV Seroconversion in a Nurse." *The Lancet* 2(8510):814.

Nemes, M., H. Carvalho, and M. Souza. (2004). "Antiretroviral Therapy Adherence Rates in Brazil." *AIDS* 18(Supp. 3):S15–20.

Nguyen, Vinh-Kim. (2005). "Antiretroviral Globalism, Biopolitics, and Therapeutic Citizenship." In *Global Assemblages: Technology, Politics, and Ethics as Anthropological Problems*, edited by Aihwa Ong and Stephen J. Collier, pp. 124–44. Malden, MA: Blackwell.

Nichter, Mark, and N. Vuckovic. (1994). "Agenda for an Anthropology of Pharmaceutical Practice." *Social Science and Medicine* 39(11):1509–25.

Ogden, J., L. Lush, and G. Walt. (2003). "The Politics of Branding in Public Health: The Case of DOTS for Tuberculosis Control." *Social Science and Medicine* 57(1):163–72.

Okie, S. (2006). "Fighting HIV—Lessons from Brazil." *New England Journal of Medicine* 354(19):1977–81.

Oldani, Michael. (2004). "Thick Prescriptions: Toward an Interpretation of Pharmaceutical Sales Practices." *Medical Anthropology Quarterly* 18(3):325–56.

Oliveira, M. A., A. F. Esher, E. M. Santos, M. A. E. Cosendey, V. L. Luiza, and J. A. Z. Bermudez. (2002). "Avaliação da Assistência Farmacêutica às Pessoas

Vivendo com HIV/AIDS no Município do Rio de Janeiro." *Cadernos de Saúde Pública* 18(5):1429–39.

Ong, Aihwa. (1999). *Flexible Citizenship*. Durham, NC: Duke University Press.

———. (2006). *Neoliberalism as Exception: Mutations in Citizenship and Sovereignty*. Durham, NC: Duke University Press.

Ong, Aihwa, and Stephen J. Collier (eds.). (2005). *Global Assemblages: Technology, Politics, and Ethics as Anthropological Problems*. Malden, MA: Blackwell.

Ostrom, Elinor. (2002). "Policy Analysis in the Future of Good Societies." *The Good Society* 11(1):42–48.

Oyugi, J., J. Byakika-Tusiime, E. Charlebois, C. Kityo, and R. Mugerwa. (2004). "Multiple Validated Measures of Adherence Indicate High Levels of Adherence to Generic HIV Antiretroviral Therapy in a Resource-limited Setting." *JAIDS* 36(5):1100–1102.

Paley, Julia. (2001). *Marketing Democracy: Power and Social Movements in Postdictatorship Chile*. Berkeley: University of California Press.

Paraguassú, L. (2001). "Governo Pode Quebrar Patente de Remédio." *Folha de São Paulo*, December 15.

Parker, Richard. (1990). *Bodies, Pleasures, and Passions*. Boston: Beacon.

———. (1994). *A Construção da Solidariedade: AIDS, Sexualidade e Política no Brasil*. Rio de Janeiro: Relume Dumará, ABIA, IMS/UERJ.

Parker, Richard (ed.). (1997). *Políticas, Instituições e AIDS: Enfrentando a Epidemia no Brasil*. Rio de Janeiro: Jorge Zahar/ABIA.

Parker, Richard. (2001). "Estado e Sociedade em Redes: Descentralização e Sustentabilidade das Ações de Prevenção das DSTs/AIDS." Paper presented at the Fourth National Conference on DST/AIDS Prevention, Cuiabá, Brazil, September 11.

Parker, Richard, Cristiana Bastos, Jane Galvão, and José Stálin Pedrosa (eds.). (1994). *A AIDS no Brasil*. Rio de Janeiro: Relume Dumará, ABIA, IMS/UERJ.

Parker, Richard, and Kenneth R. Camargo Jr. (2000). "Pobreza e HIV/AIDS: Aspectos Antropológicos e Sociológicos." *Cadernos de Saúde Pública* 16(1):89–102.

Parker, Richard, and Herbert Daniel. (1991). *AIDS: A Terceira Epidemia*. São Paulo: Editora Iglu.

Patton, Cindy. (1990). *Inventing AIDS*. New York: Routledge.

———. (1996). *Fatal Advice: How Safe-Sex Went Wrong*. Durham, NC: Duke University Press.

———. (2002). *Globalizing AIDS*. Minneapolis: University of Minnesota Press.

Pechansky, F., G. Woody, G. J. Inciardi, H. Surratt, F. Kessler, L. Von Diemen, and D. B. Bumaguin. (2006). "HIV Seroprevalence among Drug Users: An Analysis of Selected Variables Based on 10 Years of Data Collection in Porto Allegre, Brazil." *Drug and Alcohol Dependency* 82(Supp. 1):S109–S13.

Pedral-Sampaio, D., E. Netto, A. P. Alcantara, J. Souza, L. Moura, C. Brites, C. Pedroso, J. C. Bina, and R. Badaró. (1997). "Use of Standard Therapy for

Tuberculosis Is Associated with Increased Adverse Reactions in Patients with HIV." *Brazilian Journal of Infectious Diseases* 1(3):123–30.

Petersen, Alan, and Deborah Lupton. (1996). *The New Public Health: Health and Self in the Age of Risk.* St. Leonards: Allen & Unwin.

Petryna, Adriana. (2002). *Life Exposed: Biological Citizens after Chernobyl.* Princeton: Princeton University Press.

———. (2005). "Ethical Variability: Drug Development and the Globalization of Clinical Trials." *American Ethnologist* 32(2):183–97.

———. (2006). "Globalizing Human Subjects Research." In *Global Pharmaceuticals: Markets, Practices, Ethics,* edited by Adriana Petryna, Andrew Lakoff, and Arthur Kleinman, pp. 33–60. Durham, NC: Duke University Press.

Petryna, Adriana, and João Biehl. (1997). "O Estádio Clínico: A Constituição de uma Criança Inválida." *Revista da Associação Psicanalítica de Porto Alegre* 7(13):83–101.

Petryna, Adriana, Andrew Lakoff, and Arthur Kleinman (eds.). (2006). *Global Pharmaceuticals: Markets, Practices, Ethics.* Durham, NC: Duke University Press.

Phillips, K., and T. J. Coates. (1995). "HIV Counseling and Testing: Research and Policy Issues." *AIDS Care* 7(2):115–24.

Pogge, Thomas W. (2002). *World Poverty and Human Rights.* New York: Polity.

———. (2005). "Human Rights and Global Health: A Research Program." *Metaphilosophy* 36(1/2):182–209.

Prakash, Gyan. (1999). *Another Reason: Science and the Imagination of Modern India.* Princeton: Princeton University Press.

Rabinow, Paul. (1994). "Introduction: A Vital Rationalist." In *A Vital Rationalist: Selected Writings,* by Georges Canguilhem, edited by Francois Delaporte, pp. 11–22. New York: Zone Books.

———. (1996). *Essays on the Anthropology of Reason.* Princeton: Princeton University Press.

———. (1999). *French DNA: Trouble in Purgatory.* Chicago: University of Chicago Press.

———. (2003). *Anthropos Today: Reflections on Modern Equipment.* Princeton: Princeton University Press.

Rabinow, Paul, and Talia Dan-Cohen. (2005). *A Machine to Make a Future: Biotech Chronicles.* Princeton: Princeton University Press.

Rajan, Kaushik Sunder. (2006). *Biocapital: The Constitution of Postgenomic Life.* Durham, NC: Duke University Press.

Ramiah, Ilavenil, and Michael R. Reich. (2005). "Public-Private Partnerships and Antiretroviral Drugs for HIV/AIDS: Lessons from Botswana." *Health Affairs* 24(2):545–51.

Rancière, Jacques. (2004). "Who Is the Subject of the Rights of Man?" *South Atlantic Quarterly* 102(2/3):297–310.

Ranki, A., S. Valle, M. Krohn, et al. (1987). "Long Latency Precedes Overt Sero-conversion in Sexually Transmitted Human Immunodeficiency Virus Infection." *The Lancet* 2(8559):589–93.

Rapp, Rayna. (1999). *Testing Women, Testing the Fetus: The Social Impact of Amniocentesis in America.* New York: Routledge.

Raviglione, M. (2003). "The TB Epidemic from 1992 to 2002." *Tuberculosis* 83(3):4–14.

Redfield, Peter. (2005). "Doctors, Borders, and Life in Crisis." *Cultural Anthropology* 20(3):328–61.

Rheinberger, Hans-Jörg. (1997). *Toward a History of Epistemic Things: Synthesizing Proteins in the Test Tube.* Stanford: Stanford University Press.

Ribeiro, Renato Janine. (2000). *A Sociedade Contra o Social.* São Paulo: Companhia das Letras.

Riles, Annelise. (2000). *The Network Inside Out.* Ann Arbor: University of Michigan Press.

Roitman, Janet. (2005). "The Garrison-Entrepôt: A Mode of Governing in the Chad Basin." In *Global Assemblages: Technology, Politics, and Ethics as Anthropological Problems*, edited by Aihwa Ong and Stephen J. Collier, pp. 417–36. Malden, MA: Blackwell.

Rosário Costa, Nilson. (1996). "O Banco Mundial e a Política Social nos Anos 90: A Agenda para a Reforma do Setor Saúde no Brasil." In *Política de Saúde e Inovação Institucional: Uma Agenda para os Anos 90*, edited by N. Rosário Costa and J. Mendes Ribeiro. Rio de Janeiro: Secretaria de Desenvolvimento Educacional/ENSP.

Rosenberg, Charles E. (2003). "What Is Disease." *Bulletin of the History of Medicine* 77:491–505.

Rosenberg, Tina. (2001). "How To Solve the World's AIDS Crisis." *New York Times Magazine*, January 28.

———. (2006). "When a Pill Is Not Enough." *New York Times Magazine*, August 6, pp. 40–45, 52, 58–59.

Roudinesco, Elisabeth. (1997). *Jacques Lacan.* New York: Columbia University Press.

Rugg, D., R. MacGowan, K. Stark, et al. (1991). "Evaluating the CDC Program for HIV Counseling and Testing." *Public Health Reports* 106(6):708–13.

Safren, S. A., N. Kumarasamy, R. James, S. Raminani, S. Solomon, and H. M. Mayer. (2005). "ART Adherence, Demographic Variables and CD4 Outcome among HIV-positive Patients on Antiretroviral Therapy in Chennai, India." *AIDS Care* 17:853–62.

Sands, Joseph Charles. (2004). *After Decentralization: Federal Constraints, Local Oversights, and Municipal School Systems in Brazil.* Ph.D. dissertation, Princeton University.

Santoro-Lopes, G., A. M .F. de Pinho, L. H. Harrison, and M. Schechter. (2002). "Reduced Risk of Tuberculosis among Brazilian Patients with Advanced

Immunodeficiency Virus Infection Treated with Highly Active Antiretroviral Therapy." *Clinical Infectious Diseases* 34:543–46.

Saraceni, V., E. Souza, S. Passos, et al. (1996). "Importance of AIDS Anonymous Testing Site in Rio de Janeiro, Brazil." XI International Conference on AIDS, Abstracts-on-Disk.

Sassen, Saskia. (1998). *Globalization and Its Discontents.* New York: The New Press.

———. (2006). *Territory, Authority, Rights: From Medieval to Global Assemblages.* Princeton: Princeton University Press.

Scheper-Hughes, Nancy. (1991). "Reproductive Health and AIDS in Brazil" (A Preliminary Report). Department of Anthropology, University of California at Berkeley.

———. (1992). *Death Without Weeping: The Violence of Everyday Life in Brazil.* Berkeley: University of California Press.

———. (1994). "AIDS and the Social Body." *Social Science and Medicine* 39(7):991–1003.

———. (2003). "Rotten Trade: Millennial Capitalism, Human Values and Global Justice in Organs Trafficking." *Journal of Human Rights* 2(2):197–226.

Scheper-Hughes, Nancy, and Philippe Bourgois (eds). (2004). *Violence in War and Peace: An Anthology.* Malden, MA: Blackwell.

Secretaria da Saúde do Estado da Bahia. (1996). *Plano Estadual de Saúde 1996–1999.* Salvador: Coordenação de Desenvolvimento de Recursos Humanos/ Divisão de Comunicação e Documentação.

Sell, Susan. (2003). *Private Power, Public Law: The Globalization of Intellectual Property Rights.* Cambridge: Cambridge University Press.

Serra, José. (2000). *Ampliando o Possível.* São Paulo: Hucitec.

———. (2004). *The Political Economy of the Struggle Against AIDS in Brazil.* Occasional Papers (17). School of Social Science, Institute for Advanced Study.

Severe, P., P. Leger, and M. Charles, et al. (2005). "Antiretroviral Therapy in a Thousand Patients with AIDS in Haiti." *New England Journal of Medicine* 353: 2325–34.

Shapin, Steven, and Simon Schaffer. (1985). *Leviathan and the Air-Pump: Hobbes, Boyle, and the Experimental Life.* Princeton: Princeton University Press.

Sheon, Nicholas. (1996). "HIV Test Counseling, Confessional Ritual, and the Deployment of Safe Sexuality." Ph.D. dissertation prospectus, Department of Anthropology, University of California at Berkeley.

Silva, R.C.S., and J.A.Z. Bermudez. (2000). "Medicamentos Excepcionais no Âmbito da Assistência Farmacêutica no Brasil." Manuscript.

Singler, Jennifer, and Paul Farmer. (2005). "Treating HIV in Resource-poor Settings." *Journal of the American Medical Association* 288(13):1652–53.

Slaughter, Anne-Marie. (2004). *A New World Order.* Princeton: Princeton University Press.

Souza, Amaury de. (1999). "Cardoso and the Struggle for Reform in Brazil." *Journal of Democracy* 10(3):49–63.

Souza, Herbert de. (1991). "As ONGs na Década de Noventa." *Comunicações do ISER* 10(41):5–10.

Spire, B., S. Duran, et al. (2002). "Adherence to Highly Active Antiretroviral Therapies (HAART) in HIV-infected Patients: From a Predictive to a Dynamic Approach." *Social Science and Medicine* 54:1481–96.

Stepan, Nancy. (1976). *Beginnings of Brazilian Science: Oswaldo Cruz, Medical Research and Policy, 1890–1920.* New York: Science History Publications.

Sterckx, Sigrid. (2004). "Patents and Access to Drugs in Developing Countries: An Ethical Analysis." *Developing World Bioethics* 4(1):58–75.

Stiglitz, Joseph. (2002). *Globalization and Its Discontents.* New York: W. W. Norton.

Strange, Susan. (1996). *The Retreat of the State: The Diffusion of Power in the World Economy.* Cambridge: Cambridge University Press.

Styblo, K. (1984). "Some of the Main Unsolved Problems in Tuberculosis Control in Developing Countries." *Bulletin of the International Union Against Tuberculosis and Lung Disease* 59(1/2):85–87.

Szwarcwald, Celia L., E. A. Castilho, A. B. Junior, M.R.O. Gomes, E.A.M.M. Costa, B. V. Maletta, R.F.M. Carvalho, S. R. Oliveira, and P. Chequer. (2000). "Comportamento de Risco nos Conscritos do Exército Brasileiro, 1998: Uma Apreciação da Infecção pelo HIV segundo Diferenciais Sócio-econômicos." *Cadernos de Saúde Pública* 16(Supp. 1):113–28.

Teixeira, M. G., M. T. Carvalho, J. P. Dias, and J. Castro. (2002). *Boletim Epidemiológico SESAB/SUVISA/ DIVEP: Evolução Temporal das Doenças de Notificação Compulsória na Bahia de 1980 a 2000.* Salvador: Secretaria da Saúde do Estado da Bahia.

Teixeira, Paulo Roberto. (1997). "Políticas Públicas em AIDS." In *Políticas, Instituições e AIDS: Enfrentando a Epidemia no Brasil,* edited by Richard Parker. Rio de Janeiro: Jorge Zahar/ABIA.

Terra, M.C.T., and M.A.C. Bonono. (1999). "The Political Economy of Exchange Rate Policy in Brazil, 1964–1997." *Ensaios Econômicos* (Fundação Getulio Vargas), n. 341.

Thomas, Caroline. (2002). "Trade Policy and the Politics of Access to Drugs." *Third World Quarterly* 23(2):251–64.

Travis, P., S. Bennett, A. Haines, et al. (2004). "Overcoming Health-systems Constraints to Achieve the Millennium Development Goals." *The Lancet* 364:900–906.

Treichler, Paula. (1999). *How to Have Theory in an Epidemic: Cultural Chronicles of AIDS.* Durham, NC: Duke University Press.

Trevisan, João Silvério. (1986). *Devassos no Paraíso.* São Paulo: Max Limonad.

Trevisol, F. S., and M. V. da Silva. (2005). "HIV Frequency among Female Sex Workers in Imbituba, Santa Catarina, Brazil." *Journal of Infectious Diseases* 9(6):500–505.

Trouillot, Michel-Rolph. (2003). *Global Transformations: Anthropology and the Modern World*. New York: Palgrave Macmillan.

Tsing, Anna L. (2005). *Friction: An Ethnography of Global Connections*. Princeton: Princeton University Press.

Turkle, Sherry. (1991). *Psychoanalytic Politics: Jacques Lacan and Freud's French Revolution*. New York: Guilford.

———. (1997). *Life on the Screen: Identity in the Age of the Internet*. New York: Simon and Schuster.

UNAIDS. (2000a). "Brazil: Epidemiological Fact Sheets on HIV/AIDS and Sexually Transmitted Infections." Geneva: UNAIDS. Document.

———. (2000b). "UNAIDS Policy on HIV Testing and Counseling." Geneva: UNAIDS. Document.

———. (2004). *Report on the Global AIDS Epidemic*. Geneva: UNAIDS.

———. (2006). *Report on the Global AIDS Epidemic* (A UNAIDS 10th Anniversary Special Edition). Geneva: UNAIDS.

United Nations [UN]. 2001. *HIV/AIDS: Global Crisis—Global Action*. New York: United Nations. Document.

Valdiserri, R., M. Moore, R. Gerber, et al. (1993). "A Study of Clients Returning for Counseling After HIV Testing: Implications for Improving Rates of Return." *Public Health Reports* 108(1):12–18.

Varela, Drauzio. (1999). *Estação Carandiru*. São Paulo: Companhia das Letras.

Walton, D. A., P. E. Farmer, W. Lambert, F. Léandre, S. P. Koenig, and J. S. Mukherjee. (2004). "Integrated HIV Prevention and Care Strengthens Primary Health Care: Lessons from Rural Haiti." *Journal of Public Health Policy* 25(2):137–58.

Warren, Kay. (1998). *Indigenous Movements and Their Critics: Pan-Maya Activism in Guatemala*. Princeton: Princeton University Press.

Westerhaus, Michael, and Arachu Castro. (2006). "How Do Intellectual Property Law and International Trade Agreements Affect Access to Antiretroviral Theraphy?" *PLoS Medicine* 3(8):1–7.

Whyte, Susan Reynolds, Sjaak van der Geest, and Anita Hardon. (2003). *Social Lives of Medicines*. Cambridge: Cambridge University Press.

Whyte, Susan Reynolds, M. A. Whyte, L. Meinert, and B. Kyaddondo. (2006). "Treating AIDS: Dilemmas of Unequal Access in Uganda." In *Global Pharmaceuticals: Markets, Practices, Ethics*, edited by Adriana Petryna, Andrew Lakoff, and Arthur Kleinman. Durham, NC: Duke University Press.

Wilken, P.R.C., and J.A.Z. Bermudez. (1999). "Os Descaminhos da Assistência Farmacêutica Previdenciária no Brasil." Manuscript.

Williamson, John (ed.). (1990). "What Washington Means by Policy Reform." In *Latin American Adjustment: How Much Has Happened?* Washington: Institute for International Economics.

Wogart, J. P., and G. Calcagnotto. (2006). "Brazil's Fight Against AIDS and Its Implications for Global Health Governance." *World Health & Population*, January, pp. 1–16.

World Bank. (1988). *Brazil: Public Spending on Social Programs; Issues and Options, Volumes 1 and 2.* Washington, D.C.: World Bank.

———. (1993). *The Organization, Delivery and Financing of Health Care in Brazil.* Washington, D.C.: World Bank.

———. (1999). "Brazil: Partnerships in the Fight against AIDS Program Considered Model for Others." Press release, December 7 (http://www.worldbank.org/aids/).

———. (2002). "Stemming the HIV/AIDS Epidemic in Brazil." Press release, July (http://www.worldbank.org/aids/).

World Health Organization [WHO]. (1988). *From Alma-Ata to the Year 2000: Reflections at the Midpoint.* Geneva: World Health Organization.

———. (2002). *Final Report of the External Evaluation of Roll Back Malaria.* Geneva: World Health Organization.

———. (2004). *Scaling-Up Antiretroviral Therapy in Resource-limited Settings: Treatment Guidelines for a Public Health Approach, 2003 Revision.* Geneva: World Health Organization.

———. (2005). *Progress on Global Access to HIV Antiretroviral Therapy: An Update on "3 by 5."* Geneva: World Health Organization.

———. (2006). "Progress Towards Universal Access: 3 by 5 and Beyond." Geneva: World Health Organization.

World Health Organization/UNAIDS. (2003). *Treating 3 Million by 2005: Making It Happen—the WHO Strategy.* Geneva: World Health Organization/UNAIDS.

Yamey, G. (2001). "Global Campaign to Eradicate Malaria." *British Medical Journal* 322:1191–92.

Yashar, Deborah. (2005). *Contesting Citizenship in Latin America.* Cambridge: Cambridge University Press.

Young, Allan. (1995). *The Harmony of Illusions.* Princeton: Princeton University Press.

———. (2000). "Our Traumatic Neurosis and Its Brain." Unpublished manuscript.

Young, Iris. (2000). *Inclusion and Democracy.* Oxford: Oxford University Press.

Yunes, J. (1999). "Promoting Essential Drugs, Rational Drug Use and Generics: Brazil's National Drug Policy Leads the Way." *Essential Drugs Monitor* 27: 22–23.

Zhang, J., and K. F. Yu. (1998). "What's the Relative Risk? A Method of Correcting the Odds Ratio in Cohort Studies of Common Outcomes." *Journal of the American Medical Association* 280(19):1690–91.

Zizek, Slavoj. (1997). "How Did Marx Invent the Symptom?" In *Mapping Ideology.* New York: Verso.

———. (1999). *The Ticklish Subject: The Absent Centre of Political Ontology.* New York: Verso.

Zou, G. (2004). "A Modified Poisson Regression Approach to Prospective Studies with Binary Data." *American Journal of Epidemiology* 159(7):702–6.

Zweig, Stefan. (1941). *Brazil: Land of the Future.* New York: Viking.

Index

ABIA (Associação Brasileira Interdisciplinar de AIDS), 62, 169–70
accompagnateurs (local health workers), 305–6, 387
activism. *See* AIDS activism; NGOs (nongovernmental organizations)
Adair, 152*p*
Aedes aegypti mosquitoes, 90
African Comprehensive HIV/AIDS Partnership (ACHAP), 385
Agamben, Giorgio, 325
aidéticos (having AIDS), 116–17, 161, 188, 300, 303. *See also* AIDS patients; Caasah AIDS patients
AIDS & Prostitution (GAPA Bahia), 161
AIDS & Teen-agers in Public Schools (GAPA Bahia), 161
AIDS & Women (GAPA Bahia), 161
AIDS activism: AIDS program success due to, 105; background leading to, 62–68; civil space created by, 393; countering technocracy, 155–58; "dramas of inclusion" approach by, 133–34; events leading to activist state and, 68–72; facilitating invasion of HIV/AIDS science, 414n.12; "failure visibility effect" of, 107–8; loss of idealism by, 390; measuring success/undesirable realities of prevention, 160–64; micro-politics of patienthood and, 120–25; "parallel polity" of, 108; pharmaceutical form of governance and collective action by, 73–78; prevention programs through, 160–64; professionalized version of international, 105–6; as solution to AIDS crisis, 95; theatrical, 128–30; trajectories taken by, 110–14; transformations of, 105–10. *See also* civil society; NGOs (nongovernmental organizations); state-society partnerships

"AIDS citizens," 33, 302–3. *See also* citizenship
AIDS death: "acceptance of," 212; Pernalonga's, 176; providing a dignified, 318; records gathered from Caasah on, 222–25; records gathered from hospitals on, 217–22; Rose on fear of suffering and, 324
AIDS death certificates: gathering records on, 217–25; registration bias evidenced through information on, 217–22; survey regarding racial differences in, 390
AIDS epidemic: disease politics of, 248; global statistics on, 4; increased funding to fight, 377–79; institutional and medical changes due to, 405–6; pharmaceutical philanthropy and equity responses to, 384–88; rational-technical control of, 91; recognized as problem by UN (1986), 61; social mobilization as solution to, 95; transformation from fatal to lifelong disease, 375. *See also* Brazilian HIV/AIDS epidemic
AIDS Forum, 125–26
AIDS, Gender & Ethnicity (GAPA Bahia), 161
AIDS neurosis, 243
AIDS NGOs. *See* NGOs (nongovernmental organizations)
AIDS patient registrations: Bahian Epidemiological Surveillance Service on, 219*fig*; Caasah records on, 222–25; declining rates of, 235; by number of hospitalizations, 220*fig*; by reported drug use, 220*fig*; reporting bias associated with, 221–25
AIDS patients: access to care (2005) by, 397–99; *accompagnateurs* (local health workers) work with, 305–6, 387; "active partnership" between health care professionals and, 304–5; adherence

AIDS patients (*continued*)
to ARVs correlated to better outcomes
of, 303–8; "AIDS cocktail" success for,
164–69; AIDS drugs clinical testing on,
276–79; ARV rollout's coexistence with
social exclusion of, 283–86; ARV side
effects suffered by, 355–56, 358; Calixto, 191; challenges of treating homeless, 225–31; continued post-treatment
discrimination against, 354–55;
denial by, 188–89, 196; denial by rich,
189–90; disability pension paid to,
318–19; Eliane, 158, 159*p*; invisibility
of dying, 180; local economies of salvation reality of, 48, 401; micro-politics
of patienthood and, 120–25; Orlando
Lima, 226, 228; Paulinho, 126–27;
Pernalonga, 126–27*p*, 128–30, 176;
in Porto Alegre's penitentiary, 116–18,
157–59*p*; public health discourse blaming, 168–69; racial differences in mortality rates, 390; Raisa, 312; reduced
to begging, 108–10; Sebastião Santos,
183; selected characteristics of, 219*t*;
self-efficacy of, 304; social workers'
description of, 216–17; Tania Bastos
("Sheriff"), 226–28. See also *aidéticos*
(having AIDS); AIDS survivors; Caasah
AIDS patients; Winkler, Gerson
AIDS patients' invisibility: ARV rollout's
coexistence with, 283–86; author's
report on Bahian, 237; denial as factor
in, 188–89, 196; Dona Conceição on
treatment challenges related to, 150,
154–55, 185, 188–90; homelessness
contributing to, 225–31; impact on
young and impoverished, 231; infrastructure and social obstacles related to,
173–75; Lazaro de Oliveira's story on,
198–99*p*, 200; Maria Madalena's story
on, 194, 196–97*p*, 198, 200; Pelourinho
Pillory residents' stories on, 194–202;
of Porto Alegre's penitentiary patients,
116–18, 157–59*p*; registration bias
associated with, 217–25, 235; social
life of death certificates and, 217–25;
state of "apparent invisibility," 217;
system on nonintervention related to,
204–6; technologies of, 202–4. *See
also* Caasah AIDS patients; citizenship;
discrimination

AIDS post-treatment access: continued
discrimination against AIDS patients,
353–55, 404; drugs side effects and
rescue treatments, 355–58; Evangivaldo
on financial aspects of, 368–71; focus
on controlling individualized disease,
349–53; limits of quality care, 347–49;
Luis's reflections on, 358–63; revisiting
Caasah (2001) to examine, 339–49, 395;
revisiting Caasah (2005) to examine,
397–406; Rose's reflections on, 364–68,
402–3, 404. *See also* Brazilian HIV/
AIDS health care
AIDS Project I (1994–98), 65, 136
AIDS Project II (1998), 70–71, 166–67
AIDS Research and Assistance Center
(Bahia), 276, 277–79
AIDS science: ARV reverse engineering,
87–90; described, 121–22; "impure,"
414n.12
AIDS. *See* AIDS epidemic; Brazilian HIV/
AIDS epidemic
AIDS survivors: continued discrimination
against, 353–55, 404; divided by two
distinct identities, 404; drugs side effects
and rescue treatments of, 355–58; focus
on controlling individualized disease
of, 349–53; reports during 2005 visit
by, 397–406. *See also* AIDS patients;
Caasah AIDS patients
Aitken, Murray, 79
Alckmin, Geraldo, 399
Almeida-Filho, Naomar de, 262–63
Altemar, 395
Alves, Naum, 225, 348–49
Andrade, Leila, 351–52
Andrade, Tarcísio Matos de, 191, 194, 226
animal laborans concept, 325
anthropology: perspective of, 5; task of,
405; task of political, 415n.24. *See also*
ethnography; science
ANVISA (Brazil), 92–93, 383. *See also*
Brazilian Health Ministry
apadrinhamento (godfathering practice), 367
Appadurai, Arjun, 62
"apparent invisibility," 207
ARCA (Apoio Religioso Contra à AIDS),
62, 110
Arendt, Hannah, 325
Arpão (Porto Alegre penitentiary news-
letter), 158

Arruda, Lizane, 276

Artemisa (Bahian CTA study), 267–68

ARV adherence: *accompagnateurs* (local health workers) role in, 305–6, 387; "active partnership" factor of, 304–5; AIDS patient's failure to comply with, 228–29; correlated to better outcomes for patients, 303–8; institutional belonging as begetting, 306, 307; Luis on role of religion in, 361–62; pastoral vision of care facilitating, 306, 308; self-efficacy factor of, 304; subjectivity role in, 400

ARVs (antiretroviral drugs): better treatment outcomes correlated with adherence to, 303–8; Brazilian clinical trials for, 276–79, 358; Brazilian initiative to lower prices (2001) of, 77; Brazilian production of generic, 87–93; Dr. Radames's trials on, 232–34; fusion inhibitors, 382–83; global differences in prices paid for, 72; international cooperation regarding, 171; Kaletra, 383, 398; Nevirapine, 108; pharmaceuticalization of public health role of, 12; pricing impact of generic drugs on, 91–92; reverse-engineering of, 76, 93; side effects of, 355–56, 358; successful outcomes of, 164–69. *See also* Brazilian ARV rollout; drugs; Global ARV rollouts; intellectual property rights

ARVs universal access: Brazilian policy on, 3, 8–9; laws and policies preventing, 8; number of countries currently exploring, 422n.8; UNAIDS support for, 83. *See also* Brazilian ARV rollout; global ARV rollouts

ATS (alternative test site) program [U.S.], 257

Aurobindo Pharma, 385

Avila, Marlene, 245

AZT: available through SUS, 138; Brazilian law on free distribution of, 66; used early therapeutic intervention, 258; HIV positive pregnant women placed on, 275; 3 by 5 initiative on using, 72, 96, 108, 381, 388

Bahia (Brazil): AIDS incidence rates in, 181–83, 182*fig*; author's report on hidden AIDS epidemic in, 237; gross domestic product (GDP) ranking of, 412n.11; high rates of STDs in, 274; hospitals of, 207–17; report on AIDS (1996) in, 180–81; Ribeiro's report on state of AIDS in, 234–36; *A Tarde* (newspaper) report on AIDS in, 183

Bahian AIDS program: epidemiological study of, 202–4; system of nonintervention as part of, 204–6

Bahian Central Laboratory (LACEN), 250, 251*t*, 261, 272

Bahian CTA (Center for HIV Testing and Counseling): AIDS drugs clinical testing role of, 276–79; condoms distributed by, 274–75; creation of, 241; HIV testing procedure by, 241–42, 244–46; HIV window period issue and, 246–50; increasing requests for testing services of, 272–74; move to modern facility, 272; physical description of, 244; pilot study and findings on, 242–57; as prevention model, 257–59; rape victims seen as, 272–73; scientific process of objectification at, 263–67; training of personnel of, 260–61. *See also* CTA (Center for HIV Testing and Counseling); HIV testing

Bahian CTA pilot study: on Artemisa's experience, 267–68; background information on, 242–44; demographics of study group, 253–54, 255*t*; on Dog's experience, 251–52; on Equator's experience, 269; on Eyeglasses's experience, 265–66; findings of, 254–57, 270–72; on Gemini's experience, 262; on HIV testing process, 244–46; on Lion's experience, 248–49; on Love's experience, 246–47; on Mango's experience, 269; on Mulata's experience, 246; on Oxygen's experience, 270, 271; questionnaire responses, 255*t*; on retesting of subjects, 267–68; on Snake's experience, 268–69; on Star's experience, 256–57; on Sun's experience, 263; technoneurosis social reality revealed during, 270–72; on Tedania's experience, 252. *See also* CTA (Center for HIV Testing and Counseling)

Bahian Epidemiological Surveillance Service: AIDS cases recorded by, 219*fig*; work done with, 202

Bastos, Amilton, 330

Bastos, Cristiana, 68

Bastos, Tania ("Sheriff"), 226–28

Batista, Carisvaldo, 226
Batista, Luís Eduardo, 390
Batista, Valeria, 204
Bayer's Cipro patent, 77
Bear (Bahian CTA study), 242
A Bias for Hope (Hirschman), 3
Bill and Melinda Gates Foundation, 11, 86, 385
biological citizenship concept, 411n.9, 420n.6
bisexuals: AIDS patient registration of, 219*t*, 221*fig*, 222; CTA HIV testing study group reported as, 253, 254; "sexual fluidity" of, 413n.10
Bittencourt, Pedro, 121–22
Boletim Epidemiológico de AIDS report (1997), 179, 182
Botswanan African Comprehensive HIV/AIDS Partnership (ACHAP), 385
Brazil: ANVISA regulatory agency of, 92–93, 383; coexistence of ARV rollout/social exclusion mechanisms of, 283–86; development as activist state, 68–72; GDP growth rate of, 54; HIV/AIDS role in democratization of, 58–64; "moral crisis" being experienced in, 416n.9; pharmaceutical sales (2005) in, 79–84; relationship between pharmaceutical industry and, 91–93, 100; sanitation program (early 1900s), 90–91; slavery legacy for, 177; statecraft revealed by state AIDS policy of, 11; transforming state environment of, 55–58; transnational policy-space created in, 64–68; TRIPS effects on, 74–76; universal health care system struggle in, 59; WHO's AIDS program role by, 9; World Bank funding of AIDS Project I (1994), 65, 136; World Bank funding of AIDS Project II (1998), 70–71, 166–67. *See also* Brazilian HIV/AIDS policy
Brazilian ARV rollout: administrative discontinuities of, 388–90; coexistence with social exclusion mechanisms, 283–86; concept of spontaneous service demand guiding, 351–53; negotiations shaping, 376; origins and development of, 77, 87–93; power formations resulting from, 93–97; reverse-engineering of ARVs to produce, 76; 3 by 5 initiative and, 72, 96, 108, 381, 388. *See also* ARVs (anti-

retroviral drugs); ARVs universal access; Brazilian HIV/AIDS healthcare; Global ARV rollouts
Brazilian Health Ministry: *Boletim Epidemiológico de AIDS* report (1997) by, 179, 182; *casa de apoio* concept pushed by, 124, 283; national AIDS control program (1986) established by, 61; required reporting of AIDS cases to, 69; Ribeiro's report on state of AIDS to, 234–36; survey (1999) on condom use by, 274. *See also* ANVISA (Brazil)
Brazilian HIV/AIDS epidemic: activist state resulting from, 68–72; in Bahia, 180–84, 202–6; early documented cases of, 60; epidemiological study of Bahian, 202–4; evidenced in Salvador, 13–14; GAPA's framing as social problem, 163–64; impact on young and impoverished, 231; institutional and medical changes due to, 405–6; literature available on evolution of, 411n1; NGOs early response to, 7, 61–64; "official trivialization of," 399; public reports versus realities of, 179–80; racism factor of, 390; role in democratization, 58–64; social mobilization response to, 7–8; social ties recast by, 229; statistics of, 7, 207; transformation from fatal to lifelong disease, 375; turned into "drama of inclusion," 133–34; Winkler's reflections on, 390–92. *See also* HIV/AIDS epidemic; imaginary AIDS
Brazilian HIV/AIDS healthcare: "acceptance of death" role in, 212; access to (2005), 397–99; adherence to ARVs correlated to better outcomes of, 303–8; AIDS Project I (1994), 65, 136; AIDS Project II (1998), 70–71, 166–67; Caasah as venue for triage system of, 135; challenges of street patient's adherence to, 150, 154–55; DOT (Direct Observed Therapy), 304–5, 344, 401; DOT-HAART approach to, 306; homeless AIDS patients and, 225–31, 275, 352–53; inequalities (mid-1990s) of, 68; neglect of HIV-positive inmates, 157–59*p*; obstacles faced by AIDS patients, 173–75; precarious nature of, 126–27; racism affecting, 390; refiguring inequalities of, 97–101; state deferring of care to community

organizations, 236–37; successful outcomes of, 164–69; SUS (Sistema Único de Saúde) [Brazil] role in, 59, 137, 138, 388–89; "3 by 5" initiative of, 72, 96, 108, 381, 388. *See also* AIDS posttreatment access; Brazilian ARV rollout; Brazilian HIV/AIDS program; hospitals

Brazilian HIV/AIDS laws: allowing compulsory licensing of patented drugs, 57; extending disability status to people with AIDS, 62; on free distribution of AZT, 66; mandating free dispensation of ARV drugs, 68–69; on pharmaceutical dispensation (1995), 85

Brazilian HIV/AIDS policy: as contemporary form/event, 412n.4; development of public treatment, 3; flux related to, 47–49; ineffectiveness of late 1980s and early 1990s, 62–64; legal articles drafted as part of, 57–58; as model for other countries, 173; NGO activism and role in, 7, 58–64, 105–10; novel economic impacts of, 11; pharmaceuticalization of public health through, 84–87; position of deferring care to community organizations, 236–37; power formations of ARV scale-up, 93–97, 100; production of generic drugs as part of, 87–93; "psychological prophylaxis" approach of, 262–63; questions guiding research on, 5–6; recasting basic pharmacy program (mid-1990s), 84–85; state-society partnership characterization of, 10, 11, 53, 58; transforming state environment of developing, 55–58; World Bank disagreement with, 66–67. *See also* Brazil; HIV/AIDS policy; state-society partnerships

Brazilian HIV/AIDS program: AIDS Project I (1994–98), 65; correlation between three cultural processes and, 67; "Dial Health" initiative of, 170; funding distribution into, 416n.9; informal assessment of, 379–83; international cooperation with, 171; measures of success and undesirable realities of, 160–64; neglect of HIV-positive inmates in, 157–59p; nonintervention pattern of, 180; pact-making related to, 388; Programa Nacional de DST e AIDS (Brasília), 137p; providing universal access to

ARVs, 3, 8–9; structural readjustment of new national, 135–40. *See also* Brazilian HIV/AIDS healthcare; HIV/AIDS prevention programs

Brazilian HIV/AIDS transmission: changing dynamics of, 60; dominant mode of sexual, 273; gradual growth of mother-to-child, 60; heterosexual (late 1980s and early 1990s), 60; homosexual/bisexual mode (2000) of, 60

BRIC (Brazil, Russia, India, and China) markets, 79–80

Bristol-Myers Squibb, 86

Buarque de Holanda, Sérgio, 367

bubonic plague, 90

Butler, Judith, 308

Caasah AIDS patients: access to care (2005) by, 397–99; Adair, 152p; Altemar, 395; Caasah as foundational reference for, 366–67; Caminhoneiro, 145p; care baskets provided to, 420n.3; complaints about public hospitals by, 208–9, 214; "contract of responsibility" signed by, 310; disability pension paid to, 318–19; Dulce, 328–29p, 330, 341, 342p, 401; Edileusa, 20–21p, 342p; Edimilson, 300p, 313–14, 315p, 324, 330, 340–41, 342p, 400; ethnographic work with, 18–19; Evangivaldo, 33–36, 322, 324, 341, 343p, 368–69p, 370–71, 401–2, 404, 406; Evilásio Conceição, 284, 319–20, 321p, 341, 344, 345p; Fátima, 36, 322, 323p, 328, 341, 368, 370, 371, 401, 406; Francisco, 309; Jorge Antonio Santos Araújo, 146, 148, 155p; Jorge Leal, 306–7p, 340, 343p; Jorge Ramos, 26, 28, 322, 331, 332–33p, 334, 340, 342p, 395, 396; Lazaro de Oliveira, 198–99p, 200, 340, 343p; Leonardo, 300p; local economies of salvation experienced by, 48, 401; Luis Cardoso dos Santos, 22–25p, 335p, 341, 343p, 358, 359p, 360–63, 400, 404; Luzeide, 150; Manoel, 144; Marcelino, 141p; Maria Madalena, 194, 196–97p, 198, 200, 202, 343p, 395, 396; Maria Sonia, 147p; Marilda, 316, 317p, 340, 342p; Marta Damião, 309–10, 311p, 319, 320, 322, 331, 340, 343p, 351, 400; Nadia, 341, 342p, 395, 396; Naum Alves,

Caasah AIDS patients (*continued*)
225, 348–49; Nerivaldo, 30–32, 309,
340, 343*p*, 395, 396; photographs of
former, 186*p*–87*p*, 342*p*–43*p*; reports
on state of post-treatment, 340–44;
revisiting (2001) the, 339–44; Rita de
Souza, 229–31; role of religion in lives
of, 319; Romildo, 130, 131–32, 140,
142, 143–44; Rose, 26–29*p*, 322, 322*p*,
324, 331, 340, 342*p*, 402–3; selected
characteristics of, 223*t*; Sonara, 402;
Soraia, 40–42, 314, 316, 341, 343*p*, 395;
survivors and deaths among, 395–96,
399–403; Tiquinho, 43–45*p*, 288, 339,
341, 342*p*, 344, 396, 400; Valquirene,
37–39*p*, 214–15, 341, 401; Viviane, 347;
Wellington, 309. See also *aidéticos* (having AIDS); AIDS patients; AIDS patients'
invisibility; AIDS survivors
Caasah (Salvador): accomplishments and
future directions of, 334, 336; AIDS citizenship order being created by, 302–3;
AIDS death records gathered from,
222–25; Book of Proceedings of, 310;
as changed institution (2001), 339–49;
chaotic conditions leading to changes at,
288–89; conflicts surrounding, 140, 142;
controversial punishment system implemented at, 331; cooperation between
hospitals and, 349; daily logs kept at,
285, 286, 289–94, 296; death certificate records at, 218; Dona Conceição
reporting on former patients of, 184–85,
186*p*–87*p*; Dona Conceição's role in,
131, 142–44, 146, 148, 150, 154–55;
Elida (chief nurse) of, 344, 346, 356,
358; establishment of, 14–15, 130–31,
133, 284, 394; ethnographic work on,
18–19, 132–33, 285–86; as extension of
AIDS public services, 316–17; financial
and grassroots support of, 296, 298,
300; as foundational reference to patients, 366–67; functions served by, 170,
394–95; house of passage role limiting,
351–53; improved order and conditions (1996-97), 312–14; institutional
maintenance struggle of, 353–55; joint
meetings/new rules written (1995) for,
308–12; legal status attained by, 289;
Maria Luiza ousted as director of, 140,
286, 288; motto and creed followed by,

133; "natural rights" promoted by, 134;
Nerivaldo's description of life at, 30;
origins and development of, 284–86;
pastoral vision of care in, 306–8;
photographs of, 140*p*–41*p*, 151*p*–53*p*,
285*p*, 295*p*, 297*p*, 299*p*–300*p*, 301*p*;
political subjectivity of "biocommunity"
of, 324–27; Professor Carlos's work at,
28, 296, 312, 314, 316, 318, 332, 334,
354; realpolitik approach taken by, 131;
redefined as rehabilitation center (1997),
18, 327–34; religion as part of economy
of survival at, 319, 320, 322; return visit
(2005) to, 397–99; return visit to (2001),
339–49, 395; social death concept and,
142; state-society partnership role taken
by, 134–36, 142; transformed into *casa
de passgem* (house of passage), 18, 125,
230–31. See also *casas de apoio* (houses
of support)
Calcagnotto, Gilberto, 377
Calixto, 191
Caminhoneiro, 145*p*
Canguilhem, Georges, 78
Cardoso dos Santos, Luis: AIDS survivor
identity of, 404; critique of AIDS NGOs
by, 108; introduction to, 22–23; photographs of, 24*p*, 25*p*, 335*p*, 343*p*, 358,
359*p*–63; post-treatment employment
at Caasah, 341, 400; reflections on his
illness, 358, 360
Cardoso, Fernando Henrique: ARV universal access law signed by, 68–69; on
Brazilian therapeutic mobilization, 11,
87; comments on commitment to AIDS
policy by, 172; on market concept of society, 56–57, 76; on nature of the state,
55; on new "state voice," 99; on politics
of AIDS management, 87; on reaching
the neediest, 97, 98; on state-society
partnership of Brazil's AIDS response,
10, 11, 53, 58
Carlos ("Professor Carlos"): on ARV side
effects, 355–56; Caasah medicalization
developed by, 296, 312, 314; as Caasah
nurse, 28, 332; continued employment
at Caasah, 354, 401; exposed as quack
nurse, 354; on important healthcare
role by Caasah, 334; on need for doctor
for emergency services, 318; patient
advocacy by, 316; photograph of, 357*p*;

reporting on post-treatment patients, 401; seronegative boy adopted by, 314

Carmô, Dona, 183

Casa de Mãe Preta (asylum), 32

casas de apoio (houses of support): described, 63, 123; Health Ministry push for, 124, 283; origins and functions of, 123–25, 394. *See also* Caasah (Salvador)

casas de passagem (houses of passage), 18, 125, 230–31

Cassier, Maurice, 89

CDC (U.S. Centers for Disease Control), 257–58

Chequer, Pedro, 69–70, 90, 91–92, 139, 169, 171

Chile, 82–83

Cipro patent, 77

Ciro, 229

citizenship: AIDS, 33, 302–3; biological, 411n.9, 420n.6; as legitimating political order, 420n.5; prevention, 164; sexual, 413n.10. *See also* AIDS patients' invisibility

civil society: claims of representation by organized, 11; linked to political participation/execution of services, 170; new variety of AIDS mobilization created by, 109–10; public health measures jointly achieved by state and, 106; vanishing, 393–95. *See also* AIDS activism; NGOs (nongovernmental organizations); society

clinical trials: Brazilian ARV, 276–79, 358; ethical variability in global, 419n.15; judicialization of health and, 99–100

Clínicas Hospital (Porto Alegre), 99

collective action. *See* AIDS activism

Collor de Mello, Fernando, 54, 55, 63, 64

"A Commitment to Action for Expanded Access to HIV/AIDS Treatment:" (WHO), 95

Conceição, Dona: accusations made against, 225; care provided to Caasah's patients by, 142–44, 146; on challenges of treating street people, 150, 154–55, 185, 188–90; on denial by rich AIDS patients, 189–90; IBCM as new assistance foundation of, 403; interviewed during 2005 visit, 403; Jorge's praise of, 148; on reckless behavior of AIDS patients,

189; reporting on fate of different AIDS patients, 184–85, 186p–87p; Romildo's comments about, 131

Conceição, Evilásio: on Caasah as house of God, 284; Celeste on post-treatment state of, 341; on his AIDS illness, 319–20; photographs of, 321p, 343p, 345p; post-treatment interview with, 346–47

condoms: cultural/sexual codes determining use of, 128–29, 252, 265–66; free distribution by CTA, 241, 274–75; Health Ministry survey (1999) on use of, 274; HIV counseling on use of, 262; mutual trust issue of using, 251, 269–70. *See also* safe sex practices

Contrera, Wildney Freres, 120, 121, 124, 164

Correa, Marilena, 89

Costa Dourado, Maria Inês da, 207

Costa, Freire, 129

Costa, Luis Alberto, 204–5

Costa, Nilton Rosário da, 167

Costa, Roberto, 228

Coutinho, Denise, 241

Crixivan, 166

Cruz, Oswaldo, 90

CTA (Center for HIV Testing and Counseling): AIDS drugs clinical testing role of, 276–79; free distribution of condoms in, 241, 274–75; HIV counseling approach of, 262–63; installation of additional, 260–61; origins of first, 259–60; "psychological prophylaxis" function of, 262–63; scientific process of objectification at, 263–67; as World Bank funded prevention strategy, 260. *See also* Bahian CTA (Center for HIV Testing and Counseling); Bahian CTA pilot study

CTA study. *See* Bahian CTA pilot study

Cunha, Euclides da, 103

Cunha, Sérgio, 253

Damião, Marta, 309–10, 311p, 319, 320, 322, 331, 340, 343p, 351, 400

Daniel, Herbert, 60, 62, 63, 64

Das, Veena, 327

Davi, 360

death. *See* AIDS death

Declaration of Alma-Ata (1978) [WHO], 86

Deleuze, Gilles, 47, 114, 420n.1–21n.1

Delgado, Maria, 273

HIV/AIDS political economy (*continued*)
patienthood and, 120–25; TRIPS treaty,
11, 72, 74–75, 82; world trade role in,
73–78

HIV/AIDS prevention programs: CTA Bahia role in, 257–59; failure to integrate
SUS into Brazilian, 388–89; financed by
national AIDS program, 161; Larvie's
study of community-oriented, 160;
measuring success/undesirable realities
of, 160–64; study (1991) on impact of
testing and counseling, 258–59; U.S.
approach to, 257–59. *See also* Brazilian
HIV/AIDS program; GAPA Bahia HIV/
AIDS prevention programs; "prevention
citizens"

HIV counseling: CTA approach to,
262–63; as keystone to global prevention measures, 259; promoting safe sex
practices, 160–64, 262; sexual normalization goal of, 419n.9; study (1991) on
effectiveness of, 258–59. *See also* risk
behavior

HIV counseling and testing sites (CTSs)
[U.S.], 257–58

HIV Equity Initiative (Haiti), 387

HIV infection: diagnosis of, 247; incubation period of, 247; window period of,
246–50

HIV testing: AIDS neurosis associated
with, 243; conducted in Salvador
(Brazil), 191; ELISA (Enzyme-Linked
ImmunoSorbent Assay), 247, 272; fears
associated with receiving, 252–53;
increasing requests for, 272–74; at
LACEN (Bahian Central Laboratory),
250, 251*t*; required prior to employment, 230, 273–74; results used for
begging, 229, 230, 275; study (1991)
on effectiveness of, 258–59; study on
why low-risk individuals undergo, 259;
window period and, 246–50. *See also*
Bahian CTA (Center for HIV Testing
and Counseling)

homeless AIDS patients: Dona Conceição's
work with, 225, 226; failure to adhere
with ARV treatment, 228–29; HIV test
results used for begging by, 229, 230,
275; Naum Alves's service for, 225–26;
photographs of, 227*p*; remaining outside

the system (2001), 352–53; as risk
vectors, 235; Rita de Souza, 229–31;
"street tribe" of, 230, 231; Tania Bastos,
226–28

Homo Sacer (Agamben), 325

homosexuals: AIDS patient registration
of, 219*t*, 221*fig*, 222; CTA HIV testing
study group reported as, 253, 254;
GAPA Porto Alegre creating acceptance
of, 118–19

Hospital Caridade, 207, 209–13

hospitals: AIDS patient registration
information on stays at, 220*fig*; Caasah
AIDS patients' complaints about public,
208–9; cooperation (2001) between
Caasah and, 349; Dorneles Psychiatric
Hospital, 215, 316; Estadual Hospital,
316; Hospital Caridade, 207, 209–13;
State Hospital Luis Souto, 207, 208,
213–17, 218, 219*t*, 349–53; University
Hospital, 206–9. *See also* Brazilian
HIV/AIDS healthcare

Human Rights Commission, 167

imaginary AIDS: described, 250; technoneurosis social reality as, 270–72. *See
also* Brazilian HIV/AIDS epidemic

IMS Health, 79

incubation period, 247

inequalities: of Brazilian HIV/AIDS healthcare (mid-1990s), 68; public health to
refigure, 97–101

Institute for Advanced Study (Princeton),
53, 71

institutional partnerships. *See* state-society
partnership

intellectual property rights: Brazilian
production of generic ARVs violation
of, 87–93; Brazilian social mobilization against, 76–78; impact on AIDS
treatment by laws on, 8; TRIPS treaty
on, 11, 73–76, 82; world trade and role
of, 73–78. *See also* ARVs (antiretroviral
drugs)

InterAIDS, 125, 133

International Monetary Fund (IMF): Brazilian policy decisions and role of, 65;
Cardoso on sovereign power and, 56

invisibility issue. *See* AIDS patients'
invisibility

Neto, Acioly, 125–26
Nevirapine (ARV), 108
New York Times, 77
NGOs (nongovernmental organiza-
tions): ABIA (Associação Brasileira
Interdisciplinar de AIDS), 62, 169–70;
AIDS Forum, 125–26; ARCA (Apoio
Religioso Contra à AIDS), 62, 110;
Arjun Appadurai, 62; Brazilian AIDS
epidemic addressed by, 7, 61–64; Casa
de Passagem (House of Passage), 125;
Doctors Without Borders, 93, 380, 381;
dual displacement in governmentality/
resistance related to, 412n.13; Febem
(State Foundation for Youth Welfare),
127–28; GAPA Porto Alegre, 67; Global
Fund to Fight AIDS, TB, and Malaria,
377; Grupo Gay da Bahia (Gay Group of
Bahia, or GGB), 61; Grupo Pela Vidda,
62, 121; houses of support created by,
14–15; industrialization of AIDS, 106–7;
InterAIDS, 125, 133; loss of idealism by,
390; Outra Coisa, 61; Partners In Health,
305, 306, 387; Pela Vidda, 172; Somos,
61; SOS Corpo (SOS Body), 125; trajectories taken by, 110–14; transformation
of AIDS, 105–10. *See also* AIDS activism;
civil society; state-society partnership
Nogueira, Mirta, 244, 250, 251, 272, 276
Nogueira, Tais, 70

objet petit a (object cause of desire), 263–64
Oliveira, Lazaro de, 198–99*p*, 200, 340,
343*p*
"Our Traumatic Neurosis and Its Brain"
(Young), 266–67
Outeiro, Ana, 241, 242–43, 247, 249, 254
Outra Coisa (Brazil), 61
Oxygen (Bahian CTA study), 270, 271

Pan American Health Organization
(PAHO), 72, 171
Parker, Richard, 63, 64, 66, 139–40
Partners In Health, 305, 306, 387
Passarelli, Carlos, 173, 259–60
Pasteur Institute (France), 90
pastoral care: Caasah's vision of, 306–8;
facilitating, ARV adherence, 306, 308;
Foucault's insights on pastoral power,
419n.2–20n.2
patents. *See* intellectual property rights

patient citizenship. *See* citizenship
Patricia, 113*p*
Paula, Eliana de, 204
Paulinho, 126
Pela Vidda (Niterói), 172
Pelourinho, Pillory (Salvador): Andrade's
AIDS patients study conducted in, 191,
194; history of, 190; mental illness
among residents of, 416n.3–17n.3; pho-
tographs of, 192*p*–93*p*; stories of AIDS
patients from, 194–202
Pentecostal Universal Church of God's
Kingdom, 133
PEPFAR (Emergency Plan for AIDS Relief),
384–85, 422n.9–23.n9
Perin, Sandra, 157–58
Pernalonga, 126–27*p*, 128–30, 176
Petryna, Adriana, 99, 278
pharmaceutical form of public health:
Brazilian policies leading to, 84–87;
Brazilian production of generic ARVs as,
87–93; described, 11–12; interactions
leading to, 73–78; pharma-state relationship within, 91–93; role of ARVs in, 8;
short-and-long-term goals of, 84. *See
also* rational-technical controls
pharmaceutical industry: ARV adherence
education distributed by, 92; Brazilian
legislation for state activism impacting,
57–58; growing Latin American (2005)
sales, 79–84; humanitarianism and
equity responses to AIDS by, 384–88; on
Latin American demand, 80–81; pharmaceutical form of governance and role
of, 73–78; relationship between Brazilian
state and, 91–93, 100; reverse engineering impact on ARV investment by, 93;
TRIPS shaped by, 82
Picon, Paulo, 99
Pinheiro, Eloan, 88–90
Piot, Peter, 95
police search, 149*p*
political sphere: Arendt's *animal laborans*
concept of, 325; Caasah's political
subjectivity in context of, 324–25, 326–
27; Ranciere's depoliticizing approach
to, 325–26
political will notion, 80, 81
Poole, Deborah, 327
Porto Alegre (Brazil): AIDS among
prisoners in, 116–18, 157–59*p*; CTA

unit established in, 259–60; early AIDS activism in, 115–16; HIV/AIDS prevention work in gay bar in, 113*p*. *See also* GAPA Porto Alegre

Porto Alegre penitentiary: *Arpão* newsletter of, 158; declining conditions for HIV-positive inmates of, 157–58; Elaine of, 158, 159*p*; HIV-positive inmates of, 116–18; photograph of, 159*p*

post-treatment. *See* AIDS post-treatment access

"prevention citizens," 164. *See also* HIV/ AIDS prevention programs

Programa Nacional de DST e AIDS (Brasília). *See* Brazilian HIV/AIDS program

prostitutes: AIDS patients' continued work as, 229; AIDS risk and fear of, 64; inducing abortion technique used by, 257. *See also* women

PSDB (Partido da Social Democracia Brasileira) [Brazil], 399

PT. *See* Workers' Party (Partido dos Trabalhadores, PT)

public health: pharmaceutical form of, 8, 11–12, 73–78, 84–93; "psychological prophylaxis" approach of, 262–63; refiguring inequality through, 97–101; UN role in global, 96–97. *See also* Brazilian HIV/AIDS policy

racism issues, 390

Radames, Airton, 80, 232–34, 276

Raisa, 312

Ramos, Jorge, 26, 28, 322, 331, 332–33*p*, 334, 340, 342*p*, 395, 396

Ranciére, Jacques, 325–26

rape victims: CTA service for, 272–73; Oxygen's experience as, 270; two girls at stadium, 231

rational-technical controls, 91. *See also* pharmaceutical form of public health

religion: Caasah AIDS patients and role of, 319; Geertz on, 319; Luis on ARV adherence and role of, 361–62; as part of economy of survival at Caasah, 320, 322

reverse engineering: championed by Brazilian AIDS policymakers, 76; Eloan Pinheiro on, 88–90; at Farmanguinhos, 380; impact on generic ARV production by, 93

Ribeiro, Airton, 234–36, 237, 277–78

Ribeiro, Renato Janine, 97–98

Ricardo, 115

Rich, Peter, 92

Rio de Janeiro CTA, 260

risk behavior: rationales for not using condoms, 128–29, 251, 252, 265–66; "scientific" reenactment of, 268–69. *See also* HIV counseling

Roche, 164

Romildo: as Caasah founder, 130, 131–32; death of, 291; as leading Caasah takeover from Maria Luiza, 140, 288; poor physical condition of, 142, 143–44

Rose: addiction of, 322; AIDS survivor identity adopted by, 404; baby girl born to, 340; Caasah rules broken by, 331; on fear of suffering while dying, 324; introduction to, 26, 28; photographs of, 27*p*, 29*p*, 323*p*, 342*p*, 365*p*; reflections on post-treatment life, 364–68, 402–3, 404; relationship with Jorge Ramos, 332–34

safe sex practices: barriers to, 162; culture and sexual codes determining condom use, 128–29, 252, 265–66; HIV/AIDS prevention programs promoting, 160–64. *See also* condoms

Salvador (Brazil): Caasah house of support established in, 14–15; demographics of, 13; epidemiological evidence on AIDS in, 13–14; HIV testing results in, 191; Pelourinho/Pillory occupied by AIDS patients in, 190–94; State Hospital Luis Souto of, 207, 208, 213–17, 349–53

Santos Araújo, Jorge Antonio, 146, 148, 155*p*

Santos, Ribamar dos, 125

Santos, Sebastião, 183

São Paulo CTA, 260

Sarney, José, 54

Scheper-Hughes, Nancy, 62–63

science: AIDS, 87–90, 121–22, 414n.12; AIDS activism invading domain of HIV/AIDS, 414n.12; changing forms of truth in, 265; disease dynamics and role of, 249–50; generic drugs produced by public-sector, 87–93; used in reenactment of risk behavior, 268–69. *See also* anthropology; ethnography

Secure the Future Initiative (Bristol-Myers Squibb), 86
self-efficacy, 304
Serra, José: on "failure visibility effect" of AIDS activism, 107–8; on mobilization for other large-scale diseases, 86–87; presidential race between Lula and, 399; regarding state privatization measures, 76–77; regarding TRIPS agreement, 74, 75; state-market integration vision of, 78, 90, 92; on World Bank's loan to Brazil, 71
Serum Therapy Laboratory (Brazil), 90
sexual citizenship, 413n.10
"sexual fluidity" phenomenon, 413n.10
Silva, Adroaldo, 209–10
Silva, Ana, 64
Silva, Lucia, 261
Silva, Mariza, 188
Silva, Ricardo, 308, 309, 312
Silva, Zenaide, 251
Sky (Bahian CTA study), 241, 242
smallpox, 90
Snake (Bahian CTA study), 268–69
social mobilization. *See* AIDS activism
society: Cardoso on market concept of, 56–57; exclusion mechanisms of Brazilian, 283–86; social-society differences, 97–98; vanishing civil, 393–95. *See also* civil society
Somos (Brazil), 61
Sonara, 402
Sonia, Maria, 147*p*
Soraia: introduction to, 40, 42; photographs of, 41*p*, 343*p*; post-treatment state of, 341; presumed dead, 395; on Professor Carlos's care, 314, 316
SOS Corpo (SOS Body), 125
Souza, Herbert de, 59, 61
Souza, Rita de, 229–31
sovereign power concept, 55–56
Star (Bahian CTA study), 256–57
state: Brazilian development as activist, 68–72; Brazilian HIV/AIDS crisis role in democratization, 58–64; Cardoso on sovereign power of, 55–56; pharmaceutical form of governance and role of neoliberalizing, 73–78; public health measures jointly achieved by civil society and, 106; social science of a transforming, 55–58; "the temporal experience of the," 327

State Hospital Luis Souto: AIDS patients treated at, 207, 208, 213–17; death certificates records of, 218, 219*t*; visit to (2001), 349–53. *See also* Nanci, Dr. ("*Mãezona* Big Mama")
state-society partnerships: Brazilian generic drugs manifestation of, 87–93; Caasah as example of, 134–36, 142; Cardoso on Brazilian AIDS policy as, 10, 11, 53, 58; declining currency of idea of, 393–94; early development of, 122–23; to facilitate more equitable international situations, 58; new variety of AIDS mobilization created by, 109–10; political dynamics of, 172–75; in structural readjustment of national AIDS program, 135–40. *See also* AIDS activism; Brazilian HIV/AIDS policy; NGOs (nongovernmental organizations)
STDs (sexually transmitted diseases): high positive testing in Bahia, 274; testing for syphilis, 272
Stepan, Nancy, 90
"street tribe," 230, 231
"subjectivity of milieus," 47–48
Sun (Bahian CTA study), 263
SUS (Sistema Único de Saúde) [Brazil]: AZT availability through, 138; described, 59; endorsement of, 137; HIV/AIDS prevention of, 388–89
syphilis testing, 272

T-20 (Fuzeon), 382–83
The Taming of Chance (Hacking), 203
A Tarde (Bahia newspaper), 183
TB: AIDS patients suffering from, 118, 144; DOT Guidelines in Effective Tuberculosis Control (WHO) on, 305; DOTS-Plus approach to, 234; medical interventions for, 86–87, 89
technocracy countermeasures, 155–58
technologies of invisibility, 202–4
technoneurosis, 270–72
Tedania (Bahian CTA study), 252
Teenage Citizen (Adolescente Cidadão), 403
Teixeira, Paulo: on AIDS program pact-making, 388, 389–90; on changes of AIDS discourse/paradigm, 379; on events leading to breaking of patents, 61, 65, 66–67, 72, 76, 77; on impact of AIDS activism, 105–6; on international activism

increasing AIDS funding, 377; on international support of ARV rollout, 95–96; as national AIDS program coordinator, 169; on obstacles to more rapid anti-HIV treatments, 382; on pharma influence on AIDS treatment funding, 82–83; on political participation through AIDS policy, 94–96; role in WHO-UNAIDS "3 by 5" campaign, 381; on state's deferring of care to community organizations, 236–37; on state-society partnership in AIDS policy, 135–36, 138–39; on World Bank loan as social victory, 136, 138
Tenefovir, 380
theatrical AIDS activism, 128–30
3 by 5 initiative, 72, 96, 108, 381, 388
3TC: costs of, 91; "3 by 5" initiative on using, 72, 96, 108, 381, 388
Tiquinho: allowed to stay at Caasah (2001), 339, 341, 344, 396; death of, 400; introduction to, 43; photographs of, 44p, 45p, 342p; taken to Caasah by mother, 288
transmission. *See* Brazilian HIV/AIDS transmission
travestis (transvestites): Brenda Lee, 123–24, 130, 133; challenges of treating HIV-positive, 188; Don Kulick, 188; photographs of, 195p
treatment. *See* ARVs (antiretroviral drugs); Brazilian HIV/AIDS healthcare
TRIPS (Trade-Related Aspects of Intellectual Property Rights) treaty: immediate impact of, 74–75; negotiations leading to, 74; origins of, 11, 72; patent protections of, 74; pharma influence on, 82
truth: biotechnical, 243; changing forms of scientific, 265; work of literature moving away from, 420n.1-21n.1

UNAIDS (Joint United Nations Programme on HIV/AIDS): ARV treatment goal by, 4; Brazilian program praised by, 71, 83; increased budget of, 377; supporting lowering drug prices (2001), 77
United Nations (UN): AIDS problem recognized (1986) by, 61; on global public health role of, 96–97; lack of negotiating power of developing countries, 379
United Nations (UN) Conference on AIDS and Human Rights, 172

United States: economic sanctions threatened (2001) by, 77; PEPFAR (Emergency Plan for AIDS Relief) of, 384–85; prevention models used in the, 257–58
University Hospital, 206–9
U.S. ATS (alternative test site) program, 257
U.S. Centers for Disease Control (CDC), 257–58
"The Uses of Culture in the Making of AIDS Neurosis in Japan" (Miller), 243
U.S. Food and Drug Administration (FDA), 92–93, 385
U.S. HIV counseling and testing sites (CTSs), 257–58

Valquirene: complaints about public hospitals by, 214–15; introduction to, 37–38; photographs of, 39p, 342p; post-treatment state of, 341, 401
vanishing civil society, 393–95
Vasconcelos, Cristina, 162, 163
Vieira, Monica, 172
Vita (asylum), 114
Vivencial (Living) project, 127–28
Viviane, 347

Walker, Jane, 384
"Washington Consensus," 65
Wellington, 309
Wesllen, 391
Western blot (immunoblot), 247
"What Is an Author?" (Foucault lecture), 263–64, 267
Whyte, Susan Reynolds, 386–87
window period (HIV), 246–50
Winkler, Gerson: *Arpão* (Porto Alegre penitentiary newsletter) circulated by, 158; on changes in AIDS activism, 67, 106, 107, 108, 112; complaints against discourse blaming AIDS patients, 168–69; doubts on governmental activism expressed by, 123; on entering AIDS activism work, 115–16; Human Rights Commission work by, 167; on local NGOs tension with national AIDS program, 114; on loss of idealism by NGOs, 390; photograph (1995) of, 111p; reflections on his own AIDS status, 112, 114, 115; reflections on world of AIDS in Brazil by, 390–92; on resistance to working with

Winkler, Gerson (*continued*)
  state, 138; on success of AIDS cocktail
  treatment for him, 164–66, 167–68, 175;
  successful personal outcome for, 175–76;
  working with the city's AIDS program,
  156–57; working for prisoners with
  AIDS, 117–18. *See also* AIDS patients
women: rape victims, 231, 270, 272–73;
  state policy on mandatory HIV testing
  of pregnant, 275. *See also* gender differences; prostitutes
Workers' Party (Partido dos Trabalhadores,
  PT), 54, 110, 156, 379, 391, 394
World AIDS conference (Durban, 2000),
  73, 76
World Bank: accusations of exaggerated
  anti-malaria impact by, 386; accused of
  capitalizing politically on AIDS policy,
  169; AIDS Project I (1994) funded by, 65,
  136; AIDS Project II (1998) funded by,
  70–71, 166–67; Brazilian AIDS epidemic
  addressed by, 7, 65, 66, 70–71; Brazil's
  poor social welfare reported (1988)
  by, 137–38; Caasah (Salvador house
  of support) funded by, 14; Cardoso on
  sovereign power and, 56; criticized for
  negative impact on HIV spread, 71; CTA

units prevention strategy supported by,
  260; disagreement between Brazilian
  policy makers and, 66–67; Teixeira's
  comments on loan by, 136, 138;
  UNAIDS financing role by, 377
World Health Organization (WHO): ARV
  scaling-up program initiated at, 96;
  ARV therapy goal by, 4; Brazilian role in
  AIDS program of, 9; changing position
  on AIDS activism by, 72; "A Commitment to Action for Expanded Access to
  HIV/AIDS Treatment:" (2002) by, 95;
  Declaration of Alma-Ata (1978) by, 86;
  difficulties of "3 by 5" campaign by, 388;
  DOT Guidelines in Effective Tuberculosis Control of, 305; international
  classification of diseases by, 209; lack of
  negotiating power of developing countries, 379; policies preventing universal
  access to ARVs by, 8; TRIPS treaty and,
  11, 73–76, 82

yellow fever, 90
Young, Allan, 266–67
Young, Iris, 325

Zweig, Stefan, 8